Application of Nanotechnology in Drug Delivery

Application of Nanotechnology in Drug Delivery

Editor

Rodolph Donovan

Application of Nanotechnology in Drug Delivery
Edited by **Rodolph Donovan**

ISBN: 978-1-68117-238-5
Library of Congress Control Number: 2016934778

© 2017 by
SCITUS Academics LLC,
www.scitusacademics.com
Box No. 4766, 616 Corporate Way,
Suite 2, Valley Cottage,
NY 10989

Notice

Preface

Continuing improvement in the pharmacological and therapeutic properties of drugs is driving the revolution in novel drug delivery systems. In fact, a wide spectrum of therapeutic nanocarriers has been extensively investigated to address this emerging need. Nanotechnology involves the engineering of functional systems at the molecular scale. Recent years have witnessed unprecedented growth of research and applications in the area of nanoscience and nanotechnology. There is increasing optimism that nanotechnology, as applied to medicine, will bring significant advances in the diagnosis and treatment of disease. Anticipated applications in medicine include drug delivery, both in vitro and in vivo diagnostics, nutraceuticals and production of improved biocompatible materials. As such, nanomaterials are poised to take advantage of existing cellular machinery to facilitate the delivery of drugs. The use of nanotechnology in medicine and more specifically drug delivery is set to spread rapidly. Currently many substances are under investigation for drug delivery and more specifically for cancer therapy. anoparticles (NPs) containing encapsulated, dispersed, absorbed or conjugated drugs have unique characteristics that can lead to enhanced performance in a variety of dosage forms. When formulated correctly, drug particles are resistant to settling and can have higher saturation solubility, rapid dissolution and enhanced adhesion to biological surfaces, thereby providing rapid onset of therapeutic action and improved bioavailability. In addition, the vast majority of molecules in a nanostructure reside at the particle surface, which maximizes the loading and delivery of cargos, such as therapeutic drugs, proteins and polynucleotides, to targeted cells and tissues. Highly efficient drug delivery, based on nanomaterials, could potentially reduce the drug dose needed to achieve therapeutic benefit, which, in turn, would lower the cost and/or reduce the side effects associated with particular drugs. This book, Application of Nanotechnology in Drug Delivery, deals with the new and ongoing potentialities of nanotechnology application of drug delivery.In addition, it discusses advances in design, optimization, and adaptation of gene delivery systems for the treatment of cancer, cardiovascular, and infectious diseases, and considers assessment and review procedures involved in the development of gene-based pharmaceuticals.

Table of Contents

CHAPTER 1

Nanoparticles in Drug Delivery and Cancer Therapy: The Giant Rats Tail

Vinod Prabhu, Siddik Uzzaman, Viswanathan Mariammal Berlin Grace, Chandrasekharan Guruvayoorappan

Department of Biotechnology, Karunya University, Karunya Nagar, Coimbatore, Tamil Nadu, India.

ABSTRACT

Nanotechnology has the potential to offer solutions to these current obstacles in cancer therapies, because of its unique size and large surface-to-volume ratios. Nanoparticles may have properties of self-assembly, stability, specificity, drug encapsulation and biocompatibility as a result of their material composition. Nanoscale devices have impacted cancer biology at three levels: early detection, tumour imaging using radiocontrast nanoparticles or quantum dots; and drug delivery using nanovectors and hybrid nanoparticles. Other role of nanotechnology, in management of various diseases and also in drug resistance in leukemia by blocking drug efflux from cancer cells and induce efficient delivery of si RNA into lymphocytes to block apoptosis in sepsis and targeting tumors also. Nanocrystals labeling with immune cells can act as a platform technology for nanoimmunotherapy. This review addresses the advancement of nanoparticles in drug delivery and in cancer therapy.

Keywords: Nanoparticles, Cancer Therapy, Drug Delivery, Drug Targeting, Quantum Dots

1. INTRODUCTION

Nanotechnology deals with the design, production and characterization on ultra small particles which is extended to broad area in pharmaceutical, medical, chemical and engineering application due to its unique properties [1]. The development of technology occurs at the atomic, molecular or macromolecular range of approximately 1 nm - 100 nanometers (nm) to create and use structures that have novel properties [2]. Nanoparticles (NPs) are defined as a small object that behaves as a whole unit in term of its transport and properties. They can be further classified according to the size and diameter. Fine particles have the range of 100 to 2500 nm or ultrafine particles having the size of 1 to 100 nm [3]. Nanoclusters have one dimension between 1 and 10 nm and narrow size distribution and nanopowders which are agglomerates of ultrafine particles [4]. NPs research is currently an area of passionate scientific interest due to its wide variety of potential application in therapeutic and biomedical interest. The field of nanotechnology holds the promise of significant improvements in the health and well being of patients as well in manufacturing technologies [5]. Specialized nanotechnological approaches like dendrimers, quantum dots, monoclonal antibodies and intergrins which are extensively researched for diagnostic and targeted delivery of therapeutic agents [6]. Nanorobotics centers on self sufficient machines of some functionally operating at the nanoscale. These are hopes for applying nanorobots in medicine [7-9]. The advance of contemporary materials and methodologies have to be manifested with some patients granted about new nano devices which will help in establishing NPs with the use of embedded nanobioelectronics concept in future [10,11]. Nanomedicine access to drug delivery on development of nanoscale molecules which can improve drug bioavailability [12,13]. It was disclosed that this can potentially be achieved by molecules targeting by nanoengineering devices. This new method will be effective in treating a variety of illness such as neurological disorders, diabetes, osteoporosis, Alzheimer's, Parkinson's, amyotrophic lateral sclerosis, multiple sclerosis, HIV-1 associated neuro cognitive disorders, cardio vascular disorders, tuberculosis and cancer [14-22]. NPs such as lipid or polymer can be designed to improve the pharmacological and therapeutical properties of drugs [21]. Cells take up these NPs because of their size and also the ability of the drug to get into the cell cytoplasm through cell membrane. NPs have a very high surface area to volume ratio and it allows many functional group to be attached to a NPs which can bind to certain tumor cells [22]. The smaller size of the NPs facilitates them to accumulate in tumor micro environment thus facilitating newer therapeutic strategies which may replace radiation and chemotherapy.

Recent research has developed a number of NPs such as metals, semiconductor and polymeric particles used in imaging probes and delivery vehicles [23-25]. The use of NPs based drug delivery systems such as polyethleneimine liposomes (PEI), silica NPs, micelles and chitosan have effective role in drug delivery with reduced drug side effects [26,27]. Recent technology have developed many multifunctional NPs for targeting, imaging, drug delivery, sensing of anticancer agents and small interference RNA (si RNA) delivery [28,29]. Nanotechnology promises construction of artificial cells, enzymes and genes that helps in the replacement therapy of many disorders which are due to deficiency of enzymes, mutation of genes or repair in the synthesis of protein [30]. In this review, we discuss the recent emerging trends of NPs in drug delivery system and in cancer therapy.

2. MULTIFUNCTIONAL NANOPARTICLES

Liposome comprises lipid bilayer membrane surrounding an aqueous interior and it can be used as nanoparticles that have similarities with biological membrane that improves the efficacy and safety of drugs [31]. Liposomes are classified into three categories such as small unilamellar vesicles, large unilamellar vesicles and multilamellar vesicles on the basis of their size and lamellarity. The active compound can be located either in the aqueous spaces, if it is water-soluble, or in the lipid membrane, if it is lipid-soluble. The new generation of liposome called 'stealth liposomes' have the ability to evade the interception by the immune system, and have longer half-life [32]. The emulsions comprise of oil in water-type mixtures that are stabilized with surfactants to maintain size and shape. The lipophilic material can be dissolved in water organic solvent that is emulsified in an aqueous phase. Like liposomes, emulsions have been used for improving the efficacy and safety of diverse compounds [33]. Polymers such as polysaccharide chitosan NPs have been used for some time as drug delivery systems [34]. Water-soluble polymer hybrid constructs polymer–protein conjugation that reduces immunogenicity, prolongs plasma half-life and enhances protein stability. Polymer–drug conjugation promotes tumor targeting through the enhanced permeability and retention effect and at the cellular level following endocytic capture, allows lysosomotropic drug delivery [35]. Ceramic NPs are inorganic systems with porous characteristics that have emerged as drug vehicles [36]. These vehicles are biocompatible ceramic NPs such as silica, titania and alumina that can be used in cancer therapy. Gold shell NPs and other metal-based agents can serve as novel category of spherical NPs consisting of a dielectric core covered by a thin metallic shell which is typically gold. These particles possess highly favorable optical and chemical properties for biomedical imaging and therapeutic applications [37]. Carbon nanomaterials include fullerenes and nanotubes. Fullerenes are novel carbon allotropes with a polygonal structure made up exclusively by 60 carbon atoms. These NPs are characterized by having numerous points of attachment whose surfaces also can be functionalized for tissue binding [38]. Nanotubes have been one of the most extensively used types of NPs because of their high electrical conductivity and excellent strength. Carbon nanotubes can be structurally visualized as a single sheet of graphite rolled to form a seamless cylinder. There are two classes of carbon nanotubes that are single-walled (SWCNT) and multi-walled (MWCNT). MWCNT are larger and consist of many single-walled tubes stacked one inside the other compared to SWCNT. Functionalized carbon nanotubes are emerging as novel components in nanoformulations for the delivery of therapeutic molecules [39]. Quantum dots are NPs made of semiconductor materials with fluorescent properties which are mostly used in biological applications and quantum dots must be covered with other materials allowing dispersion and preventing leaking of the toxic heavy metals [40].

2.1. Synthesis of Nanoparticles

Size of the nanoparticle is very important for efficient drug delivery. Generally, 10 nm - 100 nm is considered as the optimal size for nanoparticle drug carriers. If the particle size is less than 10 nm, the NPs will be quickly eliminated by renal clearance (threshold < 6 nm) and at sizes greater than 100 nm, will have chances to be captured by the reticuloendothelial system (RES) [41]. Surface coating is essential for the stability and circulation time of NPs delivery system.

For example a sodium citrate-stabilized gold particle aggregates in phosphatebuffered saline (PBS) within several minutes but once coated with thiol-terminated polyethylene glycol (PEG) polymer provides stability not only in PBS but also under low or high pH conditions [42]. Neutral charged NPs exhibits longer circulation time and reduce the chance of nanoparticle capture by the immune system.

2.2. Gold Nanoparticles

Nanoparticles synthesis and the study of their size and its properties are fundamental importance in the advancement of recent research [43]. It is exposed that optical, electronic, magnetic and catalytic properties of metal NPs depends on their size, shape and chemical surroundings. In NP synthesis it is important to control not only particles size but also the particle shape and morphology. Colloidal gold also known as nanogold is a suspension or colloid of sub-micrometer sizes particles of gold in a fluid—usually water [44]. Gold NPs can be produced in liquid chemical method by reduction of chloroauric acid (HAuCl$_4$). After dissolving HAuCl4 the solution is then rapidly stirred followed by the addition of reducing agents. This enhanced the production of Au^{3+}ions which gets reduced to neutral gold atoms. More and more of these gold atoms from the solution turns to precipitate in the form of sub nanometer particles. Vigorously stirring of this solution results in the production of particles of uniform size [44].

2.3. Silver Nanoparticles

The uniform silver NPs can be obtained by the reduction of silver ions by ethanol at 800°C to 1000°C under atmospheric condition [45]. In this synthesis process 20 ml of aqueous solution containing silver nitrate (0.5 g of AgNO$_3$) can be treated with sodium linoleate (C$_{18}$H$_{32}$O$_2$) (1.5 g) in tubes under continuous agitation. The aqueous solution containing silver ions and sodium linoleate can be further treated with a mixture of linoleic acid and ethanol resulting in the formation of an ethanol solution phase containing silver ions. The ethanol in the liquid and solution phase reduces the silver ions into silver NPs. Linoleic acid will be absorbed on the surface of the silver NPs with alkyl chains on the outer side in a circular shape. Wang [46] demonstrated that on changing the concentration of the electrolyte a reddish brown color developed on addition of linoleic acid their by indicating 100% conversion of silver ions into silver NPs.

2.4. Copper Nanoparticles

A novel method for the preparation of copper NPs is by reducing the copper sulphate (CuSO$_4$) with hydrazine in ethylene glycol under microwave irradiation. The heating method and reaction temperature on the particle size and composition of powder have been investigated by X-ray diffractometry (XRD) and transmission electron microscopy (TEM). Well-dispersed copper nanopowder with a diameter of 15 nm can be obtained in the absence of a protective polymer [47]. In order to obtain pure-phase copper NPs using water, the reaction time of 8 hr is essential. Owing to the reduction property of ethylene glycol, the reaction rate using ethylene glycol is higher. In addition, the amount of reduction agent can be reduced largely. Polyvinylpyrrolidone (PVP) plays greater role on the size of copper particles, and increase in the (PVP) concentration that attributes to the smaller dimension particles. The mean diameter is about 4 nm when the concentration of PVP is 0.5 mmol/L. PVP acts as the polymeric capping agents in the reaction preventing the agglomeration of the copper NPs. When water is the reaction medium, the Cu^{2+} complex is reduced to Cu^+ complex and further reduction of Cu^+ will form the pure copper NPs [48].

2.5. Microencapsulation of Nanoparticles

There are various methods available for the microencapsulation but most useable methods are as follow: The particles are tumbled in a pan or other device while the coating material is applied slowly. Air-suspension coating of particles by solutions gives better control and flexibility. The particles are coated while suspended in an upward-moving air stream. They are supported by a perforated plate having different patterns of holes inside and outside a cylindrical insert. Just sufficient air is permitted to rise through the outer annular space to fluidize the settling particles. Most of the rising air (usually heated) flows inside the cylinder, causing the particles to rise rapidly. At the top, as the air stream diverges and slows, they settle back onto the outer bed and move downward to repeat the cycle. The particles pass through the inner cylinder many times in a few minutes. Spray drying serves as a microencapsulation technique when an active material is dissolved or suspended in a melt or polymersolution and becomes trapped in the dried particle. The main advantage is the ability to handle labile materials because of the short contact time in the dryer. In addition, the operation is economical. In modern spray dryers the viscosity of the solutions to be sprayed can be as high as 300 mPa·s. In chemical method the two reactants in a polycondensation meet at an interface and react rapidly. The basis of this method is the classical SchottenBaumann reaction between an acid chloride and a compound containing an active hydrogen atom, such as an amine or alcohol, polyesters, polyurea, polyurethane. Under the right conditions, thin flexible walls form rapidly at the interface. A solution of the pesticide and a diacid chloride are emulsified in water and an aqueous solution containing an amine and a polyfunctional isocyanate is added. To neutralize the acid formed during the reaction base may be added. Condensed polymer walls form instantaneously at the interface of the emulsion droplets. In a number of processes, a core material is imbedded in a polymeric matrix during formation of the particles. A simple method of this type is spray-drying, in which the particle is formed by evaporation of the solvent from the matrix material. However, the solidification of the matrix also can be caused by a chemical change [49].

2.6. Advancement of Nanoparticle Based Drug Delivery System

The important technological advantages of NPs used on drug carrier are high stability, high carrier capacity, feasibility of incorporation of both hydrophilic, hydrophobic substances and feasibility of variable routes of administration including oral application and inhalation [50]. The NPs can also be designed to allow controlled sustained drug release from the matrix. These properties will enhance to improve the drug bioavailability and reduces of dosing frequency and prevent non adherence to prescribed therapy. Micelles so called core shell structure in which the core of the micelles which is either the hydrophobic part or the ionic part of the NPs can contains small or bigger therapeutic drug [51]. The novel intracellular pH sensitive polymeric micelles drug carrier which control the systemic and sub cellular distribution of pharmacologically active drug. The micelles can be prepared from self assembling amphiphilic block copolymers, poly (ethylene glycol), poly (aspirate hydrazone adriamycin) to which the adriamycin (anticancer drug) is conjugated to the hydrophobic segments by acid sensitive hydrazone linkers. Therefore micelles can preserve drug under pH 7.4 and can release them by sensing the intracellular pH of the endosomes and lysosomes when they decreases to pH 5-6.

Nanomater sized semiconductor particles can be covalently linked with biorecognition molecules such as peptides, antibodies, nucleic acid, and small molecules ligand as biological labels [52]. The new approach of quantum dots technology with anticancer drug therapy called ZnQ Quantum dots which is loaded with anti cancer agents and encapsulated with biocompatible polymer represent a potential platform to deliver tumor targeted drugs and document the delivery process [53]. The non toxic water dispersed ZnQ quantum dots with long term fluorescence stability can be synthesized by a chemical hydrolysis method encapsulated with chitoson and loaded with anticancer drug. Chitosan enhanced the stability of quantum dots for its hydrophilicity and cationic charge of chitoson [45]. NPs are being developed as delivery vehicles for therapeutic pharmaceuticals such as liposomal NPs (LNPs), encapsulated therapeutic agents for cancer therapy, pegylated form of liposomal encapsulated doxorubicin for breast cancer, layered double hydroxide (LDHs), nanoscale polymer carrier therapy for targeting tumor cells, water soluble polymers drug conjugate to increases half life with potent antitumor effect, 5-flurorouracil loaded iron/ethylcellulose NPs for active targeting of cancer cells can be used in nanomedicine [54-58]. NPs play a vital role in developing new drugs to neural disorders [15]. It is more challenging for delivery of drugs to central nervous system (CNS) and brain but NPs and neuropeptides can over comes these problems and the drug can be delivered in the brain successfully through the carrier such as hexapeptide dalargin, dipeptide kyotorphin across blood brain barriers (BBB) through endocytosis by endothelical cell lining of the brain blood capillaries. Nimje [59] impart that NPs can be used as carrier such as mannose conjugated solid lipid NPs (SLNPs) that can be exploited for effective and targeted delivery of rifabutin. [60]. The nanosized carriers like micro nanosuspension, liposome, dendrimer, ocular inserts, hydrogels are useful in ocular drug delivery which improves the release profile and reduced toxicity.

This method of approach will also increase the efficiency of drug delivery than conventional delivery system. Ladewing [61] reveals that using of layered double hydroxides (LDHs) NPs can be used as carriers for nucleic acids and drug against the general background of bottlenecks that are encountered by cellular delivery system. Nanogels have hydrophilic or amphiphilic polymer chain which can also be used as carriers of drugs and designed spontaneously incorporated biologically active molecules by formation of salt bond, hydrogen bond or hydrophobic interaction [62]. In addition Poly electrolyte nanogels can readily incorporate oppositely changed low molecular mass drug and biomacromolecules such as oligo and polynucleotides (si RNA, DNA) and protein. A general comparison of untargeted and targeted drug delivery by using encapsulated drug system is shown in **Figure 1**.

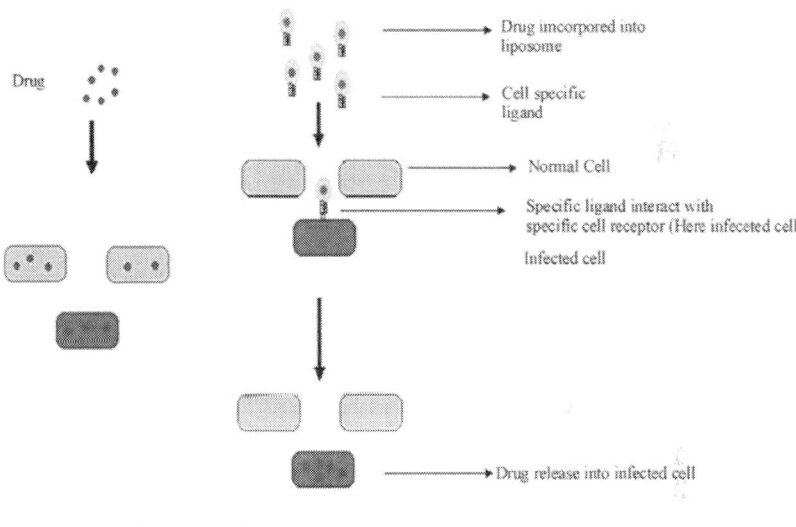

Figure 1. Drug delivery system using nanotechnology.

2.7. Role of Nanoparticles as Medicine

Nanotechnology contributes in management of lung, blood disease and also it counters multiple drug resistance in leukemia by blocking drug efflux from cancer cells and induce efficient delivery of si RNA in to lymphocytes to block apoptosis in sepsis [63]. NPs based thrombocytic agent have potential to improve effectiveness of clot removal and also used in nanodentistry in treatment like dention re-naturalization, permanent hyper sensitivity, complete orthodontic realignments and convalently bonded diamondized enamel [64]. Nanosilver which is a nanoproduct of 100 nm contains 20-15:000 silver atoms that have strong antibacterial activity which used in wounds and burn healing [65]. Nanocrysalin silver have the property of inhibiting antibiotic resistance and anti septic resistance microbes. Dendrimers is a novel polymers with well defined structure, high molecular uniformity and low polydispersity property that makes them more attractive in development of nanomedicine [66]. Dendrimers based delivery system transports drug across cellular barrier efficiently. Mesoporous silica particle (MSP), layered double hydroxide (LDHS) are used for efficient drug delivery [67,68]. NPs based drug deliveries can target intracellular infection like tuberculosis and also polymeric NPs employing polylactide co-glycolide have more potent anti-tubercular activity [22]. These NPs can be used for site specific delivery by avoiding the unwanted toxicity due to non specific distribution and improve the quality of the patients [69]. NPs can act as potent free radical scavenger and they may have anti-inflammatory activity [20]. These NP antioxidants may provide opportunities to counteract the pathogencity of Pseudonomas aeruginosa and its biofilm formation. These advances in nanotechnology research provide new set of research tools, materials, structures and application in nanomedicine in nanotherapy.

2.8. Nanoparticles Based Diagnostic in Cancer Therapy

Nanotechnology has provided an advance biomedical research tool in diagnostic imaging, therapy and targeting of NPs to individual cells and sub cellular compartment [70]. The gold NPs in cellular uptake depend upon their size and surface properties which is transported at 300 - 500 nm diameter within the cytoplasm. In advances, NPs is also used in mediating thermal therapy indicating absorption of infrared light, radiofrequency ablation and magnetically induced heating [71]. The use of radio labeled NPs tagged with radio nuclides and fluorescent NPs such as organic dye doped NPs, Quantum dots and multi function NPs which can be conjugated with several functional molecules that may promote new diagnostic tool in cancer therapy [28,72]. Nanotechnology enabled drug delivery approaches usually to reduce the systemic distribution and associated side effects typically observed with conventional chemotherapeutic molecules [73]. Recent research has developed number of NPs such as metal, semiconductor and polymeric particles to be use as imaging probes, diagnosis and as delivery vehicles in cancer therapy [28]. NPs play an important role in cancer diagnosis. The particles such as organic dye, doped polymer, liposomes, and quantum dots are used in cancer diagnosis [74-77]. Multi functional NPs have the capability to simultaneously carry therapeutic agents, squares imaging contrast agent, diamonds and targeting moieties (circle) that can be used as anti cancer agents [28]. Drug loaded NPs can also be used in treating cancer in animal model. Multi functional miscelle has been developed for in vitro cancer cell targeting, distribution imaging and anti cancer delivery [78]. Doxorubicin (DOX) drugs released from micelles have strong effects on the viability of human liver carcinoma cell line (Hep G2). In addition to DOX drugs loaded NPs have greater anti cancer activity in HER -2 over expressing human breast adenocarcinoma cell line (SK-BR-3) [79,80]. Among inorganic nanomaterials, silica or mesoporous silica materials can be used as potential delivery vehicles and imaging probes for their effective biocompatibility and easy surface fictionalization [78,81-85]. NPs play an important role by delivering drug in a targeted manner to the malignant tumor cells by reducing the systemic toxicity of the anti cancer drugs [86]. Rapamycin loaded polymeric (poly (lactide-cogyycolide) (PLGA) NPs conjugate with antibodies to epidermal growth factor receptor (EGFR) that have efficient and targeted delivery of anticancer drugs.

2.9. Nanoparticles Mediated Targeting Tumors

Nanoparticles can deliver anti cancer agent to tumor site by two strategies such as active and passive that offers significant benefits to cancer patients [85,23]. Passive targeting of tumor site will depend upon the nanoparticle size and tumor vasculature in order to enhance the efficacy of drugs. The solid tumor increase their surrounding vasculature through angiogenesis in order to grow beyond 1 mm - 2 mm diameter but during the development of blood vessels they have several abnormalities like deficiency in pericytes, aberrant basement membrane formation. These abnormalities result in leaky vessels with gap size of 200 nm to 1.2 μm between adjacent endothelial cells that allow extravasations of NPs through these gaps in to extra vascular space [87-89]. These NPs gain access to the tumor that has higher retention times than normal tissues [90]. The leaky vasculature with non effective lymphatic drainage induces the enhanced permeability and retention (EPR) effect lead to accumulation of NPs at tumor site [91-93]. Active targeting of tumor cells by conjugate targeting moieties to NPs lead to accumulation of NPs in tumor sites. The antibodies bind to an antigen on the tumor cell surface and assist NPs drug delivery system to tumor sites [94-97].

Lyp-1 nanoparticles target specific peptide of (PEG -PLGA) NPs to tumor lymph metastasis is a promising carrier to target specific drug delivery to lymphatic metastasis tumor [98]. The nanocarrier allows accumulation of melittin in murine tumor growth without toxicity. In a direct assay the molecularly targeted nanocarriers selectively delivered melittin to multiple tumor targets such as endothelial and cancer cells through a hemifusion mechanism. In animals it may causes regression of precancerous dysplastic lesions and this provides an innovative molecular design for chemotherapy with broad spectrum cytolytic peptides treated for cancer at multiple stages. Silver NPs can act as anti-angiogenic molecules by targeting the activation of P13K/AKT signaling pathway [99]. These silver (Ag) NPs have ability to inhibit angiogenesis, invasiveness, metastasis, vascular endothelial growth factor (VEGF) induced cell proliferation, migration and capillary like tube formation of bovine retinal endothelial cells. In addition silver NPs also have the property of inhibiting the formation of new blood micro vessels induced by VEGF in the mouse matrigel plug assay [99]. In addition NPs mediated targeting of phosphatidylinositol -3-kinase signaling inhibits angiogenesis [100]. NPs enabled targeting the p13K pathway result in inhibition of proliferation and induction of apoptosis of B16F-10 melanoma. Therefore NPs enabled targeting of p13K pathway resulted in inhibition of endothelial cell proliferation and tumor angiogenesis [100]. Singh [101] demonstrated that canine parvovirus NPs (CPVNPs) can be used for targeting tumor cells. The viral particles are nanostructures with nanocontainer for cellular delivery as they have naturally evolved mechanisms for binding to an entering to cells. Canine parvovirus have natural affinity for transferrin receptor (TIRs) canine and human origin and this property could be harnessed as (TIRs) are over expressed by a variety of human tumor cells. CPV based VLPs for TIRs which provides a novel nanomaterial for delivery of a therapy. Labeling nanocrystals with immune cells act as platform technology for nanoimmunotherapy. The combination of plasmid DNA encoding a multimeric soluble form of CD40L (Psp-d-CD40L) can reduced tumor growth which can be established through B16F-10 melanoma tumor [102]. The combination of Toll-like Receptor (TLR) agonists, -C-phosphate-G- (CpG) and poly (i:c) reduces the tumor growth and increase the survival rate. It is also associated with reduction of intra tumoral CD11c+ dendritic cells and an influx of CD8 T cells. The intra tumoral injection of Psp-d-Cb40L containing NPs formed form poly ethylenimine (PEI) in combination with CpG +poly (i:c) may have dramatic anti-tumor effect and it can also treat B16F-10 tumor bearing mice.

3. CONCLUSIONS

The multidisciplinary field of nanotechnology's application for discovering new molecules and manipulating those available naturally could be excited in its potential to improve health care. Nanotechnology is definitely a medical boon for diagnosis, treatment and prevention of various diseases including cancer. It supports and expands the scientific advances in genomic and proteomics and builds on our understanding of the molecular underpinnings of cancer and its treatment. We then review the current state of the art of nanoparticle-based therapeutics that have reached the clinic for its efficient advantage as drug carrier which are high stability, high carrier capacity, feasibility of incorporation of both hydrophilic and hydrophobic substances of variable routes of administration including oral application and inhalation. Multi functional NPs also have the capability to concurrently carry therapeutic agents, squares imaging contrast agent, diamonds and targeting moieties which can be used as anti-cancer agents. Interestingly pharmaceutical sciences are using NPs to reduce toxicity and side effects of drugs. The kind of hazards that are introduced by using NPs for drug delivery are beyond that posed by conventional hazards imposed by chemicals in classical delivery matrices. Predicting the future of nanotechnology in drug delivery system is not simple due to its fast developing technology and changing rapidly. Additional research is required in multifunctional NPs based drug delivery systems to overcome the problems for effective therapy without side effects which can improve the quality of life in cancer patients.

ACKNOWLEDGEMENTS

The valuable help and support of Dr. Patrick Gomez, Director, School of Biotechnology and Health Sciences, Karunya University is gratefully acknowledged.

REFERENCES

1. S. Majuru and O. Oyewumi, "Nanotechnology in Drug Development and Life Cycle Management," Nanotechnology in Drug Delivery, Vol. 10, No. 4, 2009, pp. 597-619.

2. K. K. Jain, "The Role of Nanobiotechnology in Drug Discovery," Drug Discover Today, Vol. 10, No. 21, 2005, pp. 1435-1442.

3. C. Buzea,I. I. Pacheco and K. Robbie, "Nanomaterials and Nanoparticles: Sources and Toxicity," Biointerphases, Vol. 2, No. 4, 2007, pp. 17-71.

4. B. D. Fahlman, "Materials Chemistry," Springer, Berlin, 2007, pp. 282-283.

5. W. Yang, J. I. Peters, R. O. and Williams, Eds., "Inhaled Nanoparticles—A Current Review," International Journal of Pharmceutical, Vol. 22, No. 356, 2008, pp. 239-247.

6. B. Mishra, B. B Patel and S. Tiwari, "Colloidal Nanocarriers: A Review on Formulation Technology, Types and Applications toward Targeted Drug Delivery," Nanomedicine, Vol. 6, No. 1, 2010, pp. 9-24.

7. Z. Ghalanbor, S. A. Marashi and B. Ranjbar, "Nanotechnology Helps Medicine: Nanoscale Swimmers and Their Future Applications," Medical Hypotheses, Vol. 65, No. 1, 2005, pp. 198-199.

8. T. Kubik, K. Bogunia-Kubik and M. Sugisaka, "Nanotechnology on Duty in Medical Applications," Current Pharmaceutical Biotechnology, Vol. 6, No. 1, 2005, pp. 17-33.

9. S. N. Kundra, "Toward the Emergence of Nanoneurosurgery: Part III-Nanomedicine: Targeted Nanotherapy, Nanosurgery and Progress toward the Realization of Nanoneurosurgery," Neurosurgery, Vol. 62, No. 6, 2008, p. 1384.

10. A. Cavalcanti, B. Shirinzadeh and R. A. Freitas, "Medical Nanorobot Architecture Based on Nanobioelectronics," Recent Patents on Nanotechnology, Vol. 1 No. 1, 2007, pp. 1-10.

11. M. Boukallcl, M. Gauthier, M. Dauge, E. Piat and J. Abadie, "Smart Microrobots for Mechanical Cell Characterization and Cell Convoying," IEE Transaction on Biomedical Engneering, Vol. 54, No. 8, 2007, pp. 1536-1540.

12. D. A. La Van, T. McGuire and R. Langer, "Small-Scale Systems for in vivo Drug Delivery," Nature Biotechnology, Vol. 21, No. 10, 2003, pp. 1184-1191.

13. A. Cavalcanti and B. Shirinzadeh, "Nanorobot Architecture for Medical Target Identification," Nanotechnology, Vol. 19, 2008, p. 015103.

14. R. R. Zhu, L. L. Qin, M. Wang, S. L. Wang, R.Zhang, Z. X. Liu, X. Y. Sun and S. D. Yao, "Preparation, Characterization, and Anti-tumor Property of PodophyllotoxinLoaded Solid Lipid Nanoparticles," Nanotechnology, Vol. 20, No. 5, 2009, p. 55702.

15. K. K. Jain, "Role of Nanotechnology in Development of New Therapies for Diseases of the Nervous System," Nanomedicine, Vol. 1, No. 1, 2006, pp. 9-12.

16. J. C. Pickup, Z. L. Zhi, F. Khan, T. Saxl and D. J. Birch, "Nanomedicine and Its Potential in Diabetes Research and Practice," Diabetes/Metabolism Research and Reviews, Vol. 24, No. 8, 2008, pp. 604-610.

17. L. Jianguo, Z. Li, Z. Yi, W. Huanan, L. Jidong,Z. Qin and L. Yubao, "Development of Nanohydroxyapatite/ Polycarbonate Composite for Bone Repair," Journal of Biomaterials Applications, Vol. 24, No. 1, 2009, pp. 31-45.

18. A. Nowacek, L. M. Kosloski and H. E. Gendelman, "Neurodegenerative Disorders and Nanoformulated Drug Development," Nanomedicine, Vol. 4, No. 5, 2009, pp. 541-555.

19. D. N. Patel and S. R. Bailey, "Nanotechnology in Cardiovascular Medicine," Catheterization and Cardiovascular Interventions, Vol. 69, No. 5, 2007, pp. 643-654.

20. S. F.Elswaifi, J. R. Palmieri, K. S. Hockey and B. A. Rzigalinski, "Antioxidant Nanoparticles for Control of Infectious Diseases," Infect Disorder Drug Target, Vol. 9, No. 4, 2009, pp. 445-452.

21. M. Smola, T. Vandamme and A. Sokolowski, "Nanocarriers as Pulmonary Drug Delivery Systems to Treat and to Diagnose Respiratory and Non Respiratory Diseases," International Journal of Nanomedicine, Vol. 3, No. 1, 2008, pp. 1-19.

22. R. Pandey and G. K. Khuller, "Nanotechnology Based Drug Delivery System(s) for the Management of Tuberculosis," Indian Journal of Experimental Biology, Vol. 44, No. 5, 2006, pp. 357-366.

23. Y. Y. Liu, H. Miyoshi and M. Nakamura, "Nanomedicine for Drug Delivery and Imaging: A Promising Avenue for Cancer Therapy and Diagnosis Using Targeted Functional Nanoparticles," International Journal of Cancer, Vol. 120, No. 12, 2007, pp. 2527-2537.

24. X. Wang, L. L. Yang, Z. Chen and D. M. Shin, "Application of Nanotechnology in Cancer Therapy and Imaging," A Cancer Journal for Clinicians, Vol. 58, 2008, pp. 97-110.

25. K. Riehemann, S. W. Schneider, T. A. Luger, B. Godin, M. Ferrari and H. Fuchs, "Nanomedicine-Challenge and Perspectives," Angewandte Chemie International Edition, Vol. 48. No. 5, 2009, pp. 872-897.

26. R. Sinha, G. J. Kim, S. Nie and D. M. Shin, "Nanotechnology in Cancer Therapeutics: Bioconjugated Nanoparticles for Drug Delivery," Molecular Cancer Therapeutics, Vol. 5, No. 8, 2006, pp. 1909-1917.

27. K. J. Cho, X. Wang, S. M. Nie and D. H. Shin, "Therapeutic Nanoparticles for Drug Delivery in Cancer," Clinical Cancer Research, Vol. 14, 2008, pp. 1310-1316.

28. S. Jiang, M. K. Ganesammandhan and Y. Zhang, "Optical Imaging Guided Cancer Therapy with Fluorescent Nanoparticles," Journal of the Royal Society Interface, Vol. 7, No. 42, 2010, pp. 3-18.

29. D. Castanotto and J. J. Rossi, "The Promises and Pitfalls of RNA-Interference-Based Therapeutics," Nature, Vol. 457, No. 7228, 2009, pp. 426-433.

30. S. Sandhiya, S. A. Dkhar and A. Surendiran, "Emerging Trends of Nanomedicine—An Overview," Fundamental & Clinical Pharmacology, Vol. 23, No. 3, 2009, pp. 263-269.

31. R. D. Hofheinz, S. U. Gnad-Vogt, U. Beyer and A. Hochhaus, "Liposomal Encapsulated Anti-cancer Drug," Anticancer Drugs, Vol. 16, No. 7, 2005, pp. 691-707.

32. S. M. Moghimi and J. Szebeni, "Stealth Liposomes and Long Circulating Nanoparticles: Critical Issues in Pharmacokinetics, Opsonization and Protein-Binding Properties," Progress in Lipid Research, Vol. 42, No. 6, 2003, pp. 463-478.

33. D. K. Sarker, "Engineering of Nanoemulsions for Drug Delivery," Current Drug Delivery, Vol. 2, 2005, pp. 297-310.

34. S. A. Agnihotri, N. N. Mallikarjuna and T. M. Aminabhavi, "Recent Advances on Chitosan-Based Microand Nanoparticles in Drug Delivery," Journal of Controlled Release, Vol. 100, No. 1, 2004, pp. 5-28.

35. L. J. Lee, "Polymer Nano-Engineering for Biomedical Applications," Annals of Biomedical Engineering, Vol. 34, No. 1, 2006, pp. 75-88.

36. A. K. Cherian, A. C. Rana and S. K. Jain, "Self-Assembled Carbohydratestabilized Ceramic Nano-particles for the Parenteral Delivery of Insulin," Drug Development and Industrial Pharmacy, Vol. 26, No. 4, 2000, pp. 459-463.

37. L. R. Hirsch, A. M. Gobin, A. R. Lowery, F. Tam, R. A. Drezek, N. J. Halas and J. L. West, "Metal Nanoshells," Annals of Biomedical Engineering, Vol. 34, No. 1, 2006, pp. 15-22.

38. S. Bosi, T. Da Ros, G. Spalluto and M. Prato, "Fullerenc Derivatives: An Attractive Tool for Biological Applications," European Journal of Medicinal Chemistry, Vol. 38, No. 11-12, 2003, pp. 913-923.

39. G. Pagona and N. Tagmatarchis, "Carbon Nanotubes: Materials for Medicinal Chemistry and Biotechnological Applications," European Journal of Medicinal Chemistry, Vol. 13, 2006, pp. 1789-1798.

40. J. Weng and J. Ren, "Luminescent Quantum Dots: A Very Attractive and Promising Tool in Biomedicine," Current Medicinal Chemistry, Vol. 13, 2006, pp. 97-909.

41. M. E. Davis, Z. G. Chen and D. M. Shin, "Nanoparticle Therapeutics: An Emerging Treatmentmodality for Cancer," Nature Reviews Drug Discovery, Vol. 7, 2008, pp. 771-782.

42. H. Pelicano, D. S. Martin, R. H, Xu and P. Huang, "Glycolysis Inhibition for Anticancer Treatment," Oncogene, Vol. 25, 2006, pp. 4633-4646.

43. S. S. Nath, D. Chakdar, G. Gope and D. K. Avasthi, "Characterizations of CdS and ZnS Quantum Dots Prepared by Chemical Method on SBR Latex," AZojono Journal of Nanotechnology Online, 2008.

44. V. R. Reddy, "Gold Nanoparticles: Synthesis and Applications," Thieme eJournals, Vol. 2006, No. 11, 2006, pp. 1791-1792.

45.   R. Das, S. S. Nath, D. Chakdar, G. Gope and R. Bhattacharjee, "Preparation of Silver Nanoparticles and Their Characterization," AZojono Journal of Nanotechnology Online, Vol. 5, No. 10, 2009, p. 2240.

46.   X. Wang, J. Zhuang, Q. Peng and Y. Li, "General Strategy for Nanocrystal Synthesis," Nature, Vol. 437, No. 7055, 2005, pp. 121-124.

47.   H. Zhu, C. Zhang and Y. Yin, "Novel Synthesis of Copper Nanoparticles: Influence of the Synthesis Conditions on the Particle Size," Nanotechnology, Vol. 16, No. 12, 2005, p. 3079.

48.   Y. Wei, H. Xie, L. Chen, Y. Li and C. Zhang, "Controlled Synthesis of Narrow-Dispersed Copper Nanoparticles," Journal of Dispersion Science and Technology, Vol. 31, No. 3, 2010, pp. 364-367.

49.   L. S. Jackson and K. Lee, "Microencapsulation and the Food Industry," Lebensmittel—Wissenschaft Technologie, 1991.

50.   S. Gelperina, K. Kisich, M. D. Iseman and L. Heifets, "The Potential Advantages of Nanoparticle Drug Delivery Systems in Chemotherapy of Tuberculosis," American Journal of Respiratory and Critical Care Medicine, Vol. 172, No. 12, 2005, pp. 1487-1490.

51.   Y. Bae, N. Nishiyama, S. Fukushima, H. Koyama, M. Yasuhiro and K. Kataoka, "Preparation and Biological Characterization of Polymeric Micelle Drug Carriers with Intracellular pH-Triggered Drug Release Property: Tumor Permeability, Controlled Subcellular Drug Distribution, and Enhanced in vivo Antitumor Efficacy," Bioconjugate Chemistry, Vol. 16, No. 1, 2005, pp. 122-130.

52.   E. Physica, "Quantum Dots in Biology and Medicine," Low-Dimensional Systems and Nanostructures, Vol. 25, No. 1, 2004, pp. 1-12.

53.   Q. Yuan, S. Hein and R. D. Misra, "New Generation of Chitosan-Encapsulated ZnO Quantum Dots Loaded with Drug: Synthesis, Characterization and in vitro Drug Delivery Response," Acta Biomaterialia, Vol. 6, No. 7, 2010, pp. 2732-2739.

54.   D. B. Fenske, A. Chonn and P. R. Cullis, "Liposomal Nanomedicines: An Emerging Field," Toxicology Pathology, Vol. 36, No. 1, 2008, pp. 21-29.

55.   S. Praveen and S. K. Sahoo, "Polylymeric Nanoparticles for Cancer Therapy," Journal of Drug Targeting, Vol. 16, No. 2, 2008, pp. 108-123.

56.   Y. Luo and G. D. Prestwich, "Cancer Targeted Polymeri Drug," Current Cancer Drug Targets, Vol. 2, No. 3, 2002, pp. 209-226.

57.   J. L Arias, M. López-Viota, A. V. Delgado and M. A. Ruiz, "5-Fluorouracil-Loaded Iron/Ethylcellulose (Core/ Shell) Nanoparticles for Active Targeting of Cancer," Journal of Drug Targeting, Vol. 17, No. 10, 2009, p. 813.

58.   T. Tanaka, P. Decuzzi , M.Cristofanilli, J. H. Sakamoto, E.Tasciotti, F. M. Robertson and M. Ferrari, "Nanotechnology for Breast Cancer Therapy," Biomedical Microdevices, Vol. 11, No. 1, 2009, pp. 49-63.

59. N. Nimje, A. Agarwal, G. K. Saraogi, N. Lariya, G. Rai, H. Agrawal and G. P. Agrawal, "Nanoparticulate Carriers of Rifabutin for Alveolar Targeting, Journal of Drug Targeting, Vol. 17, No. 10, 2009. pp. 777-787.

60. S. Wadhwa, R. Paliwal, S. R. Paliwal and S. P. Vyas, "Nanocarriers in Ocular Drug Delivery," Current Pharmceutical Design, Vol. 15, No. 23, 2009, pp. 2724-2750.

61. K. Ladewing, Z. P. Xu and G. C. Lu, "Layered Double Hydroxide Nanoparticle in Gene and Drug Delivery," Expert Opinion on Drug Delivery, Vol. 6, No. 9, 2009, pp. 907-922.

62. A. V. Kabanov and S. K. Vinogradov, "Nanogels as Pharmaceutical Carriers: Finite Network of Infinite Capabilities," Angewandte Chemie International Edition, Vol. 48, No. 30, 2009, pp. 5418-5429.

63. D. B.Buxton, "Nanomedicine for the Management of Lung and Blood Diseases," Nanomedicine, Vol. 4, No. 3. 2009, pp. 331-339.

64. R. A. Freitas, "Nanodentistry," Journal of American Dental Association, Vol. 131, No. 11, 2000, pp. 1559-1565.

65. J. B. Lyczak and P. J. Schechter, "Nanocrystalline Silver Inhibits Antibiotic, Antiseptic-Resistant Bacteria," Clinical Pharmacology Therapeutics, Vol. 77, 2005, p. 60.

66. A. Samad, M. I. Alam and K. Saxena, "Dendrimers: A Class Of Polymers in the Nanotechnology for the Delivery of Active Pharmaceuticals," Current Pharmceutical Design, Vol. 15, No. 25, 2009, pp. 2958-2969.

67. L. Pasqua, S. Cundari, C. Ceresa and G. Cavaletti, "Recent Development, Applications, and Perspectives of Mesoporous Silica Particles in Medicine and Biotechnology," Current Medical Chemistry, Vol. 16, No. 23, 2009, pp. 3054-3063.

68. K. Ladewig, Z. P. Xu and G. C. Lu, "Layered Double Hydroxide Nanoparticles in Gene and Drug Delivery," Expert Opinion on Drug Delivery, Vol. 6, No. 9. 2009, pp. 907-922.

69. H. Devalapally, A. Chakilam and M. M. Amiji, "Role of Nanothechnology in Pharmaceutical Product Development," Journal of Pharmaceutical Sciences, Vol. 96, No. 10, 2007, pp. 2547-2567.

70. B. D. Chithrani, J. Stewart, C. Allen and D. A. Jaffray, "Intracellullar Uptake, Transport, and Processing of Nanostructures in Cancer Cells," Nanomedicine, Vol. 5, No. 2, 2009, pp. 118-127.

71. E. S. Day, J. G. Morton and J. L. West, "Nanoparticles for Thermal Cancer Therapy," Journal of Biomechanical Engneering, Vol. 131, No. 7, 2009, p. 74001.

72. G. Ting, C. H. Chang and H. E. Wang, "Cancer Nanotargeted Radiopharmaceutical for Tumor Imaging and Therapy," Anticancer Research, Vol. 29, No. 10, 2009, pp. 4107-4108.

73. B. Thierry, "Drug Nanocarriers and Functional Nanoparticles Application in Cancer Therapy," Current Drug Delivery, Vol. 6, No. 4, 2009, pp. 391-403.

74.   S. Santra, D. Dutta, G. A. Walter and B. M. Moudgil, "Fluorescent Nanoparticle Probes for Cancer Imaging," Technology in Cancer Research and Treatment, Vol. 4, No. 6, 2005, pp. 593-602.

75.   K. Licha and C. Olbrich, "Optical Imaging in Drug Discovery and Diagnostic Applications," Advanced Drug Delivery Reviews, Vol. 57, No. 8, 2005, pp. 1087-1108.

76.   P. Grodzinski, M. Silver and L. K. Molnar, "Nano-Technology for Cancer Diagnostics: Promises and Challenges," Expert Review of Molecular Diagnostics, Vol. 6. No. 3, 2006, pp. 307-318.

77.   J. Rao, A. Dragulescu-Andrasi and H. Yao, "Fluorescence Imaging in vivo: Recent Advances," Current Opinion in Biotechnology, Vol. 18, No. 1, 2007, pp. 17-25.

78.   D. M. Huang, Y. Hung, B. S. Ko, S. C. Chien, C. P. Tsai, C. T. Kuo, J. C. Kang, C. S. Yang, C. Y. Mou and Y. C. Chen, "Highly Efficient Cellular Labeling of Mesoporous Nanoparticles in Human Mesenchymal Stem Cells: Implication for Stem Cell Tracking," Journal of the Federation of American Societies for Experimental Biology, Vol. 19, No. 14, 2005, pp. 2014-2016.

79.   J. Kim, J. E. Lee, S. H. Lee, J. H. Yu, J. H. Lee, T. G. Park and T. Hyecon. "Designed Fabrication of a Multifunctional Polymer Nanomedical Platform for Simultaneous Cancer-Targeted Imaging and Magnetically Guided Drug Delivery," Advanced Materials, Vol. 20, 2008, pp. 478-483.

80.   K. Shah, A. Jacobs, X. O. Breakefield and R. Weissleder, "Molecular Imaging of Gene Therapy for Cancer," Gene Therapy, Vol. 11, No. 15, pp. 1175-1187.

81.   Y. S. Lin, S. H. Wu, Y. Hung, Y. H. Chou, C. Chang, M. L. Lin, C. P. Tsai and C. Y. Mou, "Multifunctional Composite Nanoparticles: Magnetic, Luminescent, and Mesoporous," Chemistry of Materials, Vol. 18, No. 22, 2006, pp. 5170-5172.

82.   S. T. Selvan, P. K. Patra, C. Y. Ang and J. Y. Ying, "Synthesis of Silica-Coated Semiconductor and Magnetic Quantum Dots and Their Use in the Imaging of Live Cells," Angewandte Chemie International Edition, Vol. 46, No. 14, 2007, pp. 2448-2452.

83.   S. H. Wu, Y. S. Lin, Y. Hung, Y. H. Chou, Y. H. Hsu, C. Chang and C. Y. Mou, "Multifunctional Mesoporous Silica Nanoparticles for Intracellular Labeling and Animal Magnetic Resonance Imaging Studies," Chembiochem, Vol. 9, No. 1, 2008, pp. 53-57.

84.   C. H. Lee, S. H. Cheng, Y. J. Wang, Y. C. Chen, N. T. Chen, J. Souris, C. T. Chen, C. Y. Mou, C. H. Yang and L. W. Lo, "Near-Infrared Mesoporous Silica Nanoparticles for Optical Imaging: Characterization and in vivo Biodistribution," Advanced Functional Materials, Vol. 19, No. 2, 2009, pp. 215-222.

85.   J. D. Byrne, T. Betancourt and L. Brannon-Peppas, "Active Targeting Schemes for Nanoparticle Systems in cancer Therapeutics," Advanced Drug Delivery Reviews, Vol. 60, No. 15, 2008, pp. 1615-1626.

86. S. Acharya, F. Dilnawaz and S. K. Sahoo, "Targeted Epidermal Growth Factor Receptor Nanoparticle Bioconjugates for Breast Cancer Therapy," Biomaterials, Vol. 30, No. 29, 2009, pp. 5737-5750.

87. J. Folkman, "What Is the Evidence that Tumors are Angiogenesis Dependent?" Journal of naternational Cancer Insititution, Vol. 82, No. 1, 1990, pp. 4-6.

88. J. Folkman and Y. Shing, "Angiogenesis," Journal of Biology Chemistry, Vol. 267, No. 10, 1992, pp. 931-934.

89. D. F. Baban and L. W. Seymour, "Control of Tumour Vascular Permeability," Advanced Drug Delivery Reviews, Vol. 34, No. 1, 1998, pp. 109-119.

90. J. W. Baish, Y. Gazit, D. A. Berk, M. Nozue, L. T. Baxter and R. K. Jain, "Role of Tumor Vascular Architecture in Nutrient and Drug Delivery: An Invasion Percolation-Based Network Model," Microvascular Research, Vol. 51, No. 3, 1996, pp. 327-346.

91. R. K. Jain, "Delivery of Molecular Medicine to Solid Tumors: Lessons from in vivo Imaging of Gene Expression and Function," Journal of Controlled Release, Vol. 74, No. 1-3, 2001, pp. 7-25.

92. R. Duncan, "The Dawning Era of Polymer Therapeutics," Nature Reviews Drug Discovery, Vol. 2, No. 5, 2003, pp. 347-360.

93. L. Brannon-Peppas and J. O. Blanchette, "Nanoparticle and Targeted Systems for Cancer Therapy," Advanced Drug Delivery Reviews, Vol. 56, No. 11, 2004, pp. 1649-1659.

94. K. Ulbrich, T. Etrych, P. Chytil, M. Jelínková and B. Ríhová, "Antibody-Targeted Polymer-Doxorubicin Conjugates with pH-Controlled Activation," Journal of Drug Targeting, Vol. 12, No. 8, 2004, pp. 477-489.

95. Z. Xu, W. Gu, J. Huang, H. Sui, Z. Zhou, Y. Yang, Z, Yan and Y. I. Li, "In vitro and in vivo Evaluation of Actively Targetable Nanoparticles for Paclitaxel Delivery," International Journal of Pharmaceutics, Vol. 288, No. 2, 2005, pp. 361-368.

96. J. Cheng, B. A. Teply, I. Sherifi, J. Sung, G. Luther, F. X. Gu, E. Levy-Nissenbaum, A. F. Radovic-Moreno, R. Langer and O. C. Farokhzad, "Formulation of Functionalized PLGA-PEG Nanoparticles for in vivo Targeted Drug Delivery," Biomaterials, Vol. 28, No. 5, 2007, pp. 869-876.

97. S. Díez, G. Navarro and C. T. de ILarduya, "In vivo Targeted Gene Delivery by Cationic Nanoparticles for Treatment of Hepatocellular Carcinoma," The Journal of Gene Medicine, Vol. 11, No. 1, 2009, pp. 38-45.

98. G.Luo, X. Yu, C. Jin, F. Yang, D. Fu, J.Long, J.Xu, C.Zhan and W. Lu, "Lyp-1 Conjugated Nanoparticles for Targeting Drug Delivery to Lymphatic Metastatic Tumours," International Journal of Pharmaceutics, Vol. 385, No.1-2, 2010, pp. 150-156.

99. S. Gurunathan, K. J. Lee, K. Kalishwaralal, S. Sheikpranbabu, R. Vaidyanathan and S. H. Eom, "Antiangiogenic Properties of Silver Nanoparticles," Biomaterials, Vol. 30, No. 31, 2009, pp. 6341-6350.

100.   R. Harfouche, S. Basu, S. Soni, D. M. Hentschel, R. A. Mashelkar and S. Sengupta, "Nanoparticles Mediated Targeting of Phosphatidylinositol-3-Kinase Signaling Inhibits Angiogenesis,"Angiogenesis, Vol. 12, No. 4, 2009, pp. 325-338.

101.   P. Singh, "Tumour Targeting Using Canine Parvovirus Nanoparticles," Current Topics in Microbiology and Immunology, Vol. 327, 2009, pp. 123-141.

102.   G.W. Stone, S. Barzee, V. Snarsky, C. Santucci, B. Tran, R. Langer, G. T. Zugates, D. G. Anderson and R. S. Kornbluth, "Nanoparticle Delivered Multimeric Soluble CD40 LDNA Combined with Like Receptor Agonists as a Treatment for Melanoma," PLOS One, Vol. 4, No. 10, 2009, p. 7334.

CHAPTER 2

Liposomes as Potential Drug Carrier Systems for Drug Delivery

Melis Çağdaş[1], Ali Demir Sezer[2] and Seyda Bucak[3]

[1]Department of Chemistry, Boğaziçi University, Bebek, Istanbul, Turkey
[2]Department of Pharmaceutical Biotechnology, Faculty of Pharmacy, Marmara University, Haydarpaşa, Istanbul, Turkey
[3]Department of Chemical Engineering, Yeditepe University, Kayışdağı, Istanbul, Turkey

1. INTRODUCTION

Lipids are amphiphilic molecules, where one part of the molecule is water-loving (hydrophilic) and the other water-hating (hydrophobic). When lipids are placed in contact with water, the unfavorable interactions of the hydrophobic segments of the molecule with the solvent result in the self assembly of lipids, often in the form of liposomes. Liposomes consist of an aqueous core surrounded by a lipid bilayer, much like a membrane, separating the inner aqueous core from the bulk outside. They were first discovered by Bangham and his co-workers in 1961 [1] and described as swollen phospholipid systems [2]. In the following years, a variety of enclosed phospholipid bilayer structures were defined which were initially called bangosomes and then liposomes, which was derived by the combination of two Greek words, "lipos" meaning fat and "soma" meaning body.

Liposomes have been used to improve the therapeutic index of new or established drugs by modifying drug absorption, reducing metabolism, prolonging biological half-life or reducing toxicity. Drug distribution is then controlled primarily by properties of the carrier and no longer by physico-chemical characteristics of the drug substance only.

Lipids forming liposomes may be natural or synthetic, and liposome constituents are not exclusive of lipids, new generation liposomes can also be formed from polymers (sometimes referred to as polymersomes). Whether composed of natural or synthetic lipids or polymers, liposomes are biocompatible and biodegradable which make them suitable for biomedical research. The unique feature of liposomes is their ability to compartmentalize and solubilize both hydrophilic and hydrophobic materials by nature. This unique feature, coupled with biocompatibility and biodegradability make liposomes very attractive as drug delivery vehicles.

Hydrophobic drugs place themselves inside the bilayer of the liposome and hydrophilic drugs are entrapped within the aqueous core or at the bilayer interface. Liposomal formulations enhance the therapeutic efficiency of drugs in preclinical models and in humans compared to conventional formulations due to the alteration of biodistribution. Liposome binding drugs, into or onto their membranes, are expected to be transported without rapid degradation and minimum side effects to the recipient because generally liposomes are composed of biodegradable, biologically inert and non-immunogenic lipids. Moreover, they produce no pyrogenic or antigenic reactions and possess limited toxicity [3-5]. Consequently, all these properties as well as the ease of surface modification to bear the targetable properties make liposomes more attractive candidates for use as drug-delivery vehicles than other drug carrying systems such as nanoparticles [6, 7] and microemulsions [8, 9]. In the 1970s [1,10-13], liposomes were introduced as drug delivery vehicles but the initial clinical results were not satisfactory due to their colloidal and biological instability and their inefficient encapsulation of drug molecules.

Subsequent research on their stability and drug interactions resulted in several commercial liposome products in the market in the 1980s and early 1990s [14]. A schematic representation of liposomal drug delivery is given in Figure 1.

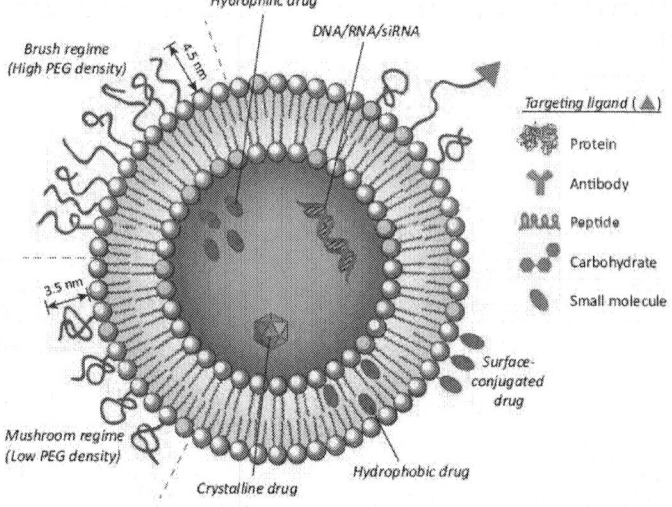

Figure 1. Structural and design considerations for liposomal drug delivery. Liposomes can be surface functionalized to endow stealth through PEGylation and to promote receptor-mediated endocytosis by using targeting ligands such as antibodies, peptides, proteins, carbohydrates, and various other small molecules. PEGylation extends liposomal circulation half-life in vivo by reducing clearance, immune recognition, and the non-specific absorption of serum proteins. Polyethylene glycol (PEG) density determines its structure at the liposome surface, with densities below 9% adopting a mushroom-like globular structure and those above 9% adopting a more rigid, extended, brush-like morphology. Chemotherapeutics or diagnostics can be encapsulated into the aqueous lumen, incorporated into the lipid bilayer, or conjugated to the liposome surface. Abbreviation: siRNA, small interfering RNA [15].

Liposomes represent versatile and advanced nanodelivery systems for a wide range of biologically active compounds [16]. The final amount of the encapsulated drug is affected by a selection of an appropriate preparation method providing a preparation of liposomes of various size, lamellarity and physicochemical properties [17]. The entrapment of the drugs, both hydrophilic and hydrophobic, into the liposomes is used to bypass the frequent generic toxicity associated with the drug as often seen in cancer drugs [18]. Thus, it represents a very effective route that enhances the drug therapeutic effect. The modification of liposomes permits a passive or active targeting of the tumor site. This effect enables an efficient drug payload into the malignant cell of tumors, while the non-malignant cells become minimally impacted.

In some of the first demonstrations of the improved in vivo activity of liposome-encapsulated drugs in animal models, the anti-cancer drug cytosine arabinoside used which showed a significant increase in the survival times of the mice bearing leukemia [19, 20] which became a popular model for testing the effects of a wide range of liposome characteristics on therapeutic outcomes. Following experiments include liposomal amphotericin B [21] and liposomal doxorubicin [22] that finally led to the first clinical trials of liposomal drugs. Nowadays; the liposomal products (as a suspension, as an aerosol or in a semi-solid form such as a gel, cream, or powder) in the market still include mostly anticancer preparations as well as antifungal and antibacterial preparations and cosmetics. In addition, liposomes are recently used as therapeutic agents to treat a disease because increased gene transfer efficiencies have been obtained via liposomal gene vectors in gene therapy.

The benefits and limitations of liposome drug carriers critically depend and based on physicochemical and colloidal characteristics such as size, composition, loading efficiency and stability, as well as their biological interaction with the cell membranes. There are four major interactions between liposomes and cells [23]. The predominant interaction among them is either simple adsorption or subsequent endocytosis. Adsorption occurs when the attractive forces exceed the repulsive ones and obviously this type of interaction depends on the surface properties of liposomes. In the delivery through endocytosis, liposome and its contents indirectly place themselves in the cytoplasm. Fusion with cell membranes, delivery of the liposomal content directly into the cell through the merge of liposome lipids into the membrane, is much rarer. The last possible interaction is the lipid exchange which is a long-range interaction that involves the exchange of bilayer constituents, such as lipids, cholesterol, and membrane bound molecules with components of cell membranes. Upon entering into the body, the delivered liposomes via one of these interaction types trigger the response of the immune system and the encapsulated material may become inactive. Therefore; substantial researches have been carried out in the development of the biocompatible and nonrecognizable liposomal surfaces.

Various types of liposomes can be prepared by different preparation methods, depending on the required application. In this chapter, these methods are summarized to give a general understanding of the relationship between structure and functionality of liposomes. As one of the advantages of liposomal formulations is the encapsulation ability of both hydrophobic and hydrophilic drugs, incorporation methods are shortly visited. Physical properties of liposomes such as stability, storage and sterilization are discussed along with the characterization techniques for size, charge, etc. Clinical applications of liposomes are a vast area of research where cancer therapy is the area of highest impact. Different clinical applications of liposomes and most recent advances in cancer therapy are summarized. New generation involving constituents other than conventional ones such as phospholipids prove to be a growing field in nanotechnology. A brief list of different types of new generation liposomes are given with short descriptions at the end of this chapter.

2. LIPOSOME PREPARATION METHODS

The manufactured liposome features are directly related to the preparation method. Although liposome formation may be spontaneous, often some mechanical agitation is required. In order to have control over the size and structure of the liposomes that are formed, increase the efficiency of entrapment of the desired molecules, and prevent subsequent leakage from the liposomes, different preparation methods have been devised.

There are a few parameters that should be considered during the method selection: 1) the physicochemical characteristics of the material to be entrapped and those of the liposomal ingredients, 2) the nature of the medium in which the liposomes are dispersed, 3) the effective concentration of the encapsulated material and its potential toxicity, 4) additional processes involved during application (delivery of the liposomes), 5) optimum size, polydispersity and shelf-life of the liposomes for the intended application and 6) batch-to-batch reproducibility and possibility of large-scale production of safe and efficient liposomal products [24-26].

Liposome size is a crucial parameter in determining the circulation half-life of liposomes in drug delivery. The amount of encapsulated drug is also related with the size and the number of bilayers of the prepared liposome. According to the desired formulation, different liposome preparation methods can be employed. The main difference in these methods is their approach to overcome the low solubility of lipids in water. Accordingly, these methods can be classified as mechanical agitation, solvent evaporation, solvent injection, and detergent solubilization. In all the above mentioned methods, drug loading is passive.

2.1. MECHANICAL AGITATION

In this method, lipids are directly solubilized in water upon application of high mechanical agitation, through the use of probe sonication. It is one of the simplest methods of liposome preparation, however, yields small liposomes that are highly unstable in terms of their size and suffers from the drawback that it is impossible to remove completely the risk of lipid degradation by contact with the hot probe, and contamination with titanium from the probe. Its advantage is the exclusion of use of organic solvents as described in the following methods. However, for drug delivery applications, liposomes prepared with mechanical agitation are not suitable due to their size instability and high leakage of encapsulated drugs [27].

2.2. Solvent Evaporation

In general this method consists of four major steps; first is the solubilization of the lipid (and a hydrophobic compound) in an organic solvent; second is solvent evaporation; third is hydration with a buffer (and the hydrophilic compound) and if need the fourth often involves obtaining unilamellar liposomes from the obtained multilamellar ones.

The aqueous volume enclosed within these lipid membranes is very small proportion of total volume used for preparation (5-10%). Consequently, large amount of water soluble drug is wasted during the preparation. On the other hand, lipid soluble drug can be encapsulated with 100% efficiency, providing that they are not present in quantities which overwhelm the structural components of the membrane [28]. The volume of entrapment can be significantly increased by the usage of negatively charged lipids in the membrane which tend to push the bilayers apart from each other. The same effect can also be achieved in the presence of neutral lipids by freezing and thawing repeatedly the obtained liposomes. 30% volume of entrapment can be achieved, which can further be increased at higher lipid concentrations [29]. The freeze-thaw protocol results in a dramatic change in liposome morphology followed by freeze-fracture electron micrographs. Before freeze-thawing, the samples exhibit the tightly packed "onions skin" arrangements of concentric bilayers normally associated with liposomal systems. After a few freeze-thaw steps, however, new structures are observed where the interlamellar spaces are much increased, and where closed lamellar systems can be intercalated between bilayers [30].

2.2.1. Solubilization of the Lipid
The starting point of this liposome preparation method is to prepare an organic solution of membrane lipids in order to ensure complete and homogenous mixing of all the components as they are required in the final membrane preparation. Compounds to be incorporated which are lipid soluble will be added to the organic solution, while compounds to be entrapped in the aqueous compartment of liposomes will be dissolved in the aqueous environment. In this method, phospholipids are first dissolved in an organic solvent along with lipid soluble compounds (if any) to be incorporated in the liposome to ensure complete and homogenous mixing.

2.2.2. Solvent Evaporation
The next step is the evaporation of the organic solvent. The simplest is to allow the solvent to evaporate in a glass container. A better method is evaporating the solvent using a rotary evaporator connected to a vacuum pump to obtain a thin

film of the lipid on the walls of a round bottom flask. In order to increase encapsulation, it is recommended to start with a large volume round bottom flask so that the lipids will be dried down onto a large surface area possible to form a very thin film. The evaporator is detached from vacuum pump and introduced to nitrogen. The container is then removed from the evaporator and fixed to a lyophilizer or exposed to high vacuum overnight to remove the residual solvent.

An alternative method of dispersing the lipids in a finely-divided form before the addition of aqueous media is to freeze-dry the dissolved lipids in an organic solvent [31]. The important concept in this method is the choice of the organic solvent which should have a freezing point above the temperature of the condenser of the freeze-drying and also be inert with regard to rubber seals of commercial lyophilizers. When these restrictions are concerned, the most suitable organic solvent happens to be tertiary buthanol.

2.2.3. Hydration

Evaporation (or freeze-drying) of the solvent is followed by hydration of lipids with the aqueous medium. Often for hydration, a suitable buffer at a temperature above the phase transition temperature of the phospholipid is employed. The solution is swirled manually or mechanically (either with a bath sonicator or vortex mixer) until all the lipids have been incorporated into the solution. The resulting product is a milky suspension of lipids which is allowed to stand for a while for the complete swelling to give MLVs [28]. Further treatment is required for the preparation of ULVs, which will be discussed later in the text.

It is possible to obtain LUVs instead of MLVs during hydration by introducing an aqueous sucrose solution down the side of the flask by inclining the flask to one side and slowly returning the flask to the upright orientation, allowing the fluid to run gently over the lipid layer on the bottom of the flask. The swelling is carried out as usual without any shaking or agitation. The suspension is then centrifuged and the layer of MLVs floating on the surface is removed, leaving LUVs in solution.

2.2.4. Obtaining Suvs from Mlvs

After preparation of MLVs by hydration of dried lipid, it is possible to continue processing the liposomes in order to modify their size and other characteristics. For many purposes, MLVs are too large of too heterogeneous population to work with. There are several methods devised to reduce their size. These include techniques such as micro-emulsification, extrusion, and ultrasonication. A second set of methods is designed to increase the entrapment volume of hydrated lipids, and/or reduce the lamellarity of the liposomes formed, and involves procedures such as freeze-drying, freeze-thawing or induction of vesiculation by ions or pH change.

Microemulsification of liposomes is performed with an equipment called micro fluidizer to prepare small vesicles from concentrated lipid suspension. This method can produce liposomes in 50-200 nm size range with the encapsulation efficiency of up to 75% [32].

Sonication [33] disrupts MLV suspensions by using sonic energy to produce SUVs with diameters in the range of 15-50 nm. There are two methods of sonication; bath sonication and probe sonication. The former method is used for large volumes of dilute lipids whereas the latter one is used for suspensions which require high energy, such as high concentration of lipid suspensions. The disadvantage of probe sonication is the contamination of preparation with metal from the tip of the probe which should be removed by centrifugation prior to use. Also, as a result of high energy, probe sonication suffers from overheating the lipid suspension causing degradation. For these reasons, bath sonicators are the most widely used instrumentation for SUV preparation.

An even gentler method of reducing the size of the liposomes is to pass through a membrane filter of defined pore size [34]. This can be at much more lower pressure and can give populations in which one can choose the upper size limit depending on the exact pore size of the filter used. This membrane extrusion technique can be used to process both LUVs and MLVs in which liposome contents are exchanged with the suspending medium during breaking and resealing of the phospholipid bilayers as they pass through the polycarbonate membrane. In order to achieve as high an entrapment as possible of water-soluble compounds, it is crucial to have these compounds present in the suspending medium during the extrusion. An almost completely unilamellar population can be produced after 5-10 repeated extrusions through two stacked membranes.

In freezing-thawing method, SUVs are rapidly frozen and thawed slowly. The short-lived sonication disperses aggregated materials to LUV. The creation of unilamellar vesicles is as a result of the fusion of SUV throughout the processes of freezing and thawing [35-37].

2.3. Solvent Injection

In this type of preparation methods, lipids are first dissolved in an organic solvent and then brought into contact with the aqueous phase containing the materials to be encapsulated within the liposome. The lipids align themselves into a monolayer at the interface between the organic and aqueous phase which is an important step to form the bilayer of the liposome [29]. There are three categories in solvent dispersion method including; (i) a miscible organic solvent with the aqueous phase, (ii) an immiscible organic solvent with the aqueous phase that is used in excess, and (iii) an immiscible organic solvent used in excess with the aqueous phase.

2.3.1. Ethanol Injection Method

In this method an ethanol solution of lipids is injected rapidly into an excess saline or other aqueous medium by a fine needle [38]. The injection force is usually sufficient to achieve complete mixing, so that ethanol is diluted in water, and lipids are dispersed evenly throughout the medium. This method yields a high proportion of SUVs. This method is extremely simple and it has a very low risk of degradation for sensitive lipids. Its major disadvantages are the limitation of solubility of lipids in ethanol and the volume of ethanol that can be introduced into the medium, which in turn limits the quantity of lipid dispersed, so that the resulting liposome solution is generally dilute. As a result, the percentage encapsulation for hydrophilic materials is very low. One last disadvantage for this method is the difficulty of the removal of ethanol from the lipid membranes.

2.3.2. Ether Injection Method

This method [39, 40] involves injecting the immiscible organic solution very slowly into an aqueous phase through a narrow needle at a temperature that the organic solvent is removed by vaporization during the process. In this method, large vesicles are formed which might be due to the slow vaporization of solvent giving rise to an ether: water gradient extending on both sides of the interfacial lipid monolayer, resulting in the eventual formation of a bilayer sheet which folds in on to itself to form a sealed vesicle [29]. Ether injection treats sensitive lipids very gently and runs very little risk of causing oxidative degradation. Since the solvent is removed at the same rate as it is introduced, there is no limit to the final concentration of lipid which can be achieved, since the process can be run continuously for a long period of time, giving rise to a high percentage of the aqueous medium encapsulated within the liposomes. The major drawbacks of this method are the long time taken to produce a batch of liposomes and the need of careful control for the introduction of lipid solution.

2.4. Surfactant (Detergent) Solubilization Method

In this method, the phospholipids are brought into contact with the aqueous phase via the intermediary of surfactants. Phospholipid molecules associate with surfactants and form mixed micelles. The basic feature of this method is the removal of the surfactant from pre-formed mixed micelles containing phospholipids, whereupon unilamellar liposomes form spontaneously. However, removal of surfactants is carried out using techniques such as, dialysis and column chromatography, inevitably remove other small water-soluble molecules, making this method not very efficient in terms of percentage encapsulation values attainable for water soluble compounds. On the other hand, surfactant solubilization method has the ability to vary the size of the liposomes by precise control of the conditions of surfactant removal and to obtain liposomes of very high size homogeneity [29].

The transfer from laboratory to industry was very important for liposomes, as it is for any biotechnological discipline. The first liposomal drug delivery experiments in humans were carried out by freshly prepared liposomes but in order to be a commercial product the liposome-drug formulation must have well-defined stability and a shelf life over a year. Of the several preparation methods described in the literature, only a few of them have the potential to be used in the large scale liposome manufacturing. The crucial problem is the presence of organic solvent residues, pyrogen control, stability, sterility, size and size distribution as well as batch to batch reproducibility.

In the parental administration the liposomes two important conditions involve being sterile and pyrogen free. In the case of animal experiments, the sufficient sterility can be obtained by the passage of the liposome preparations through the 400 nm pore size Millipore filters. In human experiments the sterilization depyrogenation techniques should be taken much more seriously starting from the raw materials, containers and working areas [41].

2.5. Loading of Drugs in Liposome Formulations

2.5.1. Encapsulation of Hydrophilic Drugs

Once lipids are hydrated in the presence of hydrophilic drugs, a portion of the drug gets entrapped inside the liposome and another portion remains in the bulk, outside the aqueous core of the liposome. As only the entrapped drug is of interest, drug in the bulk should be removed. This purification is generally done by gel filtration column chromatography (Sephadex G-50, Pharmacia LKB) and dialysis (hollow fiber dialysis cartridge) on the basis of size differences between the liposomes and the non-encapsulated material. In the cases where DNA or proteins are being encapsulated, or where there is concern that non-encapsulated material may form large aggregates, techniques such as centrifugation can be employed due to the differences in the buoyant densities of liposomes and non-encapsulated material [42, 43].

A hydrophilic drug may not be encapsulated with high efficiency because the drug molecules can diffuse in and out of the lipid membrane. Thus, the drug would be difficult to retain inside the liposomes. However, compounds with ionizable groups and those that are both water and lipid soluble can be encapsulated with high efficiency (up to 90%) by the liposomes after the formation of membranes [44] by active loading. In this technique, the pH of the interior part of the liposome is such that the unionized drug which enters the liposome by passive loading is ionized inside the liposome, and ionized drug molecules lose their ability to diffuse through the lipid membrane. Therefore, high concentration of the ionized drug is obtained inside the liposome. For example, doxorubicin and epirubicin can be entrapped in preformed SUV with high efficiency through active loading [45, 46].

The pH difference can be brought about by encapsulating a non-permeating buffer ion such as glutamate inside the liposomes at low pH and replacing the extra-liposomal buffer with one which is iso-osmolar at pH 7.0. Alternatively, charged lipids may be incorporated into the membrane at low pH, followed by adjustment of the suspending medium to neutrality. A similar approach may be adopted by using a potassium gradient, in which the membrane is made selectively permeable to potassium ions entrapped inside the liposome by incorporation of valinomycin into the lipid membrane [47, 48].

2.5.2. Encapsulation of Hydrophobic Drugs

Hydrophobic drugs are solubilized in the phospholipid bilayer of the liposomes that mainly provide a hydrophobic environment. Once trapped, they remain in the liposome bilayer as they have very low affinity towards the inner or outer aqueous regions of the liposomes. During the preparation of liposomes, hydrophobic drugs are solubilized in the organic solvent along with the phospholipids and during the subsequent hydration phase, they remain entrapped in the hydrophobic bilayer region. For example, the liposomal photosensitizer verteporfin (Visudyne) contains a hydrophobic drug that is rapidly transferred to blood proteins in vivo. Activation of the drug by targeting laser light to blood flowing though the eye causes its site-specific activity in the treatment of wet macular degeneration [49]. Amphotericin B and paclitaxel are the other most commonly investigated hydrophobic drugs in liposome formulations.

3. STABILITY OF LIPOSOMES

Liposome stability can be explained by physical, chemical and biological means which are all interrelated. Generally, chemical (degradation of phospholipids structures) and physical (uniformity of size distribution and encapsulation efficiency) stability determine the shelf-life of liposomes. Once the liposomal formulations have been obtained, maintenance of the physical properties of these preparations can be difficult. Leakage of the encapsulated material due to the permeability of the membrane, change in the size distribution and stability problems due to the hydrolytic and oxidative degradation are the general problems upon storage. Methods are devised to overcome these instability problems, those designed to minimize the degradation processes and those which help liposomes to survive in the face of conditions which encourages these processes.

Two different types of chemical degradation can affect the performance of the phospholipids bilayers; hydrolysis of the ester bonds linking the fatty acids to the glycerol backbone and oxidation of the unsaturated acyl chains, if present. The level of oxidation can be kept to a minimum by taking some precautions like starting with freshly purified lipids and freshly distilled solvents, avoiding procedures involving high temperatures, carrying out the manufacturing process in the absence of oxygen, deoxygenating the aqueous solutions by passing nitrogen, storing all liposome suspensions in an inert atmosphere and including an anti-oxidant, e.g. α-tocopherol [50], a common non-toxic dietary lipid, as a component of the lipids membrane. An alternative solution to the oxidation problem is to reduce the level of oxidizable lipids in the membrane by using saturated lipids instead of the unsaturated ones. Also, the mono-unsaturated ones have much less tendency of oxidation than the polyunsaturated ones. Thus; sphingomyelins, usually having only one double bond, are expected to degrade more slowly than other mammalian origin lipids. Entirely synthetic and saturated phospholipids; DMPC, DPPC and DSPC, can also be considered as a solution for the oxidative degradation of liposomes.

Hydrolysis type of chemical degradation of the ester linkages in the phospholipid structure occurs most slowly at pH values close to neutral. In general, the rate of hydrolysis has a "V-shaped" dependence, with a minimum at pH 6.5 and an increased rate at both higher and lower pH. In the active loading of drugs, as it is mentioned before, low pH levels are required which triggers the hydrolysis. This hydrolysis kind of chemical degradation is also very effective on the aqueous solutions of liposome due to the presence of water. Temperature also triggers the hydrolysis of the lipids which creates the need for refrigeration. In order to keep hydrolysis to a minimum during active loading, attention must be paid for the removal of residual solvent from the dried lipids. To avoid hydrolysis, instead of ester linked lipids, the usage of ether linkage containing lipids (e.g. found in the membrane of halophilic bacteria) would be an absolute solution [51]. Another chemical degradation, oxidation of the lipids in the liposome structures can be prevented by the addition of small amounts of antioxidants during the manufacturing steps.

The problems related to the lipid oxidation and hydrolysis during the shelf-life of the liposomal product can be reduced by the storage of liposomal dispersion in the dry state by freeze-drying (lyophilization), without compromising their physical state or encapsulation capacity [52]. However, freeze-drying of liposome systems without appropriate stabilizers will lead to fusion of vesicles, i.e. physical instability. To promote vesicle stability during the freeze-drying process, cycloprotectants [53-55], including saccharides (e.g. sucrose, trehalose, and lactose) and their derivatives are employed [56]. Cycloprotectants, especially sucrose because of its high glass transition temperature, are believed to be effective to protect the liposome membranes against possible fracture and rapture that might cause a change in size distribution and a loss of the encapsulated material presumably by forming glasses under the typical freezing conditions used for lyophilization [57]. Lyophilization increases the shelf-life of the finished product by preserving in a relatively more stable dry state. Some liposome products on market or clinical trials are provided as lyophilized powder. For example, AmBisomeTM, a liposomal amphotericin, is the first liposome product to be marketed in several countries is supplied as a lyophilized powder to be reconstituted with sterile water injection. Additionally, paclitaxel-liposome formulations have been developed which show good stability [58, 59]. These formulations once lyophilized can be stored at room temperature for extended time. On the other hand, once the preparation is reconstituted, it is not stable for more than a day in terms of size.

The physical degradation, leakage and fusion of liposomes, can occur as a result of the lattice defects in the membrane introduced during the manufacture, particularly in SUVs that are prepared below the membrane phase transition temperature. Annealing process, incubating the liposomes at a higher temperature than the phase transition temperature, can wipe out these defects by equalizing the differences in packing density between opposite sides of the bilayers. Even in annealed vesicles, aggregation and fusion can occur over a long period of time. In neutral liposomes, aggregation takes place because of the van der Waals interactions and because of the increased surface area it tends to be more pronounced in large liposomes. The simplest solution to overcome this aggregation is to add a small amount of negatively charged phospholipid (e.g. 10% PA or PG) to the liposome composition [29]

SUVs have much more tendency to fusion when compared to large liposomes due to the presence of stress arising from the high curvature of the membrane. Since this can occur specifically at the transition temperature of the membrane, it would be better to store these liposomes at a temperature much lower than the transition temperature of the lipids. For example, SUVs should be stored above their transition temperature for no longer than 24 hours but LUVs can be stored for a longer period of time if the temperature of the solution is kept in a range of 4-8 0C for approximately 1 week before the leakage of the encapsulated material starts due to the hydrolytic degradation on the membrane structure [60]. Also, addition of cholesterol to the phospholipid mixture would be a solution to reduce or eliminate the transition. The presence of cholesterol prevents packing and aggregation by inducing orientation and more rigidity to the phospholipids. Other than cholesterol, peptide incorporation to the lipid membrane also enables the lipid membrane to be more rigid at physiological temperature [61-63].

Permeability of liposome membranes depends highly on the membrane lipid composition, as well as on the encapsulated material. Large polar or ionic molecules will be retained much more efficiently than low molecular weight lipophilic compounds. Generally, for both type of encapsulated material, a rigid, more saturated membrane with a higher ratio of cholesterol forms the most stable lipid membrane concerning the leakage of the encapsulated material.

Many attempts have been made to enhance the physical stability of liposomes. Among these, surface modification of liposomes is an attractive method to improve liposomal stability both in vitro and in vivo. Some improvements in chemical and physical stability of polymer coated liposomes prepared with polysaccharide derivatives, such as mannan or amylopectin, have been demonstrated [64]. Several other substances also have been used for preparation of polymer coated liposomes such as poloxamer, polysorbate 80, carboxymethyl chitosan, and dextran derivatives [65-69]. While the possibility of coating liposomes with these polymers has been reported, few papers have dealt with the systematic evaluation of the physical stability of polymer coated liposomes. Moreover, contravening results have been also reported such as that polymer coated liposomes showed less stability than non-coated ones [65, 70].

In vivo stability of liposomes is also dependent on their charge. In serum, there are several proteins that are both positively and negatively charged. Liposomes with neutral charge are found to be more stable as they have much less electrostatic affinity towards proteins. [71].

Biological liposome stability plays important roles at various stages of drug delivery. However, liposomes are somewhat biologically unstable as a parenteral drug delivery system owing to their rapid uptake and clearance from circulation by cells of the mononuclear phagocytic system (MPS) located mainly in the liver and spleen [72, 73]. Biological stability of liposomes is dependent on the presence of agents such as proteins that interact with liposomes upon application to the subject and the administration route. There have been many strategies to enhance the biological stability of liposomes that improve the liposomal drug delivery in vivo and increase the circulation time in blood stream [74]. The complexation between polymers and liposomes has been studied as a way to increase the long-term stability of liposomes. Grafting hydrophilic polymers onto the head groups of phospholipids, or the addition of water soluble polymers containing several hydrophobic groups has been shown to increase the circulation time in vivo, as well as to inhibit liposome fusion [75-77]. These kinds of liposomes are called stealth liposome [78] or sterically stabilized liposomes [79]. The steric repulsion of these liposomes stabilizes the liposomes against aggregation. One of the most popular and successful methods to obtain long-circulating biologically stable liposomes is to coat the surface of the liposome with poly(ethylene glycol), PEG [80-84]. Although the PEG chemistry is successful in coating the liposome surface, alternative sterically protecting polymers are also under research. The candidate polymers should be biocompatible, soluble, hydrophilic and have highly flexible main chain for drug delivery. Some of these polymers given in the literature are synthetic polymers of vinyl series i.e. poly(vinyl pyrrolidone) (PVP) and poly(acrylamide) (PAA) [85, 86]. PVP has a similar history on pharmaceutical application to PEG [87, 88]. It shows high degree of biocompatibility and also acts as efficient steric protector for liposomes. It was found that the liposomal bilayers containing lipids with covalently attached to polyethylene glycol by which the membrane surface steric inhibits protein and cellular interactions with liposomes drastically prolonging the blood circulation time when injected in animals [89]. Doxil® is the liposomal doxorubicin available in the market which is stable for more than 18 months in the liquid state due to being stabilized by the usage of polyethylene glycol.

4. STERILIZATION OF LIPOSOMES

Pharmaceutical industry in general differentiates between two principally different approaches to ensure sterility of a parental product: terminal sterilization of the final product in its container (steam sterilization) and aseptical manufacturing. Terminal sterilization is the commonly used one because of its higher sterility assurance level achieved when compared with the aseptical methods. However, terminal sterilization is not applicable to many liposomal drug carrier formulations.

There are several sterilization methods; such as filtration, gamma irradiation, final steam sterilization, dry heat sterilization, ethylene oxide sterilization, and ultraviolet sterilization. Bearing in mind the susceptibility of liposomes to the previously mentioned physical and chemical degradation mechanisms, the conditions required in conventional sterilization techniques (except filtration) are rather concerning since they involve the usage of heat, radiation and/or chemical sterilizing agents. Therefore, identification of a suitable method for sterilization of liposome formulations is a major challenge.

TABLE 1. Summary of the Sterilization Techniques Applied on Liposomal Preparations.

Sterilization Technique	Advantage(s)	Disadvantage(s)	Convenience
Filtration	Low operation temperature	Applicable to liposomes lower than 200 nm in diameter Operation under aseptic conditions	Low
γ-irradiation	Moderate operation temperature Highest microbial death reliability	Large scale operation Risk of degradation of liposomes	High
Final steam sterilization	Low cost and convenient	Risk of degradation of liposomes	High
Dry heat	Low cost and convenient	Risk of degradation of liposomes	High
Ethylene oxide	Low operation temperature	Possible carcinogenic residues	Low
UV-sterilization	Low cost and convenient	Poor penetration into products Risk of degradation of liposomes	High

Filtration is the most suitable sterilization technique for the thermolabile liposomes since it does not include any form of heat or condition that can result in the degradation of liposomes or leakage of the encapsulated material. However, filtration has some drawbacks such as; being only applicable to the liposomes that are smaller than 200 nm in diameter and being an expensive method due to the equipment requiring to work under high pressure (25 kg/cm^2 and above). Additionally, this technique must be performed under aseptic conditions [90].

Filtration sterilization is relatively time-consuming and not efficient for the removal of viruses [91]. Studies have shown that polycarbonate membranes are less effective than hydrophobic Fluoropore membrane and cellulose acetate/surfactant-free membrane filtration units [91]. Although the limitations of filtration provoked researches on other sterilization methods, all resulted in the formation of degradation products via the previously mentioned degradation pathways. Filtration and the other methods are summarized according to their applicability on liposomal preparations in Table 1[92], given above.

5. CHARACTERIZATION OF LIPOSOMES

After preparation and before application, liposomes have to be characterized in order to ensure their in vitro and in vivo performance. Liposomal properties that are commonly discussed include lamellarity (the number of bilayers present in liposomes), diameter and size distribution, lipid composition and concentration determination, the encapsulant concentration and its encapsulation efficiency.

For the characterization of chemical properties, phospholipids can be quantitatively in terms of concentration either by Bartlett Assay or Stewart Assay. The phospholipid hydrolysis might be followed by HPLC where the column outflow can be monitored continuously by UV absorbance to obtain a quantitative record of the eluted components. Moreover, the phospholipid oxidation can also be followed by a number of techniques i.e., UV absorbance method, TBA method (2-thiobarbutiric acid) (for endoperoxides), iodometric method (for hydroperoxides) and GLC (gas-liquid chromatography) method [93].

The most direct method for determination of liposome size is the electron microscopy due to the possibility of viewing the liposomes individually and obtaining the exact information about the liposome population over the whole range of sizes [94]. As liposomes do not naturally create a contrast to be visible by electron microscopy, either cryo-TEM (Figure 2) should be used or staining of the liposome sample is required. Either way, it is a very time-consuming method and it requires equipments that may not always be immediately accessible. The other method for the determination of liposome size, dynamic light scattering [95, 96], is very simple and rapid to perform but it measures an average size of liposome bulk. More recently, atomic force microscopy is also used to determine the morphology, size and stability of liposomal structures. All these size determination methods are very expensive. If only an approximate size range is required, gel exclusion chromatography might be suitable.

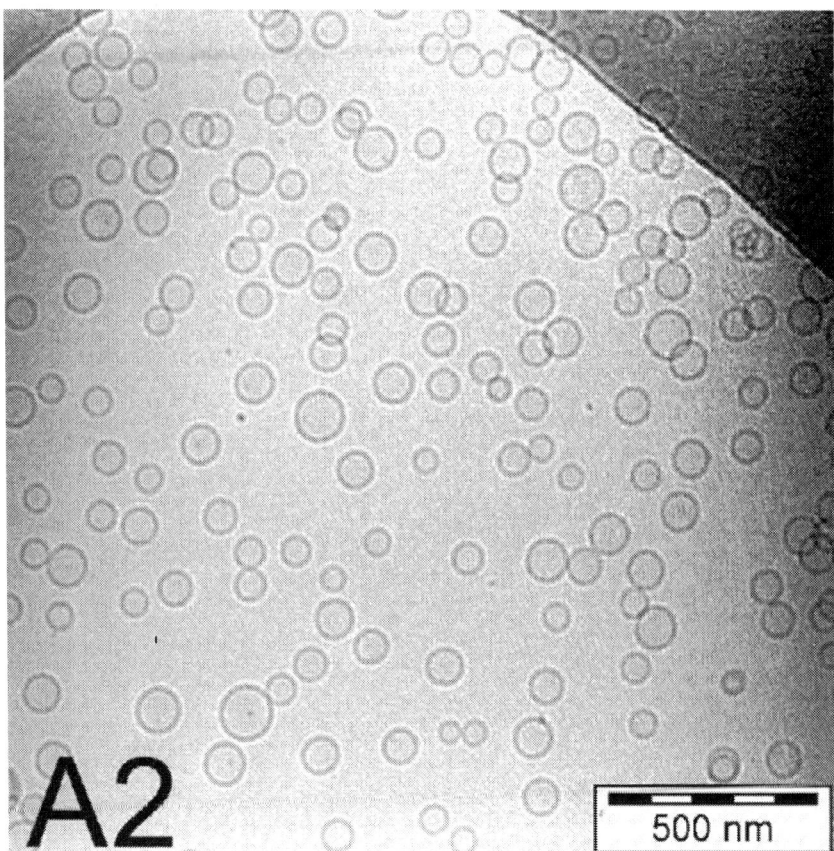

Adopted from Holzer, M., Barnert, S., Momm, J., Schubert, R., 2009. Preparative size exclusion chromatography combined with detergent removal as a versatile tool to prepare unilamellar and spherical liposomes of highly uniform size distribution. J. Chromatogr. A 1216, 5838–5848.

Figure 2. Cryo-TEM pictures of Size Exclusion Chromatography fractions eluted at 90 minutes and prepared from egg-phosphatidylcholine.

Electrostatic stabilization of liposomes may be a desirable feature to prevent fusion. The surface charge on the liposomes is measured by zeta-potential measurements [98]. These measurements are useful in determination of the in vivo behavior of liposomes. Often zeta potential values <-25 mV or >+25 mV are considered stable [99]. However, as mentioned earlier, charged liposomes have the disadvantage of being unstable in biological conditions.

Residual solvent is very unacceptable for drug delivery applications, therefore residual solvent should be kept at a minimum in the formulations. Quantification of residual solvents as a result of preparation methods is done through gas chromatography (GC) [100-101] This is a very rapid and reliable method and most analytical and organic laboratories are equipped with a GC.

An important feature of liposomes is the existence of a temperature dependant, reversible phase transition, where the hydrocarbon chains of the phospholipid structures undergo a transformation from an ordered gel state to a more disordered fluid, liquid crystalline, state. This transition temperature is important in optimizing the storage conditions (i.e Temperature) to minimize fusion and drug leakage. These changes have been monitored by freeze fracture electron microscopy and much more easily by differential scanning calorimetry (DSC) [102-104, 93].

Entrapped volume is a crucial parameter that governs the morphology of liposomes. This internal volume is defined as the aqueous entrapped volume per unit quantity of lipids. The most promising way to determine the internal volume is to measure the quantity of water by replacing external medium (water) with a spectrophotometrically inert fluid (i.e. deuterium oxide) and then measuring water signal by NMR [93].

It is essential to measure the quantity of the encapsulated material inside liposomal structures before studying the behavior of this encapsulated material physically and biologically since the effects observed experimentally will be dose related. After the removal of the non-encapsulated material by the separation techniques the quantity of material remained can be assumed as 100% encapsulated. Minicolumn centrifugation and protamine aggregation methods are the general separation procedures that are commonly used [93].

Methods for determining the amount of material encapsulated within the liposomes typically rely on the destruction of the lipid bilayer and subsequent quantification of the released material [105]. In these measurements, the signal due to intact liposomes is typically monitored prior to bilayer disruption. The techniques used for this quantification depend on the nature of the encapsulant and include spectrophotometry [106, 107], fluorescence spectroscopy [108], enzyme-based methods [109] and electrochemical techniques. If a separation technique such as HPLC of field-flow fractionation (FFF) is applied, the percent encapsulation can be expressed as the ratio of the unencapsulated peak area to that of a reference standard of the same initial concentration [110, 111]. This method can be applied if the liposomes do not undergo any purification following preparation. Either technique serves to separate liposome encapsulated materials from those that remain in the extravesicular solution and hence can also be used to monitor the storage stability in terms of leakage or the effect of various disruptive conditions o the retention of encapsulants. Some authors have combined the size distribution and encapsulation efficiency determination in one assay by using FFF-MALS (multi angled light scattering) coupled to a concentration detector suitable for the encapsulant [112].

Since techniques used to separate free materials from liposome-encapsulated contents can potentially cause leakage of contents and, in some cases, ambiguity in the extent of separation, research using methods that do not rely on separation are of interest. Reported methods have included 1H NMR where free markers exhibited pH sensitive resonance shifts in the external medium versus encapsulated markers [113]; diffusion ordered 2D NMR which relied on differences in diffusion coefficients of entrapped and free marker molecules [114]; fluorescence methods where the signal from unencapsulated fluorophores was quenched by substances present in the external solution [115]; electron pin resonance (ESR) methods which rely on the signal broadening of unencapsulated markers by the addition of a membrane-impermeable agent [116, 117].

The drug release from liposomes can be followed by the usage of a well calibrated in vitro diffusion cell in order to predict pharmacokinetics and bioavailability of drug before expensive and time-consuming in vivo studies. For the determination of pharmacokinetic performance of liposomal formulations, dilution-induced drug release in buffer and plasma was employed and for the determination of drug bioavailability, another procedure is followed which involves the liposome degradation in the presence of mouse-liver lysosome lysate [93].

6. CLINICAL APPLICATIONS OF LIPOSOMES

New drug delivery systems such as liposomes are developed when the existing formulations are not satisfactory. Among all the nanomedicine platforms, liposomes have demonstrated one of the most established nanoplatforms with several FDA-approved formulations for cancer treatment, and had the greatest impact on oncology to date, because of their size, biocompatibility, biodegradability, hydrophobic and hydrophilic character, low toxicity and immunogenicity [118]. A vast of literature describes the feasibility of encapsulation of a wide range of drugs, including anti-cancer and antimicrobial agents, peptide hormones, enzymes, other proteins, vaccines and genetic materials, in the aqueous or lipid phases of liposomes which showed enhanced therapeutic activity and/or reduced toxicity in preclinical models and in humans when compared to their non-liposomal formulations.

Liposome applications in drug delivery depend, and are based on, physicochemical and colloidal characteristics such as composition, size, loading efficiency and the stability of the carrier, as well as their biological interactions between liposomes and cells. Based on these liposome properties, several modes of drug delivery can be listed: the major ones are enhanced drug solubilization (e.g. amphotericin B, minoxidil), protection of sensitive drug molecules (e.g. cytosine arabinose, DNA, RNA, antisense olgionucleotides, ribozymes), enhanced intracellular uptake (all agents, including antineoplastic agents, antibiotics and antivirals) and altered pharmacokinetics and biodistribution of the encapsulated drug.

Although lipid based formulations have advantages as drug carriers, drug-delivery systems based on unmodified liposomes are limited by their short blood circulation time, instability in vivo and lack of target selectivity [119, 120]. To increase accumulation of liposomal formulations in the desired cells and tissues, the use of targeted liposomes including surface-attached ligands such as; antibodies, folates, peptides and transferrin that are capable of recognizing and binding to the desired cells. Despite of some improvements in targeting efficiency by these immunoliposomes, the majority of these modified liposomes were still eliminated rapidly by the reticulo endothelial system, primarily in the liver [120]. Better target accumulations are expected if liposomes can be made to remain in the circulation long enough.

Schematic drawing of cytosolic delivery and organelle-specific targeting of drug loaded nanoparticles (i.e. most frequently liposomes) via receptor-mediated endocytosis is shown in Figure 3.

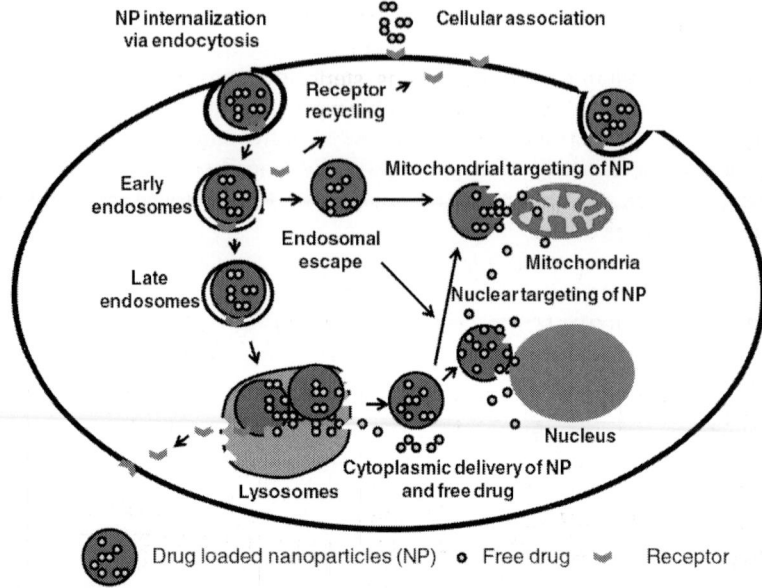

Figure 3. Schematic drawing of the cytosolic delivery and organelle-specific targeting of drug loaded nanoparticles via receptor-mediated endocytosis. After receptor mediated cell association with nanoparticles, the nanoparticles are engulfed in a vesicle known as an early endosome. Nanoparticles formulated with an endosome disrupting property disrupt the endosomes followed by cytoplasmic delivery. On the other hand, if nanoparticles are captured in early endosomes, theymaymake theirway to lysosomes as late endosomes where their degradation takes place. Only fraction of non-degraded drug released in the cytoplasm interacts with cellular organelles in a random fashion. However, cytosolic delivery of a fraction of organelle-targeted nanoparticles via endosomal escape or from lysosomes travel to the targeting organelles to deliver their therapeutic cargo [121].

Different methods have been suggested to achieve liposomes with high stability and long circulation times in vivo, including the surface coating of the liposomes with inert, biocompatible polymers such as PEG (stealth liposomes), which forms a protective layer over the liposome surface and slow down liposome recognition by opsonins and therefore subsequent clearance of liposomes [80, 84]. Long circulating liposomes are now being investigated in detail and are widely used in vitro and in vivo studies due their flexibility and also they found their place in the clinical applications. The flexibility allows a relatively small number of surface-grafted polymer molecules to create an impermeable layer over the liposome surface [122, 123]. Long-circulating liposomes demonstrate dose-dependent, non-saturable, log-linear kinetics and increased bioavailability [124].

The studies that attempt to combine the properties of long-circulating liposomes and immunoliposomes in one preparation place themselves in the literature as the further development in the liposomal formulations as drug carriers [125, 126]. In the early experiments, simple co-immobilization of an antibody and PEG on the surface of the same liposome has been performed despite the possibility of PEG creating steric hindrance for target recognition with the targeting moiety [125]. To achieve better selectivity of PEG-coated liposomes, it is advantageous to attach the targeting ligand via a PEG spacer arm, so that the ligand is extended outside the dense PEG brush which reduces steric hindrance of binding to the target [127]. The use of PEG-conjugated immunoliposomes for increasing drug carrying capacity of monoclonal antibody has been demonstrated [128]. In addition to costly monoclonal antibodies, common molecules such as folic acid, trensferrin and RGD peptides have also been studied for tumor targeting with enhanced selective uptakes [120].

Encouraging results of liposomal drugs in the treatment or prevention of a wide spectrum of diseases in experimental animals and in human, indicate that more liposome-based products for clinical and veterinary applications may be forthcoming. These could include treatment of eye and skin diseases in therapeutic applications, antimicrobial and anticancer therapy in clinical applications, metal chelation, enzyme and hormone replacement therapy, vaccine and diagnostic imaging, etc. Some of the liposome applications in terms of drug delivery are discussed below.

6.1. Ocular Applications

The eye is protected by three highly efficient mechanisms (a) an epithelial layer which is the barrier to penetration (b) tear flow (c) the blinking reflex. All these mechanisms are responsible for the poor drug penetration into the deeper layers of the cornea and the aqueous humor and for the rapid wash out of drugs from the corneal surface. Initially, in 1981 the enhanced efficiency of liposomes encapsulated idoxuridine in herpes simplex infected corneal lesions in rabbits was reported [129]. In 1985, it was concluded that ocular delivery of drugs can be either promoted or impeded by the use of liposome carriers, depending on the physicochemical properties of the drugs and the lipid mixture employed [130]. The use of mucoadhesive polymers, carbopol 934P and carbopol 1342 to retain liposomes at the cornea was proposed [131]. While precorneal retention times were indeed significantly enhanced under appropriate conditions, liposomes even in the presence of the mucoadhesive had migrated toward the conjuctival sac with very little activity remaining at the corneal surface.

6.2. Pulmonary Applications

Lung is a natural target for the delivery of therapeutic and prophylactic agents such as peptides and proteins. The past 15 years have been marked by intensive research efforts on pulmonary drug delivery not only for local therapy but also for systemic therapy as well as diagnostic purposes, primarily due to the several advantages the pulmonary route offers over other routes of drug administration. Drugs that undergo gastrointestinal degradation (such as proteins and peptides) are ideal candidates for pulmonary delivery.

Targeted drug delivery to the lungs has evolved to be one of the most widely investigated systemic or local drug delivery approaches. The use of drug delivery systems for the treatment of pulmonary diseases is increasing because of their potential for localized topical therapy in the lungs. This route also makes it possible to deposit drugs more site-specific at high concentrations within the diseased lung thereby reducing the overall amount of drug activity while reducing systemic side effects. To further exploit the other advantages presented by the lungs, as well as to overcome some challenges, scientists developed interests in particulate drug delivery systems for pulmonary administration, such as liposomes, micelles, nano-and micro-particles based on polymers.

The use of liposomes as drug carriers for pulmonary delivery has been reported for different kinds of therapeutics such as anti-microbial agents, cytotoxic drugs, antioxidants, anti-asthma compounds and recombinant genes for gene therapy in the treatment of cystic fibrosis.

Liposomes as carrier systems for pulmonary delivery offer several advantages over aerosol delivery of the corresponding non-encapsulated drug. Liposomes might be used to solubilize poorly soluble drugs, provide a pulmonary sustained release reservoir prolonging local and systemic therapeutic drug levels, facilitate intracellular delivery of drugs especially to alveolar macrophages, tumor cells or epithelial cells, prevent local irritation of lung tissue and reduce the drug's toxicity, target specific cell populations using surface bound ligands or antibodies and be absorbed across the epithelium to reach the systemic circulation intact [132].

Local delivery of medication to the lungs is highly desirable, especially in patients with specific pulmonary diseases such as cystic fibrosis, asthma, chronic pulmonary infections or lung cancer. The principal advantages include reduction of systemic side effects and application of higher doses of the medication at the site of drug action. Although simple inhalation devices and aerosols containing various drugs have been used since the early 19th century for the treatment of respiratory disorders, the past 15 years have been marked by intensive research efforts on pulmonary drug delivery not only for local therapy but also for systemic therapy as well as diagnostic purposes due to the several advantages the pulmonary route offers over other routes of drug administration. Lung is a natural target for the delivery of therapeutic and prophylactic agents such as peptides and proteins due to the large surface area available for absorption, the very thin absorption membrane and the elevated blood flow which rapidly distributes molecules throughout the body. Moreover, the lungs exhibit relatively low local metabolic activity, and unlike the oral route of drug administration, pulmonary inhalation is not subject to first pass metabolism [133].

Inhaled drug delivery devices can be divided into three principal categories: nebulizers, pressurized metered-dose inhalers and dry powder inhalers; each class presents unique strengths and weaknesses. A good delivery device has to generate an aerosol of suitable size and provide reproducible drug dosing. It must also protect the physical and chemical stability of the drug formulation.

For controlled delivery of drug to the lung, liposomes are one of the most extensively investigated systems in recent studies given that they can be prepared with phospholipids such as egg phosphatidylcholine (PC), distearoyl phosphatidylcholine (DSPC) and dipalmitoylphosphatidylcholine (DPPC) endogenous to the lung.

A significant disadvantage of many existing inhaled drugs is the relatively short duration of resultant clinical effects, which requires most medications to be inhaled at least twice daily. This often leads to poor patient compliance. A reduction in the frequency of dosing would be convenient, particularly for chronic diseases such as asthma. The advantages of such an approach include reduced dosing, increased effectiveness of rapidly cleared medicine and enhanced residence time at the target site for the treatment of infection. Many challenges exist in developing controlled release inhalation medicine, which is reflected in the fact that no commercial product exists. Cytotoxic agents, bronchodilators, anti-asthma drugs, antimicrobial and antiviral agents and drugs for systemic action, such as insulin and proteins are being investigated.

6.3. Cancer Therapy

The numerous anti-cancer agents that have a high cytotoxic effect on the tumor cells in vitro exhibit a remarkable decrease of the selective ant-tumor effect for in vivo procedures applicable in the clinical treatment. One of the significant limitations of the anti-cancer drugs is their low therapeutic index meaning that the dose required to produce an anti-tumor effect is toxic to normal tissues. The low therapeutic index of these drugs results from the inability to achieve therapeutic concentrations at the specific target sites, tumors. Further, it results from the non-specific toxicity to normal tissues such as bone marrow, renal, gastrointestinal tract, and cardiac tissue and also from the problems associated with a preparation of a suitable formulation of the drugs [134].

Many different liposome formulations of various anticancer agents were shown to be less toxic than the free drug so that most of the medical applications of liposomes that have reached the preclinical stage are in cancer treatment [135-137]. Entrapment of these drugs into liposomes resulted in increased circulation lifetime, enhanced deposition in the infected tissues, and protection from the drug metabolic degradation, altered tissue distribution of the drug, with its enhanced uptake in organs rich in mononuclear phagocytic cells (liver, spleen and bone marrow) and decreased uptake in the kidney, myocardium and brain. To target tumors, liposomes must be capable of leaving the blood and accessing the tumor. However, because of their size liposomes cannot normally undergo transcapillary passage. In spite of this, various studies have demonstrated the accumulation of liposomes in certain tumors in a higher concentration than found in normal tissues [138, 139]. Anthracyclines are drugs which stop the growth of dividing cells by intercalating into the DNA and therefore kill predominantly quickly dividing cells. These cells are not only in tumors but are also in hair, gastrointestinal mucosa, and blood cells; therefore, this class of drugs is very toxic. Many research efforts have been directed towards improving the safety profile of the anthracyclines cytotoxics, doxorubicin and daunorubicin, along with vincristine. Encapsulation of these drugs into the liposomes showed reduced cardiotoxicity, dermal toxicity and better survival of the experimental animals compared to the controls receiving free drugs [138].

Such beneficial effects of liposomal anthracyclines have been observed with a variety of liposome formulations regardless of their lipid composition and provided that lipids used high cholesterol concentration of phospholipids with high phase transition temperature are conducive to drug retention by the vesicles in the systemic circulation [45].

Active targeting of cancer drugs to the tumors is shown schematically in Figure 4.

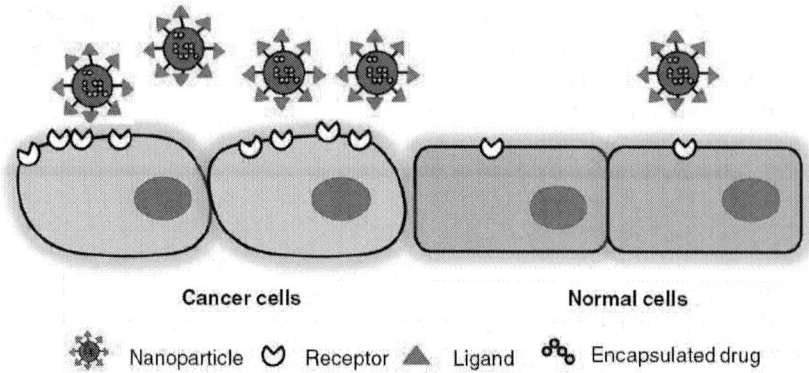

Cancer cells   Normal cells

Nanoparticle   Receptor   Ligand   Encapsulated drug

Figure 4. Representation of active targeting via receptors [121].

Currently several liposomal formulations are in the clinical practice containing different chemotherapeutics such as doxorubicin (Doxil1/Caelyx1), doxorubicin (Myocet1), daunorubicin (DaunoXome1) and cytarabine (DepoCyte1) for treating the ovarian cancer, AIDS related Kaposi's sarcoma, multiple myeloma, lymphomas, leukemia with meningeal spread. Several other liposomal chemotherapeutic drugs containing doxorubicin, annamycin, mitoxantrone, cisplatin, oxaliplatin, camptothecine, 9-nitro-20 (S)-camptothecin, irinotecan, lurtotecan, topotecan, paclitaxel, vincristine, vinorelbine and floxuridine are at the various stages of clinical trials [140].

Two liposomal formulations have been approved by the US Food and Drug Administration (FDA) and are commercially available for the treatment of AIDS-related Kaposi's sarcoma. Doxil, first liposomal drug approved by FDA and has been on the market since 1995, is a formulation of doxorubicin precipitated in sterically stabilized liposomes and has been on the market since 1995 [141], while DaunoXome, approved six months later than Doxil, is daunorubicin encapsulated in small liposomes with very strong and cohesive bilayers, which can be referred as mechanical stabilization [142].

DaunoXome is composed of small unilamellar vesicles containing distearoylphosphatidylcholine-cholesterol (2:1) with daunorubicin loaded by a pH gradient [137]. These liposomes are selectively stable in the circulation because they are small and their membrane is electrically neutral and mechanically very strong [142]. This reduces the charge-induced and hydrophobic binding of plasma components but does not protect against van der Waals adsorption. Also, uncharged liposomes are colloidally less stable than charged ones.

Doxil is a liquid suspension of 80-100 nm liposomes (2000PEG-distearoylphosphatidylethanolamine-hydrogenated-soya-bean phosphatidylcholine-cholesterol, 20 mM) loaded with doxorubicin HCl by ammonium sulfate gradient technique and additionally precipitation with encapsulated sulfate anions. These liposomes circulate in patients for several days, which increase their chances of extravasating at sites with a leaky vascular system. Their stability is due to their surface PEG coating as well as to their mechanically very stable bilayers [141, 142].

Cytarabine (Ara-C) is an effective hydrophilic chemotherapeutic agent used widely for the treatment of acute myelogenous leukaemia and lymphocytic leukeamia [143]. It has often been utilized in the combination chemotherapy, against solid tumors and leukaemias. Cytarabine is a cell cycle-dependent drug; hence, prolonged exposure of cells to cytotoxic concentrations is critical to achieve maximum cytotoxic activity. The toxicity of cytarabine is reduced if it is able to maintain an effective therapeutic level for a long period of time and, thus, it is a suitable candidate for administration in a controlled-release dosage form. Liposome encapsulated liposomes (DepoCytTM) are now commercially available.

Etoposide (VP-16-213) is another successful chemotherapeutic agents used for the treatment of human cancers. The drug is currently in its third decade of clinical use and is a front line therapy for a variety of malignancies, including leukaemias, lymphomas and several solid tumors [144]. It has a short biological half-life (3.6 h) with a terminal half-life of 1.5 h intravenously and a variable oral bioavailability ranging from 24% to 74%. Although intraperitoneal injection would result in initial high local tumor concentrations, prolonged exposure of tumor cells may not be possible [145].

The harmful and even destructive effect of cytotoxic drugs on healthy body cells makes it necessary to search for new delivery methods for drugs like cytarabine and etoposide. There are many articles describing the results of investigations of incorporation of cytarabine [146] and etoposide [147] into liposome. However, there is no information about their simultaneous incorporation, in spite of the fact that these two drugs have been used for more than 30 years.

Taxanes are complexes of diterpenoid natural products and semisynthetic analogs. Presently, these drugs belong to prominent anticancer agents used for combined chemotherapy [148]. Paclitaxel (PTX), the prototype of this class, emerges from a natural source [149]. This drug have been used for various cancers including ovarian, breast, head and neck, and non-small cell lung cancers [150].

The commercial PTX preparation (Taxol®) is formulated in the vehicle composed of Cremophor EL® (polyethoxylated castor oil used as a solubilizing surfactant) and dehydrated ethanol, which provides a homogenous preparation. However, some drawbacks have been reported for its clinical applications of this formulation such as severe hypersensitivity reactions, neurotoxicity and neutropenia [151, 152]. It was reported that these adverse effects associated with this formlation would be due to Cremophor EL rather than PTX itself [153]. PTX solubilized in Cremophor EL shows also an incompatibility with the polyvinyl chloride of the administration sets [152]. Furthermore, the short-term

stability of PTX upon dilution with aqueous media can result in possible drug precipitation [154].

Special requirements regarding a proper filter device as well as appropriate containers and infusion bags for the storage and administration of the drug have to be fulfilled in order to overcome the problems of incompatibility and instability during the clinical application of Taxol®. Hence, the development of an improved delivery system for PTX is of high importance. Current approaches are focused mainly on the development of formulations that are devoid of Cremophor EL, investigation of the possibility of a large-scale preparation and a request for a longer-term stability. There are some promising possibilities to replace Taxol® by a less irritable preparation such as micelle formulations, water-soluble prodrug preparations, enzyme-activatable prodrug preparations conjugated with antibodies or albumin, parenteral emulsions, microspheres, cyclodextrins, and nanocrytals [155-162].

The preparation of an optimal PTX formulation requires important considerations such as the optimization of the liposomal composition, the balance of the PTX amount encapsulated in the liposomes and the stability of the prepared PTX liposomes during storage in aqueous media [163]. The main characteristics of PTX molecule are asymmetry, bulkiness, hydrophobicity, low solubility and tendency to crystallization in aqueous media. All these factors affect the final design and preparation of a suitable drug formulation.

Liposomes provide suitable environment enhancing the solubility of the hydrophobic nature PTX by associating the molecule within the membrane bilayers. Commonly prepared formulations of PTX with liposomes were able to encapsulate the highest achievable content of PTX, 3-4 mol% with stability for weeks to months whereas 4-5 mol% paclitaxel was stable in the time range of just several hours to a day, and 8% paclitaxel loading only resulted in 15 minutes of liposome stability. Generally, increasing the encapsulated amount of PTX causes a reduction in the stability of the liposomal-PTX formulation due to the crystallization of the drug molecule. Thereby, to achieve a high drug/lipid ratio while retaining the long-term physical-chemical stability, a freeze-drying method is employed to obtain a dry drug-lipid powder, which is rehydrated in an aqueous solution immediately before use [58]. The encapsulation of PTX into liposomes enhances the drug therapeutic efficacy, thus, the same therapeutic effect could be reached by a decreased PTX-dose. On the other hand, the maximum tolerated dose (MTD) of liposome-encapsulated PTX increased compared with the Taxol® [Straubinger, R.M. and S.V. Balasubramanian, Preparation and Characterization of Taxane-containing Liposomes, Methods Enzymol. 391 (2005) 97-117.]. [163].

Taxane liposomes have shown slower elimination, higher antitumor activity against various murine and human tumors and lower systemic toxic effect compared to Taxol® [58]. They have also shown antitumor effect in Taxol-resistant tumor models [164]. Abraxane®, the only nonliposomal preparation of PTX, (albumin nanoparticle-based PTX preparation) and Lipusu® (liposomal PTX approved by State FDA of China) have entered the field of clinical applications. LEP-ETU (NeoPharm) and EndoTAG®-1 (Medigene) have reached the phase II of the clinical trials. Generally, liposomes and protein nanoparticles represent a promising approach to the optimization of PTX delivery. Their commercialization is at the doorstep of modern drug delivery market.

7. NEW GENERATION LIPOSOMES

Liposomes made up of commonly used ester phospholipids such as phosphatidylcholine are referred as conventional liposomes. These structures are very attractive for encapsulation and drug delivery applications to entrap both hydrophilic and hydrophobic materials due to the presence of aqueous core part as well as the lipid bilayer. Up to this date, there are many formulations in the market and also in the clinical trials. However, none of them truly overcome their chemical and physical instability problems especially during the transfer to the site of action [120]. Various attempts like modification of the liposome surface with i.e. hydrophilic polyethylene glycol polymers, using cryoprotectants or incorporation of high amount of cholesterol into the bilayer have led to only limited success. Other than instability problems, liposomal drug vehicles show extensive leakage of water-soluble drugs during the passage through the gastrointestinal tract and they are heterogeneous in terms of size distribution. Therefore, scientists have been looking for new drug delivery formulations that could address these issues about liposomes, which lead to the so-called new generation of liposomes which will be summarized in this section.

Archaeosomes are liposomal formulations that are prepared with one or more lipids, mainly containing diether and/or tetraether linkages, found in archaeobacterial membrane [165]. These archaeobacterial lipids present unique features and higher stabilities to several conditions (high or low temperatures, high salinity, acidic media, anaerobic atmosphere, high pressure) over conventional liposomes [166]. The definition of archaeosomes also includes the use of synthetically derived lipids that have the properties of archaeobacterial ether lipids, that is, regularly branched phytanyl chains attached via ether bonds at sn-2,3 glycerol carbons [167]. The surprising stability of archaeosomes can be attributed to some properties brought by the archaeobacterial lipids' structure: (i) the ether linkages that are more stable than esters over a wide range of pH, and the branching methyl groups help both to reduce crystallization and permeability; (2) the stability towards oxidative degradation of these lipid membranes are provided by the fully saturated alkyl chains in the archaeobacterial lipids; (3) the unusual stereochemistry of the glycerol backbone ensures the resistance of the membrane to enzymatic attack; (4) the bipolar lipids span the membranes and enhance their stability properties [167, 168].

Archaeosomes can be prepared by using conventional procedures (hydration of a thin film followed by sonication or extrusion, detergent dialysis) at any temperature in the physiological range or lower, thus making it possible to encapsulate thermally labile compounds. Additionally, they can be prepared and stored in the presence oxygen without any degradation. According to the clinical experiments, in vivo and in vitro, these new drug delivery vehicles are not toxic. Thus, the biocompatibility and better stability of archaeosomes in numerous conditions offer advantages over conventional liposomes for their usage in biotechnology including vaccine and drug/gene delivery [167]. Consequently, they can be considered as better carriers than conventional liposomes, especially for protein and peptide delivery due to their high stability. Li et al. showed the superiority of archaeosomes over conventional liposomes in their study in which they used insulin as a model peptide for its oral delivery [169].

Another development aiming to enhance tissue targeting is virosomes in which the liposome surface is modified with fusogenic viral envelope proteins [170]. Virosomes have been used for the intracellular delivery of drugs and DNA [171, 172] as well as the basis of the newly developed vaccines which are very effective in the delivery of protein antigens to the immune system [173]. As a result, a whole set of virosomes-based vaccines have been developed for human and animal use. Special attention has been paid to the delivery of influenza vaccine using virosomes containing the spike proteins of influenza virus. Virosome-based vaccines were found to be highly immunogenic and well tolerated in children. A similar approach was used to prepare virosomal hepatitis A vaccine that elicited high antibody titres after primary and booster vaccination of infants and young children which was also confirmed for the healthy adults and elderly patients [174-176]. In general, virosomes can provide an excellent opportunity for the efficient delivery of both various antigens and many drugs, including nucleic acids, cytotoxic drugs and toxoids [177, 178], although they might present certain problems associated with their stability, leakiness and immunogenicity.

Niosomes, exhibiting a similar behavior to liposomes, are the vesicles that are made up of nonionic surfactants (e.g. alkyl ethers and alkyl esters) and cholesterol. These structures are stable on their own and they increase the stability of the encapsulated drugs. No special conditions are needed for handling and storage of these surfactants. Niosomes improve the oral bioavailability of poorly absorbed drugs, and enhance skin penetration of drug. When compared with liposomes, their oral absorption is better due to the replacement of phospholipids with nonionic surfactants which are less susceptible to the action of bile salts, parenteral, as well as topical routes. These delivery systems are biodegradable, biocompatible and non-immunogenic. Niosomes improve the therapeutic performance of drug molecules by delaying the clearance from the circulation and protecting the drug from biological environment [179].

The transdermal delivery is one of the most important routes of drug administration. The main factor which limits the application of transdermal route for drug delivery is the permeation of drugs through the skin. Human skin has selective permeability for drugs. Lipophilic drugs can pass through the skin but the drugs which are hydrophilic in nature can not pass through. Water soluble drugs either show less or no permeation. To improve the permeation of drugs through the skin various mechanisms have been investigated, including use of chemical or physical enhancers, such as iontophoresis, sonophoresis, etc. Liposomes and niosomes are not suitable for transdermal delivery due to poor skin permeability, breaking of the system, aggregation, drug leakage, and fusion of vesicles [180].

A new type of carrier system, suitable for transdermal delivery, called transfersome has been proposed for the delivery of proteins and peptides like insulin, bovin, serum albumin, vaccines, etc. These systems are soft and malleable carriers that offer noninvasive delivery of drug into or across the deeper skin layers and/or the systemic circulation [181]. Transfersomes improve the site specificity while providing the safety of the drug. Transfersomes are the lipid supramolecular aggregates which make them very flexible. This flexibility

as well as their good penetration ability causes them to be used in the effective delivery of non-steroidal anti-inflammatory agents like ibuprofen and diclofenac [182].

Alternatively, unlike classic liposomes [183, 184], that are known mainly to deliver drugs to the outer layers of skin, ethosomes can enhance permeation through the stratum corneum barrier [185-187]. Ethosomes, developed by Touitou in 1997, are the slight modification of well established drug carrier liposome, containing phospholipids, alcohol (ethanol or isopropyl alcohol) in relatively high concentration and water [188]. The size of these soft vesicles can vary from nanometers to microns [189-193]. The high concentration of ethanol makes the ethosomes unique. The ethanol in ethosomes causes disturbance in the skin lipid bilayer organization, hence when incorporated into a vesicle membrane, it enhances the vesicle's ability to penetrate the stratum corneum. Also, because of the high concentration of ethanol the lipid membrane is packed less tightly than conventional vesicles but has equivalent stability, allowing a more malleable structure and improves drug distribution ability in stratum corneum lipids. Ethosomes can be used for many purposes in drug delivery for the treatment of many diseases such as Minoxidil for baldness, testosterone as steroidal hormone, Trihexyphenidyl hydrochloride for Parkinson's disease, Zidovudine and Lamivudine as anti-HIV, Bacitracin as antibacterial, Erythromycin as antimicrobial, DNA for genetic disorders, Cannabidol in the treatment of rheumatoid arthiritis and many others [190, 192, 194-201].

Novasomes are the modified forms of liposomes [202] or a type of niosomes prepared from the mixture of monoester of polyoxyethylene fatty acids, cholesterol and free fatty acids with the diameter of 0.1-1.0 microns. They consist of two to seven bilayer shells that surround an unstructured space occupied by a large amorphous core of hydrophilic or hydrophobic materials [203]. The inner amorphous core can be loaded up to 80-85% with a medical drug and the surfaces of novasomes can be positive, negative or neutral.

Novasomes offer several advantages to the owners of the product such as: Both hydrophilic and hydrophobic products can be incorporated in the same formulation, drugs showing interactions can be incorporated in between bilayers to prevent incompatibility, they can be made site specific due to their surface charge characteristics, they can deliver a large volume of active ingredient, thus also reducing the frequency of application, and they have the ability of adhering skin or hair shafts which makes novasomes applicable in the cosmetic formulations [204].

Novasomes have extensive utilization in fields of foods, cosmetics, personal care, chemical, agrochemical and pharmaceuticals. The technology enhances absorption rate via topical delivery of pharmaceuticals and cosmeceuticals by utilizing non-phospholipid structures. Various FDA-regulated products such as human pharmaceuticals and vaccines can be developed by this technology [205, 206]. These nonionic vesicles composed of glyceryl dilaurate with cholesterol and polyoxyethylene-10-stearyl ether have been known to deliver greater amounts of cyclosporine into and through hairless mouse skin than phosphatidyl choline or ceramide based vesicles [206]. Among various liposomal formulations, novasomes appeared more effective when delivered under non-occluded conditions from a finite dose [206]. Various vaccines based on novasomes have been licensed for the immunization of fowl against Newcastle disease virus and avian rheovirus [135]. Some of the novasome-based vaccines against bacterial and viral infections have been developed such as small pox vaccine while still many are under development [207]. Novasomes inactivate viruses such as orthomyxoviruses, paramyxoviruses, coronaviruses and retroviruses, etc., by fusing with enveloped virus and that the nucleic acid of the virus denatures shortly after the fusion [208].

Although liposomes are like biomembranes, they are still foreign objects of the body. Therefore, liposomes are known by the mononuclear phagocytic system (MPS) after contact with plasma proteins. Accordingly, liposomes are cleared from the blood stream. For more than two decades, various PEG derivatives have been used to stabilize for increasing efficiency in drug or gene delivery. Most 'stabilized' liposomes, the so-called stealth liposomes [78], or cryptosomes [84], contain a certain percentage of PEG-derivatized phospholipids, which reduce the uptake by MPS, thereof prolonging the circulation times and making available abundant time for these liposomes to leak from the circulation through the leaky endothelium. Unlike, conventional liposomes, PEG-liposomes do not show dose dependent blood clearance kinetics [209]. Vesicles containing PEG-conjugated lipids at various concentrations, molecular weights, or various sizes of PEG-containing vesicles were reported to have different circulation times [81, 84, 210-212]. These kind of liposomal systems are generally used in the ligand-mediated drug targeting [213]. This stealth principle has been used to develop the successful doxorubicin-loaded liposome product that is presently available in the market as Doxil (Janssen Biotech, Inc., Horsham, USA) or Caelyx (Schering-Plough Corporation, Kenilworth, USA) for the treatment of solid tumors.

Cryptosomes is a liposomal composition for targeted delivery of drugs. The composition comprises poloxamer molecules and liposomes encapsulating one or more delivery agents. Poloxamers are polyethylene oxide (PEO)-polypropylene oxide (PPO)-polyethylene oxide tri-block co-polymers of different molecular weights. The hydrophobic PPO group in the middle links the two hydrophilic PEO groups. The hydrophilic PEO groups of a poloxamer, on either side of the central PPO unit, can provide steric protection to a bilayer surface. The amphiphilic nature of the poloxamers makes them extremely useful in various applications as emulsifiers and stabilizers. It is considered that the central PPO unit, being hydrophobic, would tend to push into the bilayer interior serving as an anchor. Dislodging the poloxamer molecule from the bilayer is achieved by reducing its hydrophobicity which is achieved by decreasing the temperature. In an aqueous medium, poloxamers stay as individual molecules at temperatures below their critical micelle temperature (CMT), but at temperatures above the CMT, they form micelles due to their amphiphilic nature. In the presence of lipid bilayers, some poloxamer molecules would partition into the bilayers as well as forming micelles with other poloxamer units. If the temperature again goes below the CMT, the poloxamer molecules lose their amphiphilic nature and disassociate from the lipid bilayer or micelle [179].

Emulsome, having the characteristics of both liposomes and emulsions, is a novel lipoidal vesicular system with an internal solid fat core surrounded by phospholipid bilayer. Emulsomes comprise a hydrophobic core (composed of solid fates instead of oils) as in standard oil-in-water emulsions, but the core is surrounded and stabilized by one or more envelopes of phospholipid bilayers as in liposomes allowing water insoluble drugs in the solution form without requiring any surface active agent or co-solvent. Emulsomes differ from liposomes since their internal core is a lipid, whereas the internal core in liposomes is an aqueous compartment. The drug loading is generally followed by sonication to produce emulsomes of smaller size [214]. These systems are often prepared by melt expression or emulsion solvent diffusive extraction. The lipid assembly of emulsomes, stabilized by cholesterol and soya lecithin (5-10% by weight), has features that are intermediate between liposomes and oil-in-water emulsions droplets. Emulsomes provide the advantages of improved hydrophobic drug loading in the internal solid lipid core and the ability of encapsulating water-soluble medicaments in the aqueous compartments of surrounding phospholipid layers.

Beside the other vesicular formulations, emulsomes are much stabilized and nano range vesicles. It is a new emerging delivery system and therefore could play a fundamental function in the effective treatment of life-threatening viral infections and fungal infections such as hepatitis, HIV, Epstein-Barr virus, leishmaniasis, etc. appear promising for the treatment of visceral leishmaniasis specifically and hepato-splenic candidiasis [214-216]. Emulsomes could be utilized in order to improve oral controlled delivery of drug, vaccine, and biomacromolecules. It is due to the fact that they are nano sized in range and could be utilized for the intravenous route. The common application areas of emulsomes are drug targeting, anti-neoplastic treatment, leishmaniasis (a disease in which a parasite of the genus Leishmania invades the cells of the liver and spleen) treatment, and biotechnology. Moreover, emulsomes could represent a more economical alternative to current commercial lipid formulations for the treatment of viral infections and fungal infections. Emulsomes provide a controlled and sustain release of drug. In comparison to the liposomes, emulsomes provide a prolong release of drug up to 24 hours, whereas liposomes have shown release up to 6 hours [217-219]. Emulsomes are nano size range in comparison to other vesicular delivery system such as niosomes and ethosomes. Due to the reduced size (10-250 nm) they can be used to enhance bioavailability to drug and as the best carrier for the intravenous drug delivery as well as oral drug delivery.

The lipid core of emulsomes may contain one or more anti-oxidants which are generally α-tocopherol or its derivatives that are the members of Vitamin-E family. The presence of anti-oxidants reduces the formation of oxidative degradation products of unsaturated lipids such as peroxides. The need of anti-oxidant can be prevented by the usage of saturated fatty acids during the preparation of the lipid core [220]. In the formation of emulsomes, like in the case of liposomes, cholesterol is essential component for the system that influences the stability of emulsomal systems and plays an important role in the drug encapsulation [221-224].

The most important advantage of emulsomes is their ability to protect the encapsulated drug from harsh gastric environment of stomach before oral administration because the drug is inside the triglyceride lipid core which can be supported that the gastric pH and the gastric enzymes are unable to hydrolyze triglycerides. Also, they resist development of multi drug resistance, often associated with over expression of a cell membrane glycoprotein, which cause efflux of the drug from the cytoplasm and results in an ineffective drug concentration inside the cellular compartment [225].

The development of emulsomes, however, is still largely empirical, and in vitro models that are predictive of oral bioavailability enhancement are lacking. There is a need for in vitro methods for predicting the dynamic changes involving the drug in the gut in order to monitor the solubilization state of the drug in vivo. Attention also needs to be paid to the interactions between lipid systems and the pharmacologically active substance. The characteristics of various lipid formulations also need to be understood, so that guidelines can be established that allow identification of suitable candidate formulations at an early stage. Future research should involve human bioavailability studies as well as more basic studies on the mechanisms of action of this fascinating and diverse group of formulations.

Unilamellar vesicles or liposomes are commonly used as simple cell models and as drug delivery vehicles to follow the release kinetics of lipophilic drugs that require compartmental models in its therapeutics and triggers. The localization of the drug at the site of action, rate of achieving the therapeutic index and circulation lifetime are the key parameters for a liposome. Lately, their arises a need for a multi-compartment structure consisting of drug-loaded liposomes encapsulated within another bilayer, is a promising drug carrier with better retention and stability due to prevention enzymes or proteins reaching the interior bilayers. A vesosome is a more or less heterogeneous, aggregated, large lipid bilayer enclosing multiple, smaller liposomes that offer a second barrier of protection for interior compartments and can also serve as the anchor for active targeting components [226, 76]. The multi-compartment structure of vesosome can also allow for independent optimization of the interior compartments and exterior bilayer; however, just the bilayer-within-a-bilayer structure of the vesosome is sufficient to increase drug retention from minutes to hours [227, 228].

In nature, eukaryotes increased their ability to optimize their response to their surroundings by developing multiple compartments, each of which has a distinct bilayer membrane, usually of quite varied composition and physical structure. Mimicking this natural progression to nested bilayer compartments led to the development of the vesosome, or vesicles deliberately trapped within another vesicle. The vesosome has distinct inner compartments separated from the external membrane; each compartment can encapsulate different materials and have different bilayer compositions. In addition, while it has proven difficult to encapsulate anything larger than molecular solutions within lipid bilayers by conventional vesicle self-assembly, the vesosome construction process lends itself to trapping colloidal particles and biological macromolecules relatively efficiently [229, 230]. The nested bilayer compartments of the vesosome provide a degree of freedom for optimization not possible with a single membrane enclosed compartment and a more realistic approximation of higher order biological organization.

The vesosome structure could be used to deliver a cocktail of antibiotics or antimicrobials to sites at a fixed ratio; such mixtures have been shown to act synergistically when delivered in a single liposome [231]. Such multi-drug formulations may be useful to avoid inducing pathogen resistance to a single drug.

As vesosomes are simply liposomes within liposomes, it should be possible to directly translate the extensive body of research on liposome drug delivery to the vesosome with only minor changes, and perhaps significant major improvements. The vesosome is created by simply self-assembly steps very similar to those used in making conventional unilamellar liposomes [229]. An important question is whether such additional effort in developing new structures will provide a therapeutic benefit over direct injection of the free drug or drug delivery by conventional unilamellar liposomes. The most obvious potential application for the vesosome is for drugs that have already shown increased efficacy by delivery with conventional liposomes. As an example, ciprofloxacin (cipro), a synthetic bactericidal fluoroquinolone antibiotic with broad spectrum efficacy, is released much more quickly from unilamellar liposomes in serum relative to saline [232, 233]. Conventional pH-loaded liposomes can retain essentially all encapsulated ciprofloxacin when stored in buffer for 12 weeks at 21 0C and 8 weeks at 37 0C [234, 235]. Although liposomal cipro has shown increased efficacy due to prolonged residence of cipro in the blood (free cipro is cleared in minutes), the half-life of release from the liposomes was only 1 hour, yet the liposomes themselves circulated for more than 24 hours [232, 235]. A second example is vincristine, a naturally occurring dimeric catharanthus alkaloid that has been used extensively as an antitumor agent since 1960's. The therapeutic activity of vincristine is dictated by the duration of therapeutic concentrations at the tumor site [236, 238, 239]. However, conventional liposomes, while offering improved bioavailability, also cannot encapsulate vincristine for sufficient time to give optimal results [234, 236, 237]. Future work will determine if multiple compartment structures like vesosome give sufficient enhancement of small drug entrapment to lead to new therapeutics.

Genetics play an increasingly important role in medicine and is used routinely to diagnose diseases and to understand malfunctions at the molecular level. The active approach of trying to amend genetic defects or insufficiencies is a logical next step. Major elements in the successful advance of gene therapy are identification of the disease and target cells, tissues and organs as well as construction of appropriate gene vectors, effective gene transfer and expression in the targeted cells. Many inherited diseases follow the Mendelian inheritance pattern in which the cause is due to a single genetic defect. Because the existing therapeutic treatments of such diseases are in most cases very limited, it is hoped that by transfecting appropriate cells with the correct gene or by adding a missing one, the disease could be alleviated. Examples of such potential treatments are for cystic fibrosis, hemophilia, sickle cell anemia or hypercholesterimia and mutant tumor suppressor genes.

The aim of gene therapy is to deliver DNA, RNA or antisense sequences to appropriate cells in order to alleviate symptoms or prevent the occurrence of a particular disease, i.e. repair the defect and also its cause. The major approaches to gene therapy include gene replacement, addition of genes for production of natural toxins, stimulation of the immune system or over expression of highly immunogenic genes for immune self-attack and sensitization of cells to other treatments.

Recently, the studies on gene delivery into eukaryotic cells by the use of non-viral-lipid-based macromolecular delivery systems have been experiencing a growing interest owing to the appearance of clinical protocols for gene therapy. Although the efficiency and specificity of such non-viral delivery systems are not yet very high, some of the problems concerning transfection methods are being successfully solved. To date, the transfection mediators that ensure effective and directed gene delivery into various cells have been created. Transfection of plasmid DNA is closely connected to the problem of condensation of its molecule since the plasmid is too large (13-15 kb) to effectively overcome the cellular membrane barrier. Besides, free DNA has to be protected from destruction by endogenous nucleases. Lastly, it is necessary to neutralize the negative charge on DNA.

Genosomes are the artificial functional complexes for functional gene or DNA delivery to cell [238]. For the production of genosomes, cationic phospholipids were found to be more suitable because they possess high biodegradability and stability in the blood stream. Gene delivery is a vast area of research and a detailed summary of work in that field is beyond the scope of this chapter.

New generation liposomes and their features are summarized in Table 2.

Table 2. New generation liposomes and their features.

Type	Main constituent	Advantage
Liposomes	Phospholipids	
Archaesomes	One or more lipids containing diether linkages	High stability at several conditions
Niosomes	Non-ionic surfactant and cholesterol	Less prone to action of bile salts
Novasomes	Monoester of polyoxyethylene fatty acids, cholesterol and free fatty acids. Two to seven bilayer shells	High loading of drugs
Transfersomes	Lipid supramolecular aggregates	More flexible hence better transdermal delivery
Ethosomes	Phospholipids and alcohol in relatively high concentration	More distruptive in the skin lipid bilayer organization hence better transdermal delivery
Virosomes	Lipids surface modified with fusogenic viral envelope proteins	Intracellular delivery of antigens, drugs and DNA
Cryptosomes	Phospholipids and polaxamers or PEG	More stable
Emulsomes	Internal solid fat core surrounded by phospholipid bilayer	Better for encapsulation of hydrophobic drugs
Vesesomes	Multilamellar liposomes	Multidrug formulations are possible
Genosomes	Complex of cationic phospholipids and a functional gene or DNA	Suitable for gene delivery

Extensively motivated by the need to increase the stability and bioavailability of drugs, and to reduce their side effects by targeting to the site of action, research in new drug delivery vehicles has taken giant steps. Liposomes and their derivatives, so called new generation liposomes, present a vast area in this field where several advances have already been achieved as summarized in this chapter. However, still further research is required to overcome the limitations faced today in terms of prolonged stability, drug loading and active targeting.

8. CONCLUSION

In the last decade from the concept of clinical utility of liposomes to their recognized position in mainstream of drug delivery systems, the path has been long and winding. The liposome systems have been explored in the clinic for applications as diverse as sites of infection and imaging, for vaccine, gene delivery and small molecular drugs, for treatment of infections and for cancer treatment, for lung disease and for skin conditions etc. Several liposomal formulations are already on the market, while quite a few are still in the pipeline for treatment of diseases. Conventional techniques for liposome preparation and size reduction remain popular as these are simple to implement and do not require sophisticated equipment. However, not all laboratory scale techniques are easy to scale-up for industrial liposome production. Many conventional methods, for preparing small and large unilamellar vesicles, involve use of either water miscible/immiscible organic solvents or detergent molecules. The need for improvements in the design and stability of liposomal diagnostic and therapeutic systems will continue to motivate innovative and efficient routes to their production.

REFERENCE

1. Bangham A.D., editor. The Liposome Letters. Academic Press; 1983.
2. Bangham A.D, Standish M.M, Watkins J.C. Diffusion of Univalent Ions Across the Lamellae of Swollen Phospholipids. Journal of Molecular Biology 1965;13 238-252.
3. Gregoriadis G, Florene A.T. Liposomes in Drug Delivery: Clinical, Diagnostic and Opthalmic Potential. Drugs 1993;45 15-28.
4. Van Rooijen N, van Nieuwmegen R. Liposomes in Immunology: Multilammellar Phosphatidylcholine Liposome as a Simple Biodegradable and Harmless Adjuvant without any Immunogenic Activity of its Own. Immunology Community 1980;9 243-356.
5. Campbell P.I. Toxicity of Some Charged Lipids Used in Liposome Preparations. Cytobios 1983;37 (1983) 21-26.
6. Grislain L, Couvreur P, Lenaerts V, Roland M, Depreg-Decampeneere D, Speiser P. Pharmacokinetics and Distribution of a Biodegradable Drug-carrier. International Journal of Pharmacology 1983;15 335-338.

7.   Illum L, Gones P.D.E, Kreuker J, Daldwin R.W, Davis D.D. Adsorption of Monoclonal Antibodies to Polyhexylcyanoacrylate Nanoparticles and Subsequent Immunospecific Binding to Tumor Cells. International Journal of Pharmacology 1983;17 65-69.

8.   Hashida M, Takahashi Y, Muranishi S, Sezaki H. An Application of Water in Oil and Gelatin Mirosphere in Oil Emulsions to Specific Delivery of Anticancer Agents into Stomach Lymphatics. Journal of Pharmacokinetics and Biopharmacetics 1977;5 241-144.

9.   Mizushima Y, Hamano T, Yokohama K. Use of a Lipid Emulsion as a Novel Carrier for Corticosteriods. Journal of Pharmacology and Pharmacotherapeutics 1982;34 49-53.

10. Gregoriadis G, Ryman B.E. Liposomes as Carriers of Enzymes or Drugs: A New Approach to the Treatment of Storage Diseases. Biochemcal Journal 1971;124 58P.

11. Gregoriadis G. Drug Entrapment in Liposomes. FEBS Letters 1973;36 292-296.

12. Gregoradis G. The Carrier Potential of Liposomes in Biology and Medicine. Part 1,The New England Journal of Medicine 1976;295 704-710.

13. Gregoradis G. The Carrier Potential of Liposomes in Biology and Medicine. Part 2, The New England Journal of Medicine 1976;295 765-770.

14. Lasic D.D, Papahadjopoulos D. Liposomes Revisited. Science 1995;267 1275-1276.

15. Noble G. T, Stefanick, J. F, Ashley, J. D, Kiziltepe, T, Bilgicer, B. Ligand-targeted Liposome Design: Challenges and Fundamental Considerations. Trends in Biotechnology 2014;32 32-45.

16. Hofheinz R.D, Gnad-Vogt S.U, Beyer U, Hochhaus A. Liposomal Encapsulated Anticancer Drugs. Anticancer Drugs 2005;16 691-707.

17. Kulkarni S.B, Betageri G.V, Singh M. Factors Affecting Microencapsulation of Drugs in Liposomes. Journal of Microencapsulation 1995;12 229-246.

18. Koudelka S, Masek J, Neuzil J, Turanek J. Lyophilized Liposome-based Formulations of Alpha-Tocopheryl Succinate: Preparation and Physico-chemical Characterisation. Journal of Pharmaceutical Sciences 2010; 99 2434-2443.

19. Kobayashi T, Tsukagoshi S, Sakurai Y. Enhancement of the Cancer Chemotherapeutic Effect of Cytosine Arabinoside Entrapped in Liposomes on Mouse Leukemia L-1210., Gann 1975;66 719-720.

20. Mayhew E, Papahadjopoulos D, Rustum Y.M, Dave C. Inhibition of Tumor Cell Growth in vitro and in vivo by 1-β-D-arabinofuranosylcytosine Entrapped within Phospholipid Vesicles. Cancer Research 1976;36 4406-4411.

21. Lopez-Berestein G, Fainstein R, Hopfer R, Mehta K, Sullivan M.P, Keating M, Rosenblum, M.G, Mehta R, Luna M, Hersh E.M, Reuben J, Juliano R.L, Bodey G.P. Liposomal Amphotericin B for the Treatment of Systemic fungal Infections in Patients with Cancer: A Preliminary Study. Journal of Infectious Diseases 1985;151 704-710.

22. Gabizon A, Peretz T, Sulkes A, Amselem S, Ben-Yosef R, Ben-baruch N, Catane R, Biran S, Barenholz Y. Systemic Administration of Doxorubicin-containing Liposomes in Cancer Patients: A Phase I Study. European Journal of Cancer and Clinical Oncology 1989;25 1795-1803.

23. Lasic DD. Liposomes: From Physics to Applications. Elsevier; 1993.

24. Gomez-Hens A, Fernandez-Romero J. M. Analytical Methods for the Control of Liposomal Delivery Systems. Trends in Analytical Chemistry 2006;25 167-178.

25. Mozafari M.R, Johnson C, Hatziantoniou S, Demetzos C. Nanoliposomes and Their Applications in Food Nanotechnology. Journal of Liposome Research 2008;18 309-327.

26. Dua J.S, Rana A.C, Bhandari A.K. Liposome: Methods of Preparation and Applications. International Journal of Pharmaceutical Studies and Research 2012;3(2) 14-20.

27. Roger R.C. New Chapter 1 Introduction, Liposomes: A Practical Approach, Edited by R. R. C. New, IRL Press at Oxford University press, 1990.

28. Shashi K., Satinder K, Bharat P. A Complete Review on Liposomes. International Research Journal of Pharmacy 2012;3(7) 10-16.

29. Rickwood D, Hames BD. Liposomes: A Practical Approach. IRL Press; 1994.

30. Mayer L.D, Hope M.J, Cullis P.R, Janoff A.S. Solute Distributions and Trapping Efficiencies Observed in Freeze-thawed Multilamellar Vesicles. Biochimica et Biophysica Acta 1985;817 193-196.

31. Imperial Chemical Industries Ltd, 1978, Belgian Patent 866697.

32. Gregoriadis G. Liposome Technology. CRC Press, Boca Raton, Volume I, II and III, 1984.

33. Huang C. Studies on Phosphatidylcholine Vesicles. Formation and Physical Characteristics. Biochemistry 1969;8(1) 344-352.

34. Olson F, Hunt C.A, Szoka F.C, Vail W, Mayhew E, Paphadjopoulos D. Preparations of Liposomes of Defined Size Distribution by Extrusion Through Polycarbonate Membranes. Biochimica Biophysica Acta, 1980, 610, 559.

35. Pick U. Liposomes with a Large Trapping Capacity Prepared by Freezing and Thawing of Sonicated Phospholpid Mixtures. Archives of Biochemistry and Biophysics 1981;212 18194.

36. Ohsawa T, Miura H, Harada K. Improvement of Encapsulation Efficiency of Water-soluble Drugs in Liposomes Formed by Freeze-thawing Method. Chemical & Pharmaceutical Bulletin 1985;33(9) 3945-3952.

37. Liu L, Yonetaini T. Preparation and Characterization of Liposome-encapsulated hemoglobin by a Freeze-thaw Method. Journal of Microencapsulation 1994;11(4) 409-421.

38. Batzri S, Korn E.D. Single Bilayer Liposomes Prepared without Sonication. Biochimica et Biophysica Acta 1976;298 1015-1019.

39.   Deamer D.W, Bangham A.D. Large Volume Liposomes by an Ether Vaporization Method. Biochimica et Biophysica Acta 1976;443 629-634.

40.   Deamer D.W. Preparation and Properties of Ether-Injection Liposomes. Annals of the New York Academy of Sciences 1978;308 250-258.

41.   Lasic D.D. Novel Applications of Liposomes. Trends in Biotechnology 1998;16 307-321.

42.   Szoka Jr. F, Papahadjopoulos P. Proceedings of the National Academy of Sciences 1978;60 4194-4198.

43.   Jain N. K. Controlled and Novel Drug Delivery. CBS Publisher, pp. 304-326.

44.   Clerc S. Barenholz Y. Loading of Amphiphatic Weak Acids into Liposomes in Response to Transmembrane Calcium Acetate Gradients. Biochimica et Biophysica Acta 1995;1240 257-265.

45.   Mayer L.D, Tai L.C, Bally M.B, Mitilenes G.N, Ginsberg R.S, Cullis P.R. Characterization of Liposomal Systems Containing Doxorubicin Entrapped in Response to pH Gradients. Biochimica et Biophysica Acta 1990;1025 143-151.

46.   Mayhew E.G, Lasic D, Babbar S, Martin F.J. Pharmacokinetics and Antitumor Activity of Epirubicin Encapsulated in Long-circulating Liposomes Incorporating a Polyethylene Glycol-derivatized Phospholipid. International Journal of Cancer 1992;51 302-309.

47.   Mayer L.D, Bally M.B, Hope M.J, Cullis P.R. Uptake of Neoplastic Agents into Large Unilamellar Vesicles in Response to a Membrane Potential. Journal of Biological Chemistry 1985;260 802-808.

48.   Mayer L.D, Bally M.B, Hope M.J, Cullis P.R. Uptake of Antineoplastic Agents into Large Unilamellar Vesicles in Response to a Membrane Potential. Biochimica et Biophysica Acta 1985;816 294-302.

49.   Bressler N.M. In Response to: Verteporfin Therapy of Subfoveal Choroidal Neovascularization in Age-related Macular Degeneration: Two Year Results of a Randomized Clinical Trial Including Lesions with Occult with No Classic Choroidal Neovascularization-Verteporfin in Photodynamic Therapy Report 2. American Journal Of Ophthalmology 2002;133 168.

50.   Lambelet P, Löliger J. The Fate of Antioxidant Radicals During Lipid Autooxidation. I. The Tocopherol Radicals. Chemistry and Physics of Lipids 1984;35 185.

51.   Kates M, Kushwaha S.C. (1976), In Lipids. Paoletti, R. et al. (eds), Raven Press, NY, Vol. 1: Biochemistry, p. 267.

52.   Lasic D.D, Stuart M.C.A, Guol Fredeik P.M. Transmembrane Gradient Driven Phase Transitions within Vesicles: Lessons for Drug Delivery. Biochimica et Biophysica Acta 1995;12139 145-154.

53.   Anchordoquy T.J, Carpenter J.F, Kroll D.J. Maintenance of Transfection Rates and Physical Characterization of Lipid/DNA Complexes After Freeze-drying and Rehydration. Archives of Biochemistry and Biophysics 1997;348 199-206.

54. Li B, Li S, Tan Y, Stolz D.B, Watkins S.C, Block L.H, Huang L. Lyophilization of Cationic Lipid-protamine-DNA (LPD) Complexes. Journal of Pharmaceutical Sciences 2000;89 355-364.

55. Molina M.C, Armstrong T.K, Zhang Y, Patel M.M, Lentz Y.K, Anchordoquy T.J. The Stability of Lyophilized Lipid/DNA Complexes During Prolonged Storage. Journal of Pharmaceutical Science 2004;93 2259-2273.

56. Bendas G, Wilhelm F, Richter W. Synthetic Glycolipids as Membrane Bound Cryoprotectants in the Freeze-drying Process of Liposomes. Biochimica et Biophysica Acta-Biomemembranes 1988;939 327-334.

57. Molina M.C, Allison S.D, Anchordoquy T.J, Maintenance of Non-Viral Vector Particle Size During the Freezing Step of the Lyophilization Process is Insufficient for Preservation of Activity: Insight from Other Structural Indications. Journal of Pharmaceutical Science.2001;90 1445-1455.

58. Sharma A, Mayhew E, Bolcsak L, Cavanaugh C, Harmon P, Janoff A, Bernacki R.J. Activity of Paclitaxel Liposome Formulations Against Human Ovarian Tumor Xenografts. International Journal of Cancer 1997;71 103-107.

59. Straubinger R, Sharma A, Mayhew E. Taxol Formulation. US Patent 5415869, 16 May, 1995.

60. http://www.avantilipids.com/index.php?view=items&cid=5&id=8&option=com_quickfaq&Itemid=385

61. Sospedra P, Nagy I.B, Haro I, Mestres C, Hudecz F, Reig F. Physicocehmical Behavior of Polylysine-[HAV-VPS Peptide] Constructs at the Air-water Interface. Langmuir 1999;15 5111-5117.

62. Zoonens M, Reshetnyak Y.K, Engelman D.M. Bilayer Interactions of pHLIP, a Peptide that can Deliver Drugs and Target Tumors. Biophyscal Journal 2008;95 225-235.

63. Coulon A, Berkane E, Sautereau A.M, Urech K, Rouge P, Lopez A. Modes of Membrane Interaction of a Natural Cysteine-rich sPeptide:Viscotoxin A3. Biochimica et Biophysica Acta-Biomembranes 2002;1559 145-159.

64. Sunamoto J, Iwamoto K, Takada M, Yuzuriha T, Katayama K. Improved Drug Delivery to Target Specific Organs Using Liposomes Coated with Polysaccharides. Polymer Science and Technology. 1983;23 157-168.

65. Jamshaid M, Farr S.J, Kearney P, Kellaway I.W. Polyoxamer Sorption on Liposomes: Comparison with Polystyrene latex and Influence on Solute Efflux. International Journal of Pharmaceutics 1998;48 125-131.

66. Kronberg B, Dahiman A, Carifors J, Karlsson J, Artursson P. Preparation and Evaluation of Sterically Stabilized Liposomes: Colloidal Stability, Serum Stability, Macrophage Uptake, and Toxicity. Journal of Pharmaceutical Sciences 1990;79 667-671.

67. Dong C, Rogers J.A. Polymer-coated Liposomes: Stability and Release of ASA from Carboxymethyl Chitin-coated Liposomes. Journal of Controlled Release 1991;17 217-224.

68.   Alamelu S, Rao K.P. Studies on the Carboxymethyl Chitosan-containing Liposomes for Their Stability and Controlled Release of Dapson. Journal of Microencapsulation 1991;8 505-515.

69.   Elferink M.G.L, de Wit J.G, Veld J.G, Int R.A, Driessen A.J.M, Ringdorf H, Konings W.N. The Stability and Functional Properties of Proteoliposomes Mixed with Dextran Derivatives Bearing Hydrophobic Anchor Groups. Biochimica et Biophysica Acta 1992;1106 23-30.

70.   O'Connor C.J, Wallace R.G, Iwamoto K, Taguchi T, Sunamoto J. Bile Salt Damage of Egg Phosphatidylcholine Liposomes. Biochimica et Biophysica Acta 1985;817 95-102.

71.   Hernandex-Caselles T, Villalain J, Gomez-Fernandez J.C. Influence of Liposome Charge and Composition on Their Interaction with Human Blood Serum Proteins Molecular and Cellular Biochemistry 1993;120(2) 119-126.

72.   Beaumier P.L, Hwang K.J. Effects of Liposome Size on the Degradation of Bovine Brain Sphingomyelin/Cholesterol Liposomes in the Mouse Liver. Biochimica et Biophysica Acta 1983;731 23-30.

73.   Alving C.R, Wassef N.M. Complement-dependent Phagocytosis of Liposomes: Suppression by Stealth Liposomes. Journal of Liposome Research 1992;2 383-395.

74.   Simoes S, Moreira J.N, Fronseca C. On the Formulation of pH Sensitive Liposomes with Long Circulating Time. Advances Drug Delivery Reviews 2004;56 947-965.

75.   Trubetskoy V.S, Torchillin V.P. Use of Polyoxyethylene-lipid Conjugates as Long-circulating Carriers for Delivery of Therapeutic and Diagnostic Agents. Advance Drug Delivery Reviews 1995;16 311-320.

76.   Barenholtz, Y. Liposome Application: Problems and Prospects. Current Opinion in Colloid and Interface Science 2001;6 66-77.

77.   Hwang M.L, Prud'homme R.K, Kohn J, Thomas J.L. Stabilization of Phosphatidyleserine/Phosphatidylethanolamine Liposomes with Hydrophilic Polymers Having Multple 'Sticky Feet'. Langmuir 2001;17 7713-7716.

78.   Woodle M.C, Lasic D.D. Sterically Stabilized Liposomes. Biochimica et Biophysica Acta 1992;1113 171-199.

79.   Woodle M.C. Controlling Liposome Blood Clearance by Surface Grafted Polymers. Advanced Drug Delivery Reviews 1998;32 139-152.

80.   Klibanov A.L, Maruyama K, Torchillin V.P, Huang L. Amphiphatic Polyethyleneglycols Effectively Prolong the Circulation Time of Liposomes. FEBS Letters 1990;268 235-237.

81.   Mori A, Klibanov A.L, Torchillin V.P, Huang L. Influence of Steric Barrier Activity of Amphiphatic poly(ethyleneglycol) and Ganglioside GM1 on the Circulation Time of Liposomes and on the Target Binding of Immunoliposomes in vivo. FEBS Letters 1991;284 263-266.

82.   Allen T.M, Mehra T, Hansen C, Chin Y.C. Stealth Liposomes: An Improved Sustained Release System for 1-β-D-arabinofuranosylcytosine. Cancer Research 1992;52 2431-2439.

83. Gabizon A, Papahadjopoulos D. The Role of Surface Charge and Hydrophilic Head Groups on Liposome Clearance in vivo. Biochimica et Biophysica Acta 1992;1103 94-100.

84. Blume G, Cevc G. Molecular Mechanism of the Lipid Vesicle Longevity in vivo. Biochimica et Biophysica Acta 1993;1146 157-168.

85. Torchillin V.P, Shtilman M.I, Trubetskoy V.S, Whiteman K, Milstein A.M. Amphiphilic Vinyl Polymers Effectively Prolong Liposome Circulation Time in vivo. Biochimica et Biophysica Acta 1994;1195 181-184.

86. Torchillin V.P, Trubetskoy V.S, Whiteman K, Caliceti P, Ferruti P, Veronese F.M. New Synthetic Amphiphilic Polymers for Steric Protection of Liposomes in vivo. Journal of Pharmaeutical Science 1995;84 1049-1053.

87. Dreschler H. Indication to prescription of Blood and Blood Substitutes in Operative Medicine. Journal of Medical Laboratory and Diagnosis (Stuttg.) 1974;27 115-121.

88. Kramer S.A. Effect of Povidone-Iodine on Wound Healing: A Review. Journal of Vascular Nursing 1999;17 17-23.

89. Allen T.M, Hansen C, Martin F, Redemann C, yau-Young A. Liposome Containing Synthetic Lipid Derivatives of Poly(ethylene Glycol) Show Prolonged Circulation Half-lives in vivo. Biochimica et Biophysica Acta 1991; 1066 29.

90. Lukyanov A.N, Torchillin V.P. Autoclaving of Liposomes. Journal of Microencapsulation 1994;11 669-672.

91. Meure L.A, Foster N.R, Dehghani F. Conventional and Dense Gas Techniques for the Production of Liposomes: A Review. AAPS Pharm SciTech 2008;9 798-809.

92. Toh M.R, Chiu G.N.C. Liposomes as Sterile Preparations and Limitations of Sterilization Techniques in Liposomal Manufacturing. Asian Journal of Pharmaceutical Science XXX, 2013;E1-E8.

93. Jain NK. Controlled and Novel Drug Delivery. CBS Publisher, 304-326.

94. Ho R.J.Y, Rouse B.T, Huang L. Target-sensitive Immunoliposomes-Preparation and Characterization. Biochemical 1986;25 5500-5506.

95. Lodge T.P, Han C.C, Akcasu A.Z. Temperature Dependene of Dynamic Light Scattering in the Intermediate Momentum Transfer Region. Macromolecules 1983;16 1180-1183.

96. Ostrowsky N. Liposome Size Measurements by Photon Correlation Spectroscopy. Chemistry and Physics of Lipids 1993;64, 45-56.

97. Strömstedt A.A, Wessman P, Ringstad L, Edwards K, Malmsten M. Effect of Lipid Headgroup Composition on the Interaction between Melittin and Lipid Bilayers. Journal of Colloid and Interface Science 2007;311 59–69.

98. du Plessis J, Ramachandran C, Weiner N, Muller D. The Influence of Lipid Composition and Lamellarity of Liposomes on the Physical Stability of Liposomes upon Storage. International Journal of Pharmaceutics 1996;127 273-278.

99.   Labhasetwar V, Mohan M.S, Dorle A.K. A study on Zeta Potential and Dielectric Constant of Liposomes. Journal of Microencapsulation 1994;11(6)663-668.

100. Vemuri S, Rhodes C. Preparation and Characterization of Liposomes as Therapeutic Delivery Systems: A Review. Pharmacetica Acta Helvtiae 1995;70 95-111.

101. Sullivan J.C, Budge S.M, Timmins A. Rapid Method for Determination of Residual tert-butanol in Liposomes Using Solid-phase Microextraction and Gas Chromatography. Journal of Chromatographic Science 2010;48(4) 289-93.

102. Biltonen R, Lichtenberg D. The Use of Differential Scanning Calorimetry as a Tool to Characterize Liposome Preparations. Chemistry and Physics of Lipids 1993;64 129-142.

103. Ohtake S, Schebor C, Palecek S, de Pablo J. Effect of Sugar-phosphate Mixtures on the Stability of DPPC Membranes in Dehydrates Systems. Cryobiology 2004;48 81-89.

104. Schofield M, Jenski L, Dumaual A, Stillwell A. Cholesterol versus Cholesterol Sulfate-Effects on Properties of Phospholipid-bilayers Containing Docosahexaenoic Acid. Chemistry and Physcs of Lipids 1998;95 23-36.

105. Buboltz J, Feigenson G. A Novel Strategy for the Preparation of Liposomes: Rapid Solvent Exchange. Biochimica et Biophysica Acta 1999;1417 232-245.

106. Manosroi A, Kongkaneramit L, Manosroi J. Characterization of Amphotericin B Liposome Forulations. Drug Development and Industrial Pharmacy 2004;30 535-543.

107. Sezer A, Bas A, Akbuga J. Encapsulation of Enrofloxacin in Liposome Preparation and in vitro Characterization of LUV. Journal of Liposome Research 2004;14 77-86.

108. Guo W, Ahmad A, Khan S, Dahhani F, Wang Y, Ahmad I. Determination by Liquid Chromatography with Fluorescence Detection of Total 7-ethyl-10-hydroxy-camptothecin (SN-38) in Beagle Dog Plasma After Intravenous Administration of Liposome-based SN-38 (LE-SN38). Journal of Chromatography B 2003;791 85-92.

109. Ramaldes G, Deverre J, Grognet J, Puisieux F, Fattal E. Use of an Enzyme Immunoassay for the Evaluation of Entrapment Efficiency and in vitro Stability in Intestinal Fluids of Liposomal Bovine Serum Albumin. International Journal of Pharmaceutics 1996;143 1-11.

110. Moon M, Giddings J. Size Distributions of Liposomes by Flow Field-Flow Fractionation. Journal of Pharmaceutical and Biomedical Analysis 1993;11 911-920.

111. Teshima M, Kawakami S, Nishida K, Nakamura J, Skaeda T, Terazono H, Kitahara T, Nakashima M, Sasaki H. Prednisolone Retention in Integrated Liposomes by Chemical Approach and Pharmaceutical Approach. Journal of Controlled Release 2004;97 211-218.

112. Arifin D, Palmer A. Determination of Size Distribution and Encapsulation Efficiency of Liposome Encapsulated Hemoglobin Blood Substitutes Using Asymmetric Flow Field-Flow Fractionation

Coupled with Multi-angle Light Scattering. Biotechnology Progress 2003;19 1798-1811.

113. Zhang X, Patel A, de Graaf R, Behar K. Determination of Liposomal Encapsulation Efficiency Using Proton NMR Spectroscopy. Chemistry and Physics of Lipids 2004;127 113-120.

114. Hinton D, Johnson C. Simultaneous Measurement of Vesicle Diffusion Coefficients and Trapping Efficiencies by Means of Diffusion Ordered 2D NMR Spectroscopy. Chemistry and Physics of Lipids 1994;69 175-178.

115. Oku N, Kendall D, MacDonald R. A Simple Procedure for the Determination of the Trapped Volume of Liposomes. Biochimica et Biophysica Acta 1982;691 332-340.

116. Mehlhorn R, Candau P, Packer L. Measurement of Volumes and Electrochemical Gradients with Spin Probes in Membrane Vesicles. Methods in Enzymology 1982;88 751-772.

117. Vistnes A, Puskin J. A Spin Label Method for Measuring Internal Volumes in Liposomes or Cells, Applied to Ca-dependent Fusion of Negatively Charged Vesicles. Biochimica et Biophysica Acta 1981;644 244-250.

118. Muthu M.S, Singh S. Targeted Nanomedicines: Effective Treatment Modalities for Cancer, AIDS and Brain Disorders. Nanomedicine-UK 2009;4 105-118.

119. Wang X, Wang Y, Chen Z.G, Shin D.M. Advances of Cancer Therapy by Nanotechnology. Cancer Research and Treatment 2009;41 1-11.

120. Torchilin V.P. Recent Advances with Liposomes as Pharmaceutical Carriers. Nature Reviews Drug Discovery 2005;4 145-160.

121. Biswas S, Torchilin V. P. Nanopreparations for Organelle-specific Delivery in Cancer. Advanced Drug Delivery Reviews 2013;66C 26-41.

122. Torchilin V.P, Omelyanenko V.G, Papisov M.I, Bogdanov A.A, Trubetskoy V.S, Herron J.N, Gentry C.A. Poly(ethylene glycol) on the Liposome Surface: on the Mechanism of Polymer-coated Liposome Longevity. Biochimica et Biophysica Acta 1994;1195 11-20.

123. Torchilin V.P, Trubetskoy V.S. Which Polymers can Make Nanoparticulate Drug Carriers Long-circulating? Advance Drug Delivery Reviews 1995;16 141-155.

124. Allen T.M, Hansen C. Pharmacokinetics of Stealth versus Conventional Liposomes: Effect of Dose. Biochimica et Biophysica Acta 1991;1068 133-141.

125. Torchilin V.P, Klibanov A.L, Huang L, O'Donnell S, Nossiff N.D, Khaw B.A. Targeted Accumulation of Polyethylene glycol-coated Immunoliposomes in Infracted Rabbit Myocardium. FASEB Journal 1992;6 2716-2719.

126. Blume G, Cevc G, Crommelin M.D, Bakker-Woudenberg I.A, Kluft C, Storm G. Specific Targeting with Poly(ethylene glycol)-modified Liposomes: Coupling of Homing Devices to the Ends of Polymeric Chains Combines Effective Target Binding with Long Circulation Times. Biochimica et Biophysica Acta 1993;1149 180-184.

127.   Torchilin V.P, Levchenco T.S, Lukyanov A.N, Khaw B.A, Klibanov A.L, Rammohan R, Samokhin G.P, Whiteman K.R. p-Nitrophenylcarbonyl-PEG-PE-liposomes: Fast and Simple Attachment of Specific Ligands, Including Monoclonal Antibodies, to Distal Ends of PEG Chains via p-Nitrophenylcarbonyl Groups. Bochimica et Biophysica Acta 2001;1511 397-411.

128.   Huwyler J, Wu D.F, Pardridge W.M. Brain Drug Delivery of Small Molecules Using Immunoliposomes. Proceeding of the National Academy of Sciences USA 1996;93 14164-14169.

129.   Smolin G, Okumoto M, Feller S, Condon D. Idoxuridine Liposome Therapy for Herpes Simplex Keratisis. American Journal of Opthalmology 1981; 91 220-225.

130.   Lee V.H, Urrea P.T, Smith R.E, Schanzlin D.J. Ocular Drug Bioavailability from Topically Applied Liposomes. Survey of Opthalmology 1985;29 335-348.

131.   Davies N.M, Fair S.J, Hadgraft J, Kellaway I.W. Evaluation of Mucoadhesive Polymers in Ocular Drug Delivery. II. Polymer Coated Vesicles. Pharmaceutical Research 1992;9 1137-1142.

132.   Gaspar M.M, Bakowsky U, Ehrhardt C. Inhaled Liposomes-Current Strategies and Future Challenges. Journal of Biomedical Nanotechnology 2008;4 1-13.

133.   Adjei AL, Gupta PK. Inhalation Delivery of Therapeutic Peptides and Proteins. Marcel Dekker Inc., New York, Basel, Hong Kong;1997.

134.   Sharma A, Sharma U.S. Liposomes in Drug Delivery: Progress and Limitations. International Journal of Pharmaceutics 1997;154 123-140.

135.   Gregoriadis G. Engineering Liposomes for Drug Delivery: Progress and Problems. Trends in Biotechnology 1995;13 527-537.

136.   Gregoriadis G., editor. Liposomes as Drug Carriers. J. Wiley;1988.

137.   Lasic, DD. and Papahadjopoulos D. Medical Applications of Liposomes. Elsevier;1988.

138.   Gregoriadis G. Overview of Liposomes. Journal of Antimicrobial Chemotherapy 1991; 28 39-48.

139.   Gregoriadis G, Florence A.T. Liposomes and Cancer Therapy. Cancer Cells 1991;4 144-146.

140.   Slingerland M, Guchelaar H.J, Gelderblom H. Liposomal Drug Formulations in Cancer Therapy: 15 Years along the Road. Drug Discovery Today 2012;17 160-6.

141.   Lasic D.D. Doxorubicin in Sterically Stabilized Liposomes. Nature 1996;380 561-562.

142.   Lasic D.D, Nedham D. The Stealth Liposome: A Prototypical Biomaterial. Chemical Reviews 1995;95 2601-2628.

143.   Gahrton G. Treatment of Acute Leukemia—Advances in Chemotherapy, Immunotherapy, and Bone Marrow Transplantation. Advances in cancer Research 1983;40 255-329.

144.   Hande K.R. Etoposide: Four Decades of Development of a Topoisomerase II Inhibitor. European Journal of Cancer 1998;34 1514-1521.

145. Reddy L.H, Adhikari J.S, Dwarakanath B.S.R, Sharma R.K. Tumoricidal Effects of Etoposide Incorporated into Solid Lipid Nanoparticles After Intraperitoneal Administration in Dalton's Lymphoma Bearing Mice. AAPS Journal 2006;8 E254-E262.

146. Hamada A, Kawaguchi T, Nakano M. Clinical Pharmacokinetics of Cytarabine Formulations. Clinical Pharmacokinetics 2002;41 705-718.

147. Sengupta S, Tyagi P. Encapsulation in Cationic Liposomes Enhances Antitumor Efficacy and Reduces the Toxicity of Etoposide, a Topoisomerase Inhibitor II. Pharmacology 2001;62 163-171.

148. Michaud L.B, Valero V, Hortobagyi G. Risks and Benefits of Taxanes in Breast and Ovarian Cancer. Drug Safety 2000;23 401-428.

149. Wani M.C, Taylor H.L, Wall M.E, Coggon P, McPhail A.T. Plant Antitumor Agents 6. Isolation and Structure of Taxol, a Novel Antileukemic and Antitumor Agent from Taxus Brevifolia. Journal of American Chemical Society 1971;93 2325-2327.

150. Rowinsky E.K, Donehower R.C. Drug Therapy-Paclitaxel (Taxol). The New England Journal of Medicine 1995;332 1004-1014.

151. Lorenz W, Reimann H.J, Schmal A, Schult H, Lang S, Ohmann C, Weber D, Kapp B, Luben L, Doenicke A. Histamine Release in Dogs by Cremophor EL and its derivatives: Oxyethylated Oleic Acid is the Most Effective Constituent. Agent Actions 1977;7 63-67.

152. Dye D, Watkins J. Suspected Anaphylactic Reaction to Cremophor EL. British Medical Journal 1980;280, 1353.

153. Weiss R.B, Donehower R.C, Wiernik P.II, Ohnuma T, Gralla R.J, Trump D.L, Jr. Baker J.R, Van Echo D.A, Von Hoff D.D, Leyland-Jones B. Hypersensitivity reactions from Taxol. Journal of Clinical Oncology 1990;8 1263-1268.

154. Xu Q.Y, Trissel L.A, Martinez J.F. Stability of Paclitaxel in 5% Dextrose Injection or 0.9% Sodium Chloride Injection at 4, 22 or 32 Degrees C. American Journal of Hospital Pharmacy 1994;51 3058-3060.

155. Lee S.C, Kim C, Kwon I.C, Chung H, Jeong S.Y. Polymeric Micelles of Poly(2-ethyl-2-oxazoline)-block-poly(epsilon-caprolactone) Copolymer as a Carrier of Paclitaxel. Journal of Controlled Release 2003;89 437-446.

156. Feng X, Yuan Y.J, Wu J.C. Synthesis and Evaluation of Water-soluble Paclitaxel Prodrugs. Bioorganic and Medicinal Chemistry Letters 2002;12 3301-3303.

157. Dosio F, Arpicco S, Brusa P, Stella B, Catel L. Poly(ethylene glycol) Human Serum Albumin-Paclitaxel Conjugates: Preparation, Characterization and Pharmacokinetics. Journal of Controlled Release 2001;76 107-117.

158. Liu C.H, Strobl J.S, Bane S, Schilling J.K, McCracken M, Chatterjee S.K, Rahim-Bata R, Kingston D.G.I. Design, Synthesis and Bioactivities of Steroid-linked Taxol Analogues as Potential Targeted Drugs for Prostate and Breast Cancer. Journal of Natural Products 2004;67 152-159.

159.   Lundberg B.B, Risovic V, Ramaswamy M, Wasan K.M. A Lipophilic Paclitaxel Derivative Incorporated in a Lipid Emulsion for Parenteral Administration. Journal of Controlled Release 2003;86 93-100.

160.   Ruan G, Feng S.S. Preparation and Characterization of Poly(lactic acid)-poly(ethylene glycol)-poly(lactic acid) (PLA-PEG-PLA) Microspheres for Controlled Release of Paclitaxel. Biomaterials 2003;24 5037-5044.

161.   Sharma U.S, Balasubramanian S.V, Straubinger R.M. Pharmaceutical and Physical Properties of Paclitaxel (Taxol) Complexes with Cyclodextrins. Journal of Pharmaceutical Sciences 1995;84 1223-1230.

162.   Merisko-Liversidge E, Sarpotdar P, Bruno J, Hajj S, Wei L, Peltier N, Rake J, Shaw J.M, Pugh S, Polin L, Jones J, Corbett T, Cooper E, Liversidge G.G. Formulation and Antitumor Activity Evaluation of Nanocrystalline Suspensions of Poorly Soluble Anticancer Drugs. Pharmaceutical Research 1996;13 272-278.

163.   Straubinger R.M, Balasubramanian S.V. Preparation and Characterization of Taxane-containing Liposomes. Methods in Enzymology 2005;391 97-117.

164.   Sharma A, Mayhew E, Straubinger R.M. Antitumor effect of Taxol-containing liposomes in a Taxol-resistant murine tumor model. Cancer Research 1993;53 5877–5881.

165.   Patel G.B, Agnew B.J, Deschatelets L, Fleming L.P, Sprott G.D. In vitro Assessment of Archaeosome Stability for Developing Oral Delivery Systems. International Journal of Pharmaceutics 2000;194 39-49, 2000.

166.   Sprott G.D. Structures of Archaeobacterial Membrane Lipids. Journal of Bioenergetics and Biomembranes 1992;24 555-566.

167.   Patel G. B, Sprott G.D. Archaeobacterial Ether Lipid Liposomes (Archaeosomes) as Novel Vaccine and Drug Delivery Systems. Critical Reviews in Biotechnology 1999;19 317-357.

168.   Jacquemet A, Barbeau J, Lemiegre L, Benvegnu T. Archaeal Tetraether Bipolar Lipids: Structure, Function and Applications. Biochimie 2009;91 711-717.

169.   Li Z, Chen J, Sun W, Xu Y. Investigation of Archaeosomes as Carriers for Oral Delivery of Peptides. Biochemical and Biophysical Research Communications 2010;394 412-417.

170.   Kaneda Y. Virosomes: Evolution of the Liposome as a Targeted Drug Delivery System. Advances Drug Delivery Reviews 2000;43 197-205.

171.   Sarkar D.P, Ramani K, Tyagi S.K. Targeted Gene Delivery by Virosomes. Methods in Molecular Biology 2002;199 163-173.

172.   Cusi M.G, Terrosi C, Savellini G.G, Genova G.D, Zurbriggen R, Correale P. Efficient Delivery of DNA to Dendritic Cells Mediated by Influenza Virosomes. Vaccine 2004;22 735-739.

173.   Bungener L, Huckriede A, Wilschut J, Daemen T. Delivery of Protein Antigens to the Immune System by Fusion-active Virosomes: A Comparison with Liposomes and ISCOMS. Bioscience Reports 2002;22 323-338.

174. Huckriede A, Bungener L, Daemen T, Wilschut J. Influenza Virosomes in Vaccine Development. Methods in Enzymology 2003;373 74-91.
175. Herzog C, Metcalfe I.C, Schaad U.B. Virosome Influenza Vaccine in Children. Vaccine 20 2002;5 B24-B28.
176. Usonis V, Bakasenas V, Valentelis R, Katiliene G, Vidzeniene D, Herzog C. Antibody Titres after Primary and Booster Vaccination of Infants and Young Children with a Virosomal Hepatitis A. Vaccine Infection 2004;32 191-198.
177. Gluck R, Moser C, Metcalfe I.C. Influenza Virosomes as an Efficient System for Adjuvanted Vaccine Delivery. Expert Opinion in Biological Therapy 2004; 4 1139-1145.
178. Moser C, Metcalfe I.C, Viret J.F. Virosomal Adjuvanted Antigen Delivery Systems. Expert Reviews of Vaccines 2003;2 189-196.
179. Gangwar M, Singh R, Goel R.K, Nath G. Recent Advances in Various Emerging Vesicular Systems: An Overview. Asian Pacific Journal of Tropical Biomedicine 2012 1176-1188.
180. Garg B.J, Saraswat A, Bhatia A, Katare O.P. Topical Treatment in Vitiligo and the Potential Uses of New Drug Delivery Systems. Indian Journal of Dermatology Venereology and Leprology 2010;76 231-238.
181. Jain S, Mishra D, Kuksal A, Tiwary A.K, Jain N.K. Vesicular Approach for Drug Delivery into or across the Skin: Current Status and Future Prospects" Available from: http://priory.com/pharmol/Manuscript-Jain.htm
182. Jain S, Jaio N, Bhadra D, Tiwary A.K, Jain N.K. Delivery of Nonsteroidal Anti-inflammatory Agents like Diclofenac. Current Drug Delivery 2005;2 223-227.
183. Size determination of liposomes, Liposomes-A practical approach, edited by RRC New (Oxford University Press, NewYork) 1990, 154.
184. Heeremans J.L.M, Gerristen H.R, Meusen S.P, Mijnheer F.W, Gangaram R.S, Panday G, Prevost R, Kluft C, Crommelin D.J.A. The Preparation of Tissue Type Plasminogen Activator (t- PA) Containing Liposomes: Entrapment Efficacy and Ultracentrifugation Damage. Journal of Drug Targeting 1995; 3 301-306.
185. Preparation of liposomes and size determination, Liposomes-A practical approach, edited by RRC New (Oxford University Press, New York) 1990, 36.
186. Touitou E, Dayan N, Levi-Schaffer F, Piliponsky A.Novel lipid vesicular system for enhanced delivery. Journal of Lipid Research 1998; 8 113-114.
187. Asbill C.S, El-Kattan A.F, Michniak B. Enhancement of Transdermal Drug Delivery: Chemical and Physical Approaches. Critical Reviews in Therapeutic Drug Carrier Systems 2000;17 621-658.
188. Touitou E. Inventor. "Composition of Applying Active Substance to or through the Skin" U.S. Patent 5 540 934. July 20, 1996.
189. Patel S. Ethosomes: A promising tool for transdermal delivery of drug. Pharma Info.Net 2007;5(3).

190.   Touitou E, Dayan N, Bergelson L, Godin B, Eliaz M. Ethosomes Novel Vesicular Carriers for Enhanced Delivery: Characterization and Skin Penetration Properties, Journal of Controlled Release 2000;65 403-418.

191.   Touitou E, inventor. Composition of applying active substance to or through the skin. US patent 5 716 638. October 2, 1998.

192.   Touitou E, Godin B, Dayan N, Weiss C, Piliponsky A, Levi-Schaffer F. Intracellular Delivery Mediated by an Ethosomal Carrier. Biomaterials 2001;22 3053-3059.

193.   Jain S, Umamaheshwari R.B, Bhadra D, Jain N.K. Ethosomes: a Novel Vesicular Carrier for Enhanced Transdermal Delivery of an anti-HIV Agent, Indian Journal of Pharmaceutical Science 2004;66 72-81.

194.   Bhalaria M.K, Naik S, Misra A.N. Ethosomes:A Novel Delivery System for Antifungal Drugs in the Treatment of Topical Fungal Diseases, Indian Journal of Experimental Biology 2009;47 368-375.

195.   Guo J, Ping Q, Sun G, Jiao C. Lecithin Vesicular Carriers for Transdermal Delivery ofCyclosporine A. International Journal of Pharmaceutics 2000;194(2) 201-207.

196.   Touitou E, Godin B, Dayan N, Vaisman B. Intracellular Delivery by Cationic Ethosoms not Containing Positively Charged Phospholipids: Characterization and Intracellular Delivery Properties. Proceedings of the 6th PAT & 4th ICRS InternationalSymposium, Eilat, Israel.2001.

197.   Lodzki M, Godin B, Rakou L, Mechoulam R, Gallily R, Touitou E. Cannabidiol- Transdermal Delivery and Anti-inflammatory Effects in a Murine Model. Journal of Controlled Release 2003;93 377-387.

198.   Horwitz E, Pisanty S,Czerninsky R, Helser M, Eliav E, Touitou E.A Clinical Evaluation of a Novel Liposomal Carrier for Acyclovir in the Topical Treatments of Recurrent Herpes Labialis.Oral Surgery Oral Pathology Oral Radiology Endodontology 1999;88 700-05.

199.   Dkeidek I, Touitou E. Transdermal Absorption of Polypeptides. AAPS Pharmaceutical Science 1999;1 202-210.

200.   Paolino D, Lucania G, Mardente D, Alhaique F, Fresta M. Ethosomes for Skin Delivery of Ammonium Glycyrrhizinate: In vitro Percutaneous Permeation through Human Skin and in vivo Anti-inflammatory Activity on Human Volunteers. Journal of Controlled Release 2005;106 99-110

201.   Dubey V, Mishra D, Dutta T, Nahar M, Saraf D.K, Jain N.K. Dermal and Transdermal Delivery of an Anti-psoriatic Agent via Ethanolic Liposomes, Journal of Controlled Release 2007;123 148–154.

202.   Lasic, DD. Liposomes and Niosomes. In: Surfactants in Cosmetics, Eds., M.M. Rieger and L.D. Rhein, Marcel Dekker: New York, 1997. p: 263-283.

203.   Pinsky, M.A, 2010. "Materials and Methods for Delivering Antioxidants into the Skin" U.S. Patent 20,100,098,752.

204.   Mills, R, Mathur R, Lawrence N. 2007.Mahonia aquifolim extract, Extraction process and pharmaceutical composition containing the same.U.S. Patent 20,070,148,226.

205.   Wallach, D.F.H, Mathur R, Redziniak G.J.M, Tranchant J.F. Some Properties of N-acyl Sarcosinate Lipid Vesicles. Journal of the Society of Cosmetic Chemists 1992;43(2) 113-118.

206.   Williams AC.Transdermal and Topical Drug Delivery from Theory to Clinical Practice. Pharmaceutical Press: United Kingdom;2003

207.   Frézard F. Liposomes: From Biophysics to the Design of Peptide Vaccines. Brazilian Journal of Medical and Biological Research 1999;32(2) 181-189.

208.   Varanelli C, Kumar S, Wallach D.F.H. 1996. Method of Inhibiting Viral Reproduction Using Non-phospholipid, Paucilamellar Liposomes.U.S. Patent 5,561,062.

209.   Desai N.P, Hubbell J.A. Solution Technique to Incorporate Polyethylene Oxide and Other Water-Soluble Polymers into Surfaces of Polymeric Biomaterials. Biomaterails 1991;12 144-153.

210.   Klibanov A.L, Maruyama, Beckerleg A.M, Torchilin V.P, Huang L. Activity of Amphipathic Polyethyleneglycol 5000 to Prolong the Circulation Time of Liposomes Depends on the Liposome Size and is Unfavorable for Immunoliposome Binding to Target. Biochimica et BiophysicaActa 1991;1062 142–148.

211.   Maruyama K., Yuda T, Okamoto A, Ishikura C, Kojima S, Iwatsuru M. Effect of Molecular Weight in Amphiphatic Polyethyleneglycol on Prolonging the Circulation Time of Large Unilamellar Liposomes. Chemical and Pharmaceutical Bulletin 1991;39(6) 1620-1622.

212.   Allen T.M, Hansen C, Martin F, Redemenn C, Yau-Young A. Liposomes Containing Synthetic Lipid Derivatives of Poly(ethylene glycol) Show Prolonged Circulation Half-lives in vivo. Biochimica et Biophysica Acta 1991;1066 29-36.

213.   Cevc G.G. Drug-carrier and Stability Properties of the Long-lived Lipid Vesicles Cryptosomes, in vitro and in vivo. Journal of Liposome Research 1992;2 355-68.

214.   Vyas S.P, Subhedar R, Jain S. Development and Characterization of Emulsomes for Sustained and Targeted Delivery of an Antiviral Agent to Liver. Journal of Pharmacology and Pharmacotherapeutics 2006;58 321-6.

215.   Paliwal R, Paliwal S.R, Mishra N, Mehta A, Vyas S.P. Engineered Chylomicron Mimicking Carrier Emulsome for Lymph Targeted Oral Delivery of Methotrexat. International Journal of Pharmaceutics 2009;380 181-8.

216.   VanCott T.C, Kaminski R.W, Mascola J.R, Kalyanaraman V.S, Wassef N.M, Alving C.R. HIV-1 Neutralizing Antibodies in the Genital and Respiratory Tracts of Mice Intranasally Immunized with Oligomeric gp160. Journal of Immunology 1998;160 2000-12.

217.   Schwarz C, Mehnert W, Lucks J.S, Muller R.H. Solid Lipid Nanoparticles (SLNs) for Controlled Drug Delivery: I. Production, Characterization and Sterilization. Journal of Controlled Release 1994;30 83-96.

218. Vyas S.P, Gupta S, Dube A. Antileishmanial Efficacy of Amphotericin B Bearing Emulsomes Against Experimental Visceral Leishmaniasis. Journal of Drug Targeting 2007;15 437-44.

219. Gupta S, Vyas S.P. Development and Characterization of Amphotericin B Bearing Emulsomes for Passive and Active Macrophage Targeting. Journal of Drug Targeting 2006;15 206-17.

220. Solid Fat Nanoemulsions As Drug Delivery Vehicles-Patent 5576016. Available from: http://www.docstoc.com/docs/48874511/Solid-Fat-Nanoemulsions-As-Drug-Delivery-Vehicles-Patent-5576016.

221. Barry B.W. Novel Mechanisms and Devices to Enable Successful Transdermal Drug Delivery. European Journal of Pharmaceutical Sciences 2001;14 101-14.

222. Devaraj G.N, Parakh S.R, Devraj R, Apte S.S, Rao B.R, Rambhau D. Release Studies on Niosomes Containing Fatty Alcohols as Bilayer Stabilizers Instead of Cholesterol. Journal of Colloid and Interface Science 2002;251 360-365.

223. Kirby C, Clarke J, Gregoriadis G. Effect of Cholesterol Content of Small Unilamellar Liposomes on Their Stability in vivo and in vitro. Biochemical Journal 1980;186 591-8.

224. Harasym T.O, Cullis P.R, Bally M.B. Intratumor Distribution of Doxorubicin Following I.V. Administration of Drug Encapsulated in Egg Phosphatidylcholine Cholesterol Liposomes. Cancer Chemotherapy Pharmacology 1997;40 309-17.

225. Gill B, Singh J, Sharma V, Hari Kumar S.L. Emulsomes: An Emerging Vesicular Drug Delivery System. Asian Journal of Pharmaceutics2012;6 87-94.

226. Bangham A.D, Standish M.M, Watkins J.C. Diffusion of Univalent Ions across the Lamellae of Swollen Phospholipids. Journal of Molecular Biology 1965;13 238-252.

227. Allen T.M. Liposomal Drug Delivery. Current Opinion in Colloid and Interface Science 1996;1: 645-652.

228. Gabizon A, Papahadjopoulos D. Liposome Formulations with Prolonged Circulation Time in Blood and Enhanced Uptake by Tumors. Proceedings of the National Academy of Sciences 1988;85 6949-6953.

229. Kisak E, Coldren B, Zasadzinski J.A. Nanocompartments Enclosing Vesicles, Colloids, and Macromolecules via Interdigitated Lipid Bilayers. Langmuir 2002;18 284-288.

230. Evans, C.A, Zasadzinski J.A. Encapsulating Vesicles and Colloids in Cochelate Cylinders. Langmuir 2003;193109-3113.

231. Schiffelers R.M, Storm G, ten Kate M.T, Stearne-Cullen L.E.T., Hollander J.G., Verbrugh H.A, Bakker-Woudenberg I. Journal of Liposome Research 2002;12 122-129.

232. Maurer N, Wong K.F, Hope M.J, Cullis P.R. Anomalous Solubility Behavior of the Antibiotic Ciprofloxacin Encapsulated in Liposomes: a H-NMR Study. Biochimica Biophysica Acta 1998;1374 9-20.

233. Maurer-Spurej E, Wong K.F, Maurer N, Fenske D.B, Cullis P.R. Factor Influencing Uptake and Retention of Amino-containing Drugs in Large

Unilamellar Vesicles Exhibiting Transmembrane pH Gradient. Biochimica Biophysica Acta 1999;1416 1-10.

234. Webb, M.S., N.L. Boman, D.J. Wiseman, D. Saxon, K. Sutton, K.F. Wong, P. Logan and M.J. Hope. Antibacteria Efficacy Against an in vivo Salmonella Typhimurium Infection Model and Pharmacokinetics of a Liposomal Ciprofloxacin Formulation. Journal of Antimicrobial Agents and Chemotherapy 1998;42 45-52.

235. Bakker-Woudenberg I, ten Kate M.T, Guo L, Working P, Mouton J.W. Improved Efficacy of Ciprofloxacin Administered in Polyethylene Glycol-coated Liposomes for Treatment of Klebsiella Pneumoniae Pneumonia in Rats. Journal of Antimicrobial Agents and Chemotherapy 2001;45 1487-1492.

236. Embree, L, Gelmon K, Tolcher A, Hudon N, Heggie J, Dedhar C, Logan P, Bally M.B, Mayer L.D. Pharmacokinetic Behavior of Vincristine Sulfate Following Administration of Vincristine Sulfate Liposome Injection. Cancer Chemotherapy and Pharmacology 1998;41 347-352.

237. Krishna R, Webb M.S, St. Onge G, Mayer L.D. Liposomal and Nonliposomal Drug Pharmacokinetics After Administration of Liposome-Encapsulated Vincristine and Their Contribution to Drug Tissue Distribution Properties. Journal of Pharmacology and Experimental Therapeutics 2001;298 1206-1212.

238. Alatorre-Meda M, Gonzales –Perez A, Rodriguez J.R. DNA Metafectene Pro-Complexation: A Physical Chemistry Study. Physical Chemistry Chemical Physics 2010;12 7464-7472.

CHAPTER 3

Nanotechnology-Based Drug Delivery Systems for Melanoma Antitumoral Therapy: A Review

Roberta Balansin Rigon, Márcia Helena Oyafuso, Andressa Terumi Fujimura, Maíra Lima Gonçalez, Alice Haddad do Prado, Maria Palmira Daflon Gremião, and Marlus Chorilli

School of Pharmaceutical Sciences, Department of Drug and Medicines, São Paulo State University, 14801-902 Araraquara, SP, Brazil

ABSTRACT

Melanoma (MEL) is a less common type of skin cancer, but it is more aggressive with a high mortality rate. The World Cancer Research Fund International (GLOBOCAN 2012) estimates that there were 230,000 new cases of MEL in the world in 2012. Conventional MEL treatment includes surgery and chemotherapy, but many of the chemotherapeutic agents used present undesirable properties. Drug delivery systems are an alternative strategy by which to carry antineoplastic agents. Encapsulated drugs are advantageous due to such properties as high stability, better bioavailability, controlled drug release, a long blood circulation time, selective organ or tissue distribution, a lower total required dose, and minimal toxic side effects. This review of scientific research supports applying a nanotechnology-based drug delivery system for MEL therapy.

1. INTRODUCTION

Malignant melanoma (MEL) are tumors that mainly affect adult and elderly patients; the highest incidence is at approximately 60 years of age [1]. However, currently, MEL recurrence has increased in young adults and can be observed in children and adolescents [2].

The World Cancer Research Fund International (GLOBOCAN 2012) estimates that there were 230,000 new cases of MEL in the world in 2012; MEL incidence rates are much higher in the White population than in the Black population, and it is uncommon in the Asian population, likely due to better protection from their skin pigment and different sun exposure habits; African and Asian societies consider fair skin beautiful [3]. In addition, rich populations have a high rate of MEL with a relatively low rate of mortality from this disease, potentially because MEL is diagnosed in early stages for this social class [4–6].

Pathogenesis of Melanoma. Melanocytic skin tumors include a wide variety of benign and malignant skin lesions with distinct clinical, morphological, and genetic profiles [7–9].

Melanoma describes melanocyte malignance; a melanocyte is a melanin-producing cell located in the basal layer of the epidermis [10]. When it functions normally, the melanocyte provides basic skin pigmentation and protects against UV radiation damage [11–13].

In summary, the most significant causes of MEL development are at personal history of MEL in the family, advanced age, the presence an atypical nevus, intense exposure to sunlight, sunburn during childhood [14], and chronic immunosuppression [15]; it is especially observed in posttransplant patients and patients with acquired immunodeficiency syndrome (AIDS) or a prior cancer diagnosis [11].

Genetic predisposition plays an important role in MEL development due to the relative risk of people with a family history of MEL developing this cancer, which is 2-3 times greater than in people without such a family history; several genes (CDKN2A; BRAFV600E; N-Ras codon 61; CKIT; GNAQ/GNA11; BRCA2; OCA1 and MC1R) related to this predisposition have been identified [2, 16–20].

UV radiation also has a profound influence on MEL development. Sunscreens use, which protect the skin against this radiation, does not prevent MEL development, because the UV radiation spectrum that causes erythema (UVB) and that traditional sunscreens protect against differ from the spectrum that promotes MEL (UVA). Thus, users of sunscreens are relatively unprotected from UVA radiation [11]. An alternative theory suggests that vitamin D, which inhibits the signaling pathway involved in MEL development [21] (i.e., the MAP kinase pathway that promotes cell proliferation), is synthesized upon UV radiation, and when radiation is blocked by sunscreen, vitamin D synthesis stops [22, 23].

The cutaneous MEL is manifested in different regions of the body through lesions on the head and neck and is associated with chronic sun exposure and lesions on the trunk related to the presence of numerous melanocytic nevi [24].

Almost all MEL lesions are pigmented and flat; malignant melanocytes growth is restricted to the epidermis ("MEL in situ"), and the cells are characterized by a relatively homogeneous brown pigmentation with slightly irregular edges [25, 26]. Over time, likely many years, these lesions present with irregular edges and pigmentation. In late stages, this neoplastic growth is vertical, and the tumor cells infiltrate through collagen fibers in the reticular dermis [27]. The subcutaneous tissue is then infiltrated by the tumor, which forms papules and nodules, and is typically confined to the lesion area [28]. Partial regression of the lesion is common, which functions through an immune mediated phenomenon that promotes malignant melanocyte elimination by cytotoxic lymphocytes [29]. However, complete MEL regression may be associated with the spread of metastasis, which is a negative, not positive prognostic sign [2, 30].

For melanocyte transformation in MEL, resistance to apoptosis is necessary [31], and MELs escape from apoptosis stimulation through overexpressing apoptosis-inhibiting genes (e.g., inhibitor of apoptosis proteins (IAPs), especially survivin) or decreasing apoptosis-inducing gene expression, which results in apoptosis dysfunction and an increased risk of metastasis [32]. The serine/threonine kinase Akt/protein kinase B and transcription factor nuclear factor- (NF-) participate in the cell proliferation control, apoptosis, and oncogenesis [33], and certain studies suggest that Akt activation can facilitate MEL progression by increasing cells survival through NF- regulation with a consequent reduction in apoptosis [20].

Classification of Melanoma and Diagnoses. MEL is clinically classified into four main groups [34]. The first group is lentigo maligna MEL, which is characterized by an invasive tumor in the head, neck, or forearms regions [35]. Another group is superficial-spreading MEL, which is characterized by a lesion with irregular edges and pigmentation that grows laterally and slowly before promoting vertical invasion [36]. The next group is nodular MEL, which is a more aggressive type that appears in the body following high levels of sunlight exposure [37]. The final group is acral lentiginous MEL, which are pigmented lesions that appear on the palms of the hands, soles of the feet, and above the nose [38, 39]. Other classifications include amelanotic MEL, mucosal MEL, and subungual MEL [2].

For MEL diagnosis, five main characteristics of the lesion are analyzed: asymmetry, border-color, diameter, and elevation; MEL diagnoses are more accurate where dermatoscopy is used [11]. However, for many people, the first area that metastasizes is the lymph node (sentinel); the next most common site of metastasis is distant skin. The organs more frequently affected are the lungs and the liver; the central nervous system and bones can also be metastasis sites [40, 41].

The MEL stage can be determined through a complete clinical examination [42], including sonography [43] of the superficial lymph nodes and the abdomen, radiography of the thorax [44], and evaluation of serum markers, such as lactate dehydrogenase (LDH) [45], S-100-beta, sialic acid, enolase, 5-S-cyseinyldopa, 6-hydroxy-5-methyoxy-indole-2-carboxylic acid, 3,4-dihydroxy-L-phenylalanine (DOPA), L-tyrosine [46], computer tomography scan [47], magnetic resonance imaging [48], bone scintigraphy [49], and positron emission tomography, which are useful for evaluating patients with metastatic disease [50].

Moreover, an immunohistochemical technique can also be used to diagnose metastasis because antigens are expressed on malignant cells' membrane and cytoplasm surface, which can be immunohistochemically detected using antibodies that are specific to these antigens [51]. The antibodies commonly used are anti-S100, HMB-45, and MART-1 e NK1/C3.

Conventional Therapeutic Strategies against Melanoma. Chemoprevention can be used to avoid MEL development; chemoprevention was originally proposed by Sporn et al. (1976) [52] and refers to using synthetic or natural agents to reverse, suppress, or prevent molecular and histological premalignant lesions that occur with invasive cancer progression [20]. Reactive oxygen species play a role in MEL progression because they lead to uncontrolled overregulation and compartmentalization of melanosomes [53], a diet rich in antioxidants, particularly carotenoids and vitamins C and E, which can be used for chemoprevention [54,55].

The conventional treatment for primary MEL is surgical; the lesion is removed, and the tissue is analyzed to determine the MEL stage, which depend on the lesion thickness and location (epidermis or dermis). The lesion is removed with a certain safety margin; however, where lesion excision is inappropriate, such as for MELs in the nasopharyngeal, sinonasal, and oral regions, radiotherapy is a way to eliminate the lesion. For patients who present risk of metastasis, the above indicated laboratory tests are also used, such as radiography of the thorax [11, 56, 57].

The conventional MEL chemotherapy treatment is performed using dacarbazine, temozolomide (dacarbazine analogue), nitrosoureas (carmustine, lomustine), vinca alkaloids (vincristine, vinblastine), platinum compounds (cisplatin, carboplatin), and taxanes (Taxol, docetaxel), but these single agents are not an improvement over dacarbazine [32, 58]. Immunotherapy has also been applied for MEL therapy; immunotherapy employs cytokines that stimulate the patient's immune system to fight cancer, such as interleukin (IL), IL-2, IL-5, IL-7, and IL-21, interferon-α (INF-α), and granulocyte macrophage colony–stimulating factor (GM-CSF) [59]. These cytokines have side effects, such as diarrhea, nausea, constipation, abdominal pain, vomiting, vitiligo, dermatitis, enterocolitis, hepatitis, toxic epidermal necrolysis, neuropathy, and endocrinopathy [11].

The benefits of therapy with interferon alfa-2b are directly related to the MEL stage [60]. However, high interferon alfa-2b doses have many side effects, such as chronic fatigue, headaches, weight loss, myelosuppression, and depression [61].

Vemurafenib and dabrafenib are BRAF inhibitors approved for use in MEL metastases that express BRAFV600E and lead to dramatic shrinkage of tumors. However, they are short-lived and resistance to treatment eventually emerges in most melanomas. In addition, treatment with BRAF inhibitors can lead to the induction of second primary cancers, including squamous cell carcinomas of the skin and new primary BRAF wild-type melanomas; other side effects are nausea, diarrhea, arthralgias, nonspecific skin rashes, fatigue, alopecia, and photosensitivity [62–64].

Tremelimumab is an antibody against the cytotoxic T lymphocyte-associated antigen 4 and is well-tolerated, but it does not offer many benefits over conventional chemotherapy [65].

Ipilimumab is a humanized antibody against CTLA-4, a negative regulatory checkpoint protein that is expressed on T cells surface after activation; the ipilimumab specifically blocks the CTLA-4 inhibitory signal, resulting in activation of T cells and tumour infiltrating lymphocytes; this is an indirect mechanism that enhances the immune response mediated by T cells. The adverse effects are colitis, dermatitis, hepatitis, endocrinopathy, and neuritis [63, 66].

Nivolumab and pembrolizumab are anti-PD-1 antibodies; PD-1, like CTLA-4, is expressed on the surface of activated T cells and has a function to turn off the T-cell response to prevent an excessive immune reaction. Anti-PD-1 antibodies may have higher response and lower toxicity rates than ipilimumab, as well as improved overall survival compared to chemotherapy [63, 66, 67].

Studies demonstrated that a combination therapy with ipilimumab and nivolumad was responsible for more adverse effects than monotherapy; on the other hand the patient's median survival was higher when patients were treated with combination therapy [66]. Therefore, several studies are being realized to evaluate the potential survival benefits of immunotherapy combination [68].

2. NANOTECHNOLOGY-BASED DRUG DELIVERY SYSTEMS

Many active ingredients used in MEL therapy present undesirable properties and, thus, have been discarded [69]. Introducing a new active ingredient on the market takes several years of research and involves high costs. The alternative employed to circumvent these high costs and reintroduce the active ingredients that were previously discarded is the development of delivery systems that increase efficiency [70].

Drug delivery systems represent an alternative strategy to carrier antineoplastic agents. Encapsulated drug could result in advantages such as high stability, better bioavailability, controlled drug release, long circulation time in blood, selective organs or tissue distribution, a reduction of the total dose required, and minimizing the toxic side effects [71–73]. Nanotechnology-based drug delivery systems are widely used to improve the effectiveness of antineoplastic agent; the most common nanosystems are hydrogel, cyclodextrins, liquid crystalline phase, and nanoparticulate pharmaceutical drug delivery systems (NDDSs), as classified by Torchilin (2014), that include liposomes; polymeric nanoparticles; polymeric micelles; silica, gold, silver, and other metal nanoparticles; carbon nanotubes; solid lipid nanoparticles; niosomes; and dendrimers [74]. This review of scientifically based research supports the application of nanotechnology-based drug delivery system for MEL therapy.

2.1. Hydrogels

Polymeric systems can be classified by their physical forms such as (i) linear polymer chain in solution, (ii) physically or covalently cross-linked reversible gels, and (iii) polymer chains grafting or adsorption on the surface of micro- and nanoparticles [75]. A hydrogel is a network of polymer chains that are hydrophilic and promote the drug release through the spaces formed in the network via dissolution or disintegration of the polymeric matrix. Swelling in certain non-water-soluble polymer demonstrates a high water absorption capacity in the reticular structure (>20%) [76, 77].

An increased interest in hydrogels as a drug delivery system has been demonstrated as a result of their easy handling and similar physical properties to animal tissue, which depend on the polymer employed [78–80]. The release rate depends on the hydrogel properties, initial drug concentration, drug solubility, and drug-polymer interaction [81].

A wide variety of polymeric materials with different properties have been used to form hydrogels. The required polymer is selected based on the tissue of interest and the specific application [80, 82]. For example, poly(vinyl alcohol) tetrahydroxyborate (PVA-THB) hydrogels have shown therapeutic potential for topically treating acute and chronic wounds due to many benefits, such as controlled release, bioadhesion, and low toxicity [83,84]. Moreover, chitosan-based hydrogels have an additional advantage as a drug delivery system because the drug can be released under various environmental stimuli; thus, these hydrogels provide an anchor for delivering therapeutic payloads to the site of action [85].

Hydrogels have been employed as a drug delivery system in MEL therapy because they may act as an intratumoral chemotherapy depot by promoting accumulation or maintenance of high intracellular levels of the chemotherapeutic agent. In recent years, a hydrogel composed of a cyclodextrin-containing linear polymer and decorated with PEG as well as transferrin was approved for commercial use in MEL therapy [86].

Hydrogels are classified as stimuli-sensitive swelling-controlled release systems because they can respond to various environmental conditions, such as pH, the surrounding fluid ionic strength, temperature, an applied electrical or magnetic field, or glucose level changes. These changes promote altered network structure, swelling, mechanical strength and permeability [83]. Thus, hydrogels may be used to improve drug delivery [87–90]. However, little evidence supports using hydrogels for topical treatment of MEL.

Certain studies suggest using topical hydrogels; topical ibuprofen-releasing hydrogels promote lower metastatic spread of primary MEL through significantly lower tumor necrosis factor (TNF)-α levels, which is the major proinflammatory cytokine that induces MEL cell migrations [91].

Injectable hydrogels have been widely explored for cancer therapy [85, 92]. Interleukin-2 was given as pulse in cancer immunotherapy because it is a potent immunomodulator that can induce antitumor activity [93]. Recombinant human interleukin-2 (rhIL-2) loaded, in situ gelling, and physically cross-linked dextran hydrogels slowly release rhIL-2 and maintain the rhIL-2 protein in an intact form that is both biologically and therapeutically active, which greatly enhances the clinical applicability of rhIL-2 immunotherapy [94].

Subcutaneous injection of a doxorubicin-loaded hydrogel composed of sugar beet pectin (SBP) associated with biodegradable gelatin (SBP/gelatin) successfully suppressed mouse MEL B16F1 cell tumor growth in nude mice [95].

The human MEL cell line Me665/2/21 derived from a cutaneous metastasis was treated for 48 h with a cisplatin-loaded hydrogel and it showed similar and, in certain cases, higher cytotoxic activity towards the MEL cell line compared with free cisplatin at the same concentration [79].

A novel system for incorporating paclitaxel has been investigated to lower toxicity and improve efficacy [96]. Tumor activity upon using a paclitaxel (PTX) loaded hydrogel composed of a pH- and temperature-sensitive block copolymer, the poly(ε-caprolactone-co-lactide)–poly(ethylene glycol)–poly(ε-caprolactone-co-lactide) (PCLA–PEG–PCLA) block copolymer, was analyzed in vivo using B16F10 MEL cells. After 2 weeks of subcutaneously injecting the B16F10 MEL cells into male mice, the tumors were allowed to grow, and the results demonstrate that saline-treated mice produced a tumor volume of approximately $17 \, cm^3$, while the PTX-treated mice tumors were smaller than $7 \, cm^3$, which demonstrates that a PTX-loaded block copolymer hydrogel can effectively suppress tumor development [97].

2.2. Liposomes and Micelles

Recently, research has described the importance of lipids in drug carrier systems such as liposomes [98]. According to Fahy and coworkers (2005) [99], lipids are hydrophobic molecules that are soluble in organic solvents. However, certain lipids have amphiphilic characteristics due to both hydrophilic and hydrophobic segments [100]. These compounds can form carrier systems, such as micelles and liposomes, because they can self-assemble in the presence of water [101]. Micelles are formed when amphiphilic components concentration exceeds a certain threshold concentration. The micelles' size and shape depend on pH, temperature, constituent geometry, and intermolecular interactions [98, 99, 102, 103].

The liposomes are microscopic spherical nanostructured with a well-defined shape and size, which varies from 10 nm to several micrometers, depending on the technique used to create them [104]. These vesicles are formed by an external phase with double phospholipid membranes and an internal phase formed by an aqueous medium [105]. These components provide an amphiphilic character due to the organized double phospholipid layer that surrounds the aqueous compartment. Thus, they can encapsulate both hydrophobic and hydrophilic compounds [106, 107]. Figures 1, 2, and 3 schematically show the micelles and liposomes structures.

Liposomes have high versatility because they can be modified based on pharmacological and pharmaceutical needs. Thus, the size, surface, lamellarity, lipid composition, volume, and inner aqueous medium composition can be modified in these vesicles [108].

Liposomes can be formed with natural lipids, such as sphingomyelins, as well as lecithins and synthetic lipids, such as dimyristoyl, distearoyl, dipalmitoyl, and dioleoyl [109]. Currently, there are several methods to obtain liposomes. Under certain conditions, they can undergo spontaneous rearrangement and be derived from preformed micelles by changing the solution or applying external energy, such as by extrusion through filter membranes, sonication, or agitation [98, 110, 111].

The extrusion technique forces a lipid suspension to pass through a polycarbonate membrane with a well-defined pore size [112]. This method can produce vesicles with a diameter near the membrane pore size used to prepare the liposomes. Over time, studies have shown several advantages from this technique; for example, the average size of the vesicles formed is reproducible due to the physical process involved in liposome formation, residual organic solvent removal at the end of the technique is unnecessary, and a large variety of lipids can be used to prepare the vesicles [111, 113, 114].

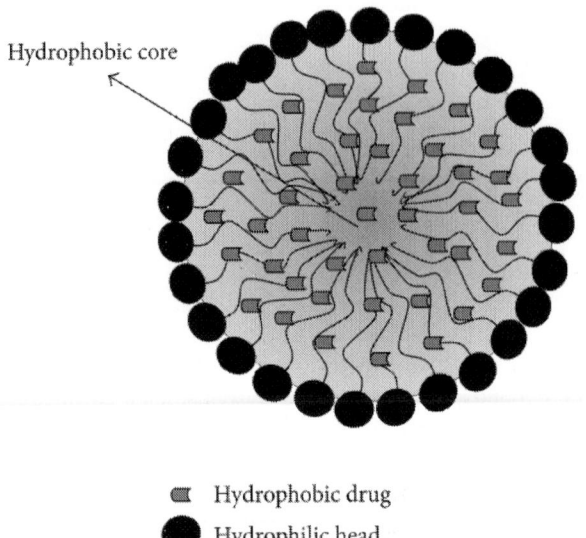

Figure 1. Micelle with hydrophobic compounds.

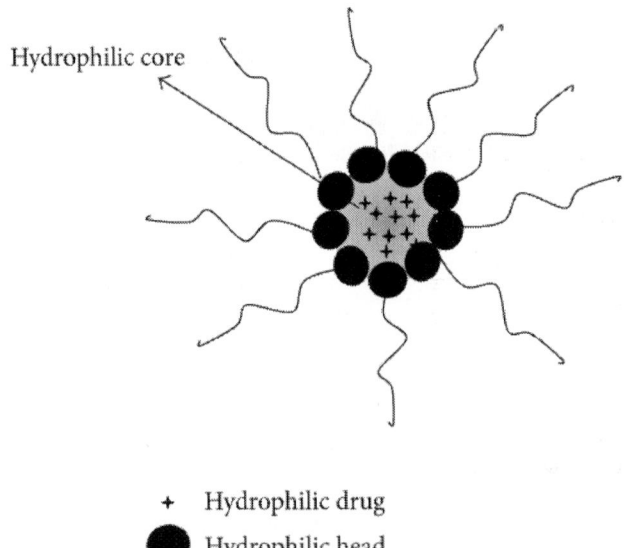

Figure 2. Micelle with hydrophilic compounds.

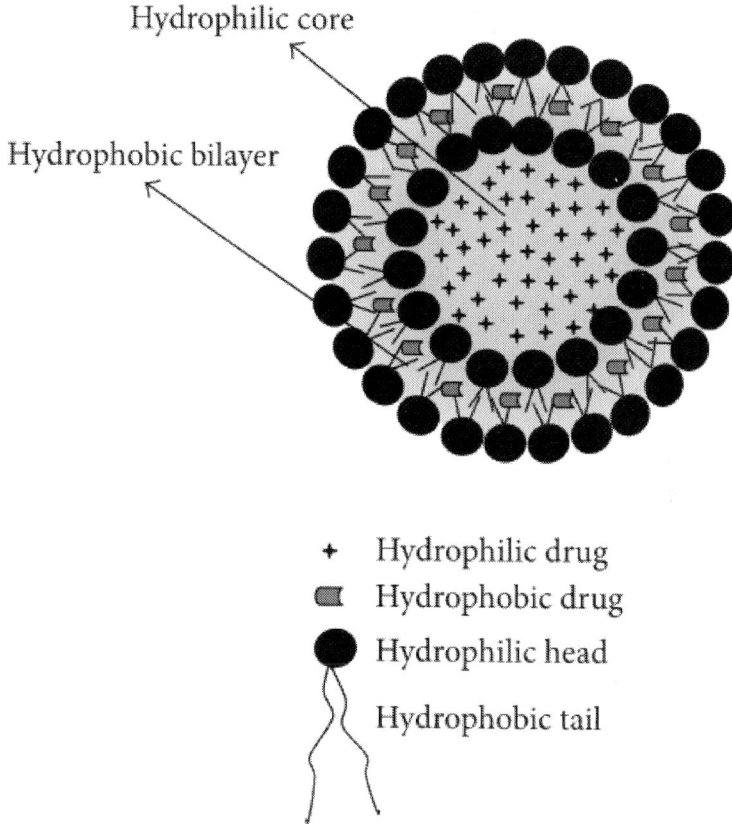

Figure 3. Liposome encapsulated hydrophobic and hydrophilic compounds.

For the sonication technique, liposomes are prepared using a sonicator to mix the lipid suspension. The pressure exerted by the sonicator stirring causes a decrease in the larger vesicles sizes. Thus, the stirring time is decisive for liposome size formed. The main advantage of this technique is less time in liposome preparation [111, 115].

Liposomes have attracted the attention of the scientific community due to their high versatility. Liposomes have greater therapeutic efficacy than conventional pharmaceutical system because they promote slow drug release at the target site [107, 116, 117]. Furthermore, liposomes are less toxic, nonimmunogenic, and biocompatible with organic tissues. They can decrease systemic toxicity and improve drug efficacy, especially for antibiotics, antifungals, and anticancer drugs [107, 118, 119]. Thus, using liposomes as a delivery system for chemotherapeutic agents offers great prospects for cancer treatment [108].

Wolf et al. (2000) [120] incorporated the DNA repair enzyme, T4 endonuclease V, into liposomes composed of phosphatidylcholine, phosphatidylethanolamine, oleic acid, and cholesteryl hemisuccinate (2 : 2 : 1 : 5

molar ratio) and applied it to human patients with a previous history of skin cancer after ultraviolet exposure. These liposomes were developed by encapsulating a purified recombinant T4 endonuclease V. The researchers observed that the enzyme was present in skin cells, which, in the skin, tends to improve DNA repair. Moreover, they reported that the T4 endonuclease V liposome prevented ultraviolet-induced upregulation of tumor necrosis factor-alpha and interleukin-10 mRNAs as well as interleukin-10 protein.

Another study using T4 endonuclease V in liposomes was conducted by Yarosh et al. (2001) [121]. They observed 30 patients with xeroderma pigmentosum in a double-blind study. The patients were randomly assigned to use either the T4N5 liposome or a placebo liposome lotion, daily for 1 year. To produce the T4 endonuclease V liposome lotion, they used 1 mg/L T4 endonuclease V encapsulated in liposomes in a 1% hydrogel lotion. The placebo lotion was prepared with the same liposomes in a 1% hydrogel solution but without the enzyme T4 endonuclease V. Patients with xeroderma pigmentosum have a genetic error in a DNA repair enzyme. Thus, the incidence of skin cancer in these patients is more frequent than in other people. The researchers noted a decrease in the xeroderma pigmentosum incidence rate and skin cancer in the groups treated with the T4 endonuclease V liposome. Furthermore, no significant adverse effects were reported.

Pierre et al. (2001) [122] proposed a topical delivery system for 5-aminolevulinic acid (5-ALA) based on liposomes with a similar composition to the stratum corneum to treat skin cancer. They prepared these liposomes using a reverse phase evaporation technique and the following components: ceramide, cholesterol, palmitic acid, cholesteryl sulfate, and α-tocopherol. 5-ALA is used in photodynamic, which had been shown effective in topical treatment for a variety of skin diseases. 5-ALA liposomal delivery system targeted and delivered 5 ALA to skin layers (viable epidermis and dermis) compared with the aqueous solutions typically applied in a 5-ALA-PDT clinical procedure. Thus, liposomes were a suitable delivery system for 5 ALA.

Chen et al. (2012) [123] developed a transdermal drug delivery system for curcumin-loaded liposomes and investigated in vitro skin permeation and the antineoplastic effect in vivo. Soybean phospholipids, hydrogenated soybean phospholipids, and egg yolk phospholipids were used to obtain the liposomes. Curcumin-loaded liposomes composed of soybean phospholipids promoted greater in vitro drug permeation. Moreover, the liposomes were effective against MEL and the liposomes composed of soybean phospholipids showed a higher capacity to inhibit MEL cells growth.

Nobayashi et al. (2002) [124] evaluated the efficiency of cationic multilamellar liposome-mediated gene transfer in murine MEL cell lines and an experimental gene therapy for subcutaneous MEL. They used B16F10, which is a murine MEL cell line and cationic liposomes composed of N-(α-trimethylammonioacetyl)-didodecyl-D-glutamate chloride (TMAG), dilauroyl phosphatidylcholine (DLPC), and dioleoyl phosphatidylethanolamine (DOPE) at molar ratio 1 : 2 : 2. They observed that repeated exposure to liposomes increased the transduction efficiency in murine MEL cells and experimental subcutaneous MEL tissue; thus, the therapy was effective for the intended purpose.

Liu and colleagues (2013) [125] developed liposomes with quercetin to improve its delivery into human skin and evaluate the potential anti-UVB effect. The liposomes were composed of soybean phosphatidylcholine, cholesterol, tween 80, and span 20. The researchers prepared liposomes with high entrapment efficiencies and a prolonged drug release. The group yielded good results; the quercetin liposomes enhanced cell viability compared with a quercetin solution in UVB-irradiated HaCaT cells, the reactive oxygen species levels decreased, the edema and inflammation were alleviated and 3.8-fold more quercetin liposomes permeated the skin compared with quercetin solution.

2.3. Cyclodextrins

Cyclodextrins (CDs) are a family of natural cyclic oligosaccharides with α-(1-4) linked glucopyranose subunits bonds [126–128]. They are produced from starch via enzymatic conversion using cyclodextrin glycosyl transferases (CGTases) [129, 130]. CDs have received more attention as a pharmaceutical excipient, because they can form drug complexes [130–132]. Furthermore, CDs are biocompatible and can be used to reduce in vitro and in vivo toxicity and the delivery profile can be modulated with great flexibility by changing the guest components [133].

β-cyclodextrin (βCD) is a CD that comprises seven α-(1,4)-linked α-d-glucopyranose units and is used extensively due to its ready availability and because its cavity size is suitable for a varied drug range [131]. Many hydrophilic, hydrophobic, and ionic CD derivatives have been developed to increase CDs versatility and decrease undesirable drug properties [134, 135].

Recent studies have demonstrated that CDs are efficient drug delivery systems for targeting cancer cells [136–138]. A complex formed between CD and a gemini surfactant (CDgemini) was used to carry curcumin analogue, and the cytotoxic effect of this system in MEL cells was analyzed. The results indicate that the drug-loaded CDgemini showed higher caspase 3/7 activity levels compared with drugs dissolved in DMSO, which enhance their ability to trigger apoptosis. Further, the researchers demonstrated that this treatment was more specific for MEL cells than for healthy keratinocytes [139].

In general, the pH surrounding tumor tissues tends to be more acidic (i.e., ~ 5.5 to 6.5) than normal tissue (i.e., 7.4) [140, 141]. Thus, pH-triggered drug release systems are promising for intracellular delivery of anticancer drugs [142]. Certain substances exhibit pH-sensitive host-guest interactions with cyclodextrin and may be used as pH-triggered drug release systems [143]. He and coworkers (2013) [144] synthesized a pH-responsive material through acetonating α-CD for PTX delivery. Results from in vitro drug release studies show a more rapid release profile in pH 5.0 buffer comparison with physiological conditions (pH 7.4). Moreover, this system can be effectively internalized by tumor cells; it demonstrates a superior cytotoxic activity and a longer incubation time results in higher efficiency. In addition, treatment with PTX loaded pH-sensitive α-CD inhibited tumor growth even at the lower PTX dose (1.1 mg/kg) [144].

Polypseudorotaxanes are inclusion complexes formed between cyclodextrins and linear macromolecules such as polymers [145]. Doxorubicin (DOX) loaded polypseudorotaxanes were developed by Chang and colleagues (2013) [146] and in vitro antitumor studies including cellular uptake and inhibition efficiency were analyzed for B16 MEL cells. The results indicate that doxorubicin-loaded polypseudorotaxanes inhibited MEL cells proliferation. The loaded doxorubicin showed slower endocytosis than doxorubicin hydrochloride, perhaps due to the larger system size. The cellular uptake of loaded doxorubicin was greater upon increasing the incubation time. Polypseudorotaxanes may be a promising carrier for DOX as antitumor MEL therapy.

4-Hydroxynonenal (4-HNE) is the end product of lipid peroxidation, which has been broadly used to inducer oxidative stress, and it produces a cytotoxic effect in cancer cells [147, 148]. The 4-HNE inclusion complex with the derivative βCD (PACM-βCD) was developed by Pizzimenti and coworkers (2013) to enhance 4-HNE stability [149]. The results demonstrate that the inclusion complex HNE/PACM-βCD was stable and significantly reduced more viable cells among the several cell lines tested, including human MEL A375 cells, than untreated control cells and cells treated with 10 μM free HNE.

Disrupting the lipid rafts' integrity, which are plasma membrane microdomains rich in cholesterol, may modify tumorigenic processes by altering the functionality of CD44, which is a cell surface receptor involved in cell migration and tumor metastasis [150, 151]. Murai and colleagues (2011) [152] showed that cholesterol reduction might be effective for preventing and treating malignant tumors progressions. Methyl-β-cyclodextrin (MβCD) forms soluble inclusion complexes with cholesterol and depletes cholesterol in plasma membranes [153]. A study conducted by Onodera et al. (2013) [154] investigated the potential of MβCD to cause apoptotic cell-death in a highly pigmented human MEL cell line. The results demonstrate that MβCD induced apoptosis through cholesterol depletion in lipid rafts, which activated caspase-3/7 and promoted cancer cell apoptosis. Thus, MβCD provides a potential strategy for treating MEL via lipid rafts modulation.

Mazzaglia and coworkers (2013) [155] developed an amphiphilic cyclodextrin (ACD) system for incorporating porphyrin derivatives to improve their water solubility and their selectivity towards MEL cells. The complexes formed showed higher cytotoxic activity in MEL cells than the free porphyrin derivative in water; thus, apoptotic cell death was observed at lower concentrations, and both cell proliferation and changes in cellular morphology were inhibited.

Mistletoe extract is often used in complementary cancer therapy [156]; it has been shown to stimulate cytokine production, modify intracellular protein synthesis, induce cell necrosis, and inhibit tumor colonization [157]. Strüh et al. (2013) [158] solubilized mistletoe triterpenoids with cyclodextrins and observed lower tumor growth, tumor necrosis, apoptotic cells, and prolonged survival in mice. These results indicate that solubilized mistletoe triterpenoids enhanced the antitumor effect of mistletoe extract.

Betulin (BET) is found in Betula sp. and has been used to treat skin diseases due to its therapeutic properties, including antitumor activity [159, 160]. Complexes formed between BET and a novel CD derivative, octakis-[6-deoxy-6-(2-sulfanyl ethanesulfonate)]-γ-CD (GCDG) were developed, and in vitro and in vivo experimental animal model experiments were conducted to verify antineoplasic activity in system. The results showed that BET complexation with CD improved BET solubility, which was an important property for enhancing BET antitumor activity. Moreover, BET promoted a lower MEL size, which was attributed to its antiangiogenic effect [160].

Interleukin-2 (IL-2) promotes immune recognition of MEL, while sparing normal cells [161]. However, secretion of certain immunosuppressive factors, such as TGF-β (transforming growth factor-β), can decrease ability of the immune system to identify the tumor as being composed of foreign cells [162]. A system composed of methacrylate-f-CD to solubilize the TGF-β inhibitor and liposomes loaded with a biodegradable crosslinking polymer and IL-2 cytokine was developed to sustain cytokines release to the tumor microenvironment and induces antitumor immune responses in a B16/B6 mouse. The results show that the TGF-β inhibitor and IL-2 reduced tumor growth. Furthermore, the natural killer cells' activity increased [163].

Cancer photodynamic therapy (PDT) combines a photosensitizer or photosensitizing drug with a specific type of light source to treat cancers [164]. A nontoxic carrier was prepared using 2-hydroxypropyl-cyclodextrins (hpCDs) and metallocomplex meso-tetrakis(4-sulfonatophenyl)porphyrin (ZnTPPS4) as the photosensitizer. The results demonstrate that low irradiation doses do not promote a substantial damaging effect on MEL cells, whereas higher irradiation doses induce cell death. Cell apoptosis or tissue necrosis depends on the radiation intensity. ZnTPPS4 complexation was an efficient sensitizer in human MEL cells [165].

2.4. Liquid Crystalline Phases

Pharmaceutical companies have shown an interest in developing nanostructured systems, such as liquid crystals, which have advantages that are mainly related to controlled drug release, and protect the active ingredients from thermal degradation or photobleaching [166, 167].

Liquid crystalline systems can compartmentalize drugs in the inner phase droplets, which have different physicochemical properties than the dispersing medium, and induce changes in the biological properties of the incorporated substances [168, 169].

Lehmann described an intermediate state in the thermal transformation from solid to liquid, which became known as liquid crystals (CLs) [170–172].

Liquid crystals are classified as lyotropic and thermotropic. When these systems are formed through adding solvents, they are lyotropic; thermotropic formation is temperature-dependent. As the surfactant concentration changes occur, different liquid-crystalline forms can be generated, such as lamellar, hexagonal (hexasomes), and cubic (cubosomes) forms. The lamellar phase is formed by parallel, planar layers of surfactant bilayers separated by a solvent layer, which form a one-dimensional network. Beginning in the hexagonal phase, aggregates are formed through an arrangement of long cylinders that form two-dimensional structures. In the cubic phase systems, the molecules are arranged in a three-dimensional system that consists of two corresponding water channel networks surrounded by lipid bilayers or surfactant [169].

Polarized light microscopy is an important tool to identify and classify liquid crystalline materials. Photomicrographs are used to demonstrate the observed textures, typically using polarized light [173]. Under polarized light plane, the sample is anisotropic if it can divert the plane of incident light that is isotropic and does not deflect light. The lamellar and hexagonal mesophases are anisotropic, while the cubic mesophase is isotropic [169, 174].

Figures 4 and 5 show microscopy systems in lamellar and hexagonal phases, respectively.

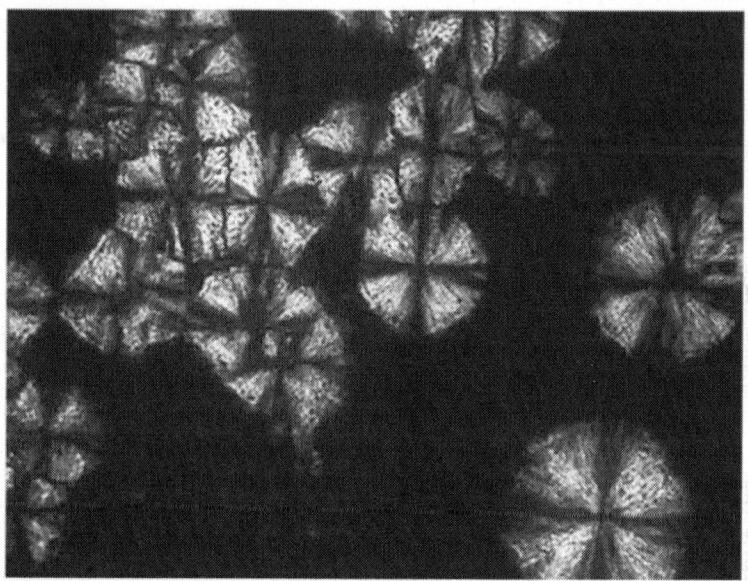

Figure 4. Polarized light microscopy of the lamellar phase (anisotropic system).

Liquid crystals have increasingly been used as delivery systems; Bitan-Cherbakovsky and colleagues (2013) [175] evaluated the release of gallic acid in cancer treatments. Liquid crystalline systems were studied as a dermal delivery system with ascorbyl palmitate to prevent skin aging [176].

Cubosomes present potential utility as a drug delivery system in skin cancer therapy, such as for MEL, due to their bioadhesion properties and enhancer penetration [177]. Bei and coworkers (2010) formulated dacarbazine-loaded cubosomes composed of glycerol monooleate RYLO MG 19 (GMO), poloxamer 407 (F127), phosphate buffer saline (PBS), and DTIC (5-(3, 3-dimethyl-1-triazeno) imidazole-4-carboxamide) and characterized their physicochemical properties. Currently, dacarbazine is a first-line chemotherapy agent against MEL. Due to the material's bioadhesion properties, it presents a potential for use in MEL therapy [178].

Figure 5. Polarized light microscopy of the hexagonal phase (anisotropic system).

5-FC phytanyl (5-FCPhy) is an amphiphile prodrug, carried in a lyotropic liquid crystalline system, and its in vivo efficacy as a chemotherapy agent against breast cancer has been investigated. The results show that the 5-FCPhy-loaded lyotropic liquid crystalline system reduced tumor size in a dose-dependent manner; the smallest average tumor volumes were observed with highest 5-FCPhy doses. Thus, a 5-FCPhy-loaded lyotropic liquid crystalline system may be used as a controlled drug delivery system for chemotherapeutic treatments such as for MEL [179].

von Eckardstein et al. (2005) developed an intracavitary carrier system composed of cubosomes that encapsuled carboplatin and paclitaxel; the release kinetics, the antitumor activity against glioma, and the prolonged survival were analyzed. The results show a significantly smaller tumor in animals treated with paclitaxel/carboplastin compared with the control group although survival did not differ among the groups studied. Both the drugs carried in the crystalline cubic phases presented cytotoxic activity in tumor cells, which indicates that they play an important role in cancer therapy [180]. The same researchers clinically observed 12 patients with a recurrent glioblastoma multiforme, who received an intracavitary application of paclitaxel and carboplatin cubosomes in different doses. The results indicate that this system is feasible and safe [181].

Many studies show the advantages of liquid crystals as a drug delivery system. However, most studies conducted using liquid crystals as a chemotherapy drug delivery systems remain at an early development stage. Several studies have been executed to characterize certain systems without efficacy trials [178, 182–186]. However, more studies are necessary to better understand the role of liquid crystals as a drug delivery system in MEL therapy.

2.5. Nanoparticles

The Food and Drug Administration (FDA) defines a nanoparticle as any material with a dimensional range of approximately 1 to 100 nm or end products with a dimension up to 1 μm that exhibit properties or biological phenomena (chemical, physical, and biological effects) [187–192]. Nanotechnology-based drug delivery systems have gained scientific notoriety due to variety of applications and many benefits; these systems may include polymeric and lipid-based nanoparticles.

In 1996, Müller and Lucks introduced the term solid lipid nanoparticle (SLN) to patent a manufacturing process using high pressure homogenization [193]. SLNs are the first generation of lipid nanoparticles (LN), which can be constructed by only using solid lipids (i.e., lipids that are solid at room temperature) [194].

Subsequent modifications to SLNs have been described, which are nanostructured lipid carriers (NLC) and are the second generation of LN [194]. Both SLN and NLC are constructed from lipid solid. However, they can be distinguished by their internal structures. The internal SLN structures only have solid lipids and NLCs are constructed using a blend of solid and liquid lipids, which produces imperfections in the crystal lattice [195,196], as shown in Figure 6. These imperfections have also been observed for SLNs because SLNs that contain multiple solid lipid components with distinct structural features may improve the drug entrapment efficiency [195, 197, 198].

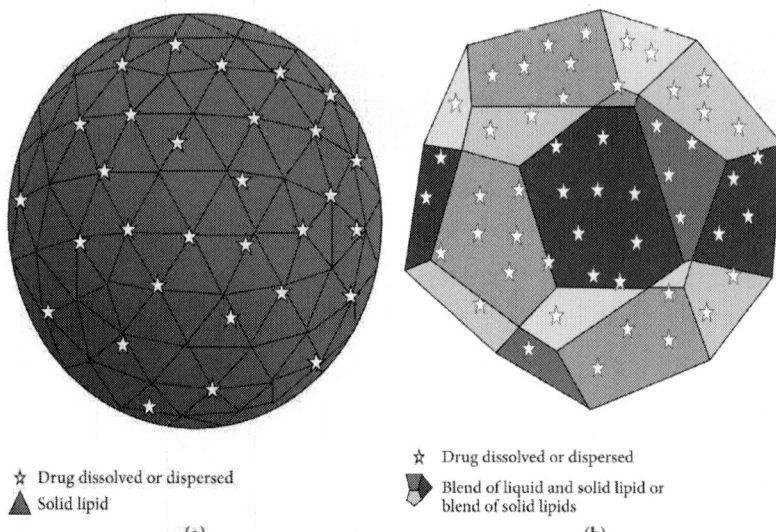

☆ Drug dissolved or dispersed
▲ Solid lipid

(a)

☆ Drug dissolved or dispersed
◆ Blend of liquid and solid lipid or blend of solid lipids

(b)

Figure 6. The image shows an SLN-organized lipid matrix composed of only solid lipids (a) and imperfections in the crystal lattice (b) on NLC or SLN that are composed of multiple solid lipid components with distinct structural features that are distorted upon forming a perfect crystal.

In addition to LN, polymeric nanoparticles (PN) may be constructed from organic polymers or inorganic materials, such as silica [199]. Polymers or lipids form solid NP nuclei, which promotes more stable systems, sustained drug release, and a uniform particle size distribution [200].

PN can be referred to as nanocapsules or nanospheres depending on their composition [166], as shown in Figure 7. The presence of oil promotes a vesicular structure in nanocapsules that forms reservoir-based drug delivery systems [201, 202], while nanospheres form matrix organized polymeric chains in the absence of oil [203, 204].

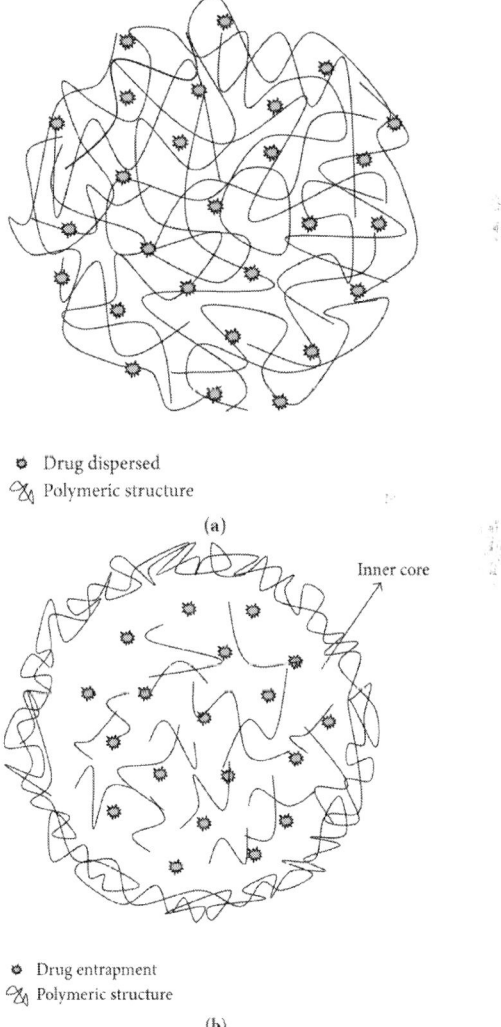

Figure 7. Polymeric nanoparticles schematics: nanospheres (a) and nanocapsules (b).

Drugs are entrapped in PN throughmixing the drug and polymer solution. Drug compounds are physically entrapped in the nanoparticle through polymer self-assembly [200]. PNs using different encapsulation mechanisms, such as dissolving it, disperse it or chemically adsorbs it in the constituents of the polymer matrix [166, 205, 206].

Both types of nanoparticles (lipid and polymeric nanoparticles) can be used as a drug delivery system with antitumor properties in MEL therapy.

Identifying tumor microenvironment properties is critically important for accumulating the most nanoparticles at the site of action, which decreased drug toxicity and adverse effects. Pathological systems' metabolism, cell morphology, and microenvironment have peculiar characteristics [207]. Through knowing these characteristics, specific biomarkers (antibodies, aptamers, peptides, and small molecules) can be identified, and molecules can be attached to the nanoparticles surface for successful targeted drug delivery to the site of action [208, 209].

Nonspecific interactions may appear in addition to specific biomarkers such as van der Waals bonds and electrostatic and steric affinities that can be used to predict the propensity for nanoparticle adhesion and uptake [210]. Thus, the particular nanoparticles structural components (lipids, surfactants, and polymers: Table 1) may improve drug targeting to the tumor tissue [198, 211], prevent opsonization and a consequent decrease in nanoparticles degradation by the immune system [212–214], improve the interactions between the surface nanoparticles and tumor cell membrane [215–217], alter the normal function of organelles, and induce apoptosis, which would increase tissue-specific cytotoxicity in cancer cells [197, 218, 219].

2.5.1. Benefits of Using Nanoparticles for MEL-Targeted Drug Delivery

Recent studies have shown improved SLN uptake and accumulation in tumor tissue [220], due to physiological tumor tissue characteristics, such as abnormalities and a dysfunctional tumor vasculature, which allow SLN in the range 30–100 nm to easily permeate the tumor. Moreover, higher SLN concentrations are maintained in the tumor for long periods of time due to low venous return and lymphatic drainage [221–223].

Xu and coworkers (2009) development a docetaxel-loaded SLN composed by egg phosphatidylcholine, dioleoylphosphatidylethanolamine, trimyristin, and lactobionic acid that showed 2.4-fold greater accumulation in tumors compared with the nonencapsulated drug 6 hours after intravenous administration. Galactosylation of the nanoparticle surfaces enhanced the cellular uptake of docetaxel and promoted passive targeting of the drug into the tumor cell, which reduced systemic toxicity [224]. Higher cationic nanoparticle uptake in HeLa cells compared with anionic nanoparticles was observed [225], which demonstrates that nanoparticle uptake is influenced by the nanoparticle surface molecules [226, 227].

Guo et al. (2010) investigated the antitumor effects of resveratrol (RES) bovine serum albumin nanoparticles. The results showed that the concentration of RES was greatly increased in the target tissue when the RES-loaded nanoparticle was injected. High levels of RES were observed in bloodstream for long periods of time after the RES suspension was administered

Table 1: Particular nanocarrier structural components for improving drug targeting to the tumor tissue.

Components for successful targeted drug delivery in antitumor	Benefits in anticancer therapy	References
	Active targeting	
Cholesterol	Cancer cells take up 100-fold more low density lipoprotein (LDL) than normal tissue due to upregulated LDL receptors in cancer cells for membrane synthesis during cell division associated with malignant transformation processes. Thus, LDL has been proposed as a drug carrier for anticancer agents.	[208, 239–245]
Polyunsaturated fatty acids (α-linolenic acid; linoleic acid; arachidonic acid; eicosapentaenoic acid; and docosahexaenoic acid).	They can be attached to the tumor cell membrane more easily, which results in disruption and fluidity of the cell membranes. Tumor progression is reduced by modulating p53, p16, and p27 expression and cell cycle regulation, as well as by inducing cell death by apoptosis and necrosis.	[246–248]
Hyaluronic acid	Hyaluronic acid is an extracellular matrix compound that specifically binds CD44, which is an extracellular membrane protein that regulates various cellular responses. CD44 is overexpressed in cancer cells, while normal cells underexpress this protein. Thus, CD44 is a good candidate biomarker for cancer cells.	[249–251]
Folic acid	Folate is important for producing and maintaining new cells because it can participate in nucleotide synthesis. Folates receptors are highly overexpressed in cancer cells. In addition, only the malignant cells, not normal cells, transport folate-conjugates; thus, the folate-drug conjugation can improve tumor-targeted drug delivery.	[248, 252–254]
	Passive targeting	
Polysaccharides; polyacrylamide; polyvinyl alcohol; polyvinylpyrrolidone; PEG; PEG-containing copolymers (poloxamers; poloxamines; polysorbates; and PEG copolymer).	They prevent the opsonin binding to the nanoparticle surfaces and, consequently, recognition as well as phagocytosis of the nanoparticles by the mononuclear phagocytic system, which enhances the blood circulation time.	[212, 255–258]
Cationic surfactants	The positive charge of a cationic surfactant interacts through electrostatics with the negatively-charged phospholipids that are preferentially expressed on the cancer cell surface.	[252, 259–261]

(nonencapsulated RES), which illustrated incomplete RES distribution. Moreover, the results show that RES-loaded nanoparticles promoted greater tumor growth inhibition [228]. Teskač and Kristl (2010) showed that where NLC (Compritol 888ATO and Phospolipon 80H as the oil phase and Lutrol as the steric stabilizer) was used to incorporate RES, it crossed the cell membrane, was delivered throughout the cytosol, and was located in the perinuclear region without inducing cytotoxicity. They found that RES solubility, stability, and intracellular release were also enhanced in the RES-loaded nanoparticle [229].

The drug release profile was modulated using drug-loaded LN. A recent study demonstrates that the camptothecin release rate can be modified by changing lipid nanoparticle inner phases. The SLN composed of precirol as the solid lipid showed the most sustained release (45% of the total drug was released after 30 hours) while the NLC composed of precirol as the solid lipid and squalene as the liquid lipid showed more rapid release (65% of the total drug was released after 30 hours). Drug mobility decreased when solid or crystalline substances were incorporated into the nanoparticles, which decreased the drug levels released as a function of time. Greater inhibition of MEL cell proliferation was observed when the cells were treated with nanoparticles, which may be because MEL cells exhibit excellent uptake endocytosis [230].

Docetaxel-loaded NLC (DTX-NLC) composed of stearic acid, glyceryl monostearate, soya lecithin, and oleic acid showed as sustained-release drug profile (77% of the total drug after 24 hours), while Duopafei (docetaxel injection provided by Qilu Pharmaceutical Co., Ltd. in China) showed 100% release after 24 hours. In addition, DTX-NLC showed greater cytotoxicity against MEL cells compared with Duopafei through enhanced apoptosis. Moreover, DTX-NLC showed low cytotoxicity for healthy cells because the drug is only released after endocytosis by a target cell [231].

Camptothecin was encapsulated into NLC, which was composed of cetyl palmitate, coconut oil, and Myverol associated with a quantum dot (metallic compounds at the core of the semiconductor NLC) as oil phase and Pluronic 68 solution as water phase. Camptothecin-loaded NLC presented superior cytotoxicity against MEL cells and the greater cell uptake compared with other carriers. Cellular endocytosis was essential for cell viability and the quantum dots showed a minimal capacity to influence proliferation. In addition, camptothecin accumulation in the MEL increased by approximately 6.4-fold following administration of the camptothecin-loaded nanoparticle. In vivo real-time monitoring showed that a camptothecin-loaded nanoparticle associated with a quantum dot was strongly localized at tumors with a persistent signal for 24 hours. The drug-loaded NLC was directed using quantum dot to efficiently transmit sustained tumor bioimaging, in addition to promoting drug release. This system offers the potential for diagnosing or monitoring evolution of the tumor through bioimaging and for drug delivery through nanocarrier [232].

Multiple synthetic and natural biodegradable polymers may be used in antitumor drug delivery systems, such as polyesters (e.g., polylactic acid, PLA), polyamino acids (e.g., polyaspartic acid), and polyoxypropylenes (e.g., poloxamers) [233].

Interleukin-2 was delivered by a polymeric nanoparticle composed of a low molecular weight polyethylenimine linked by β-cyclodextrin and conjugated with folate; this molecule was analyzed as a potential MEL antitumor therapy. High levels of tumor suppression were observed after peritumoral injection of the polymeric nanoparticle with interleukin-2, which prolonged survival in mice; thus, it is a promising gene-based therapeutic strategy for MEL. The antitumor effect can be attributed mainly to activation, proliferation, and infiltration of effector T cells and NK cells into the tumor; the therapeutic efficiency was dose-dependent and presented low cytotoxicity [234].

Polymeric nanoparticle using polylactic-co-glycolic acid (PLGA) as a polymer to incorporate coumarin increased the cellular uptake rate 2-fold versus nonencapsulated coumarin. In addition, molecular signals for mRNA expression were used to demonstrate that the coumarin-loaded nanoparticle downregulated cyclin-D1, proliferating cell nuclear antigen (PCNA), survivin, and Stat-3, and it upregulated p53 and caspase-3, promoting enhanced apoptosis of MEL tumor cells compared with nonencapsulated coumarin [235].

As a drug delivery system for apigenin, PLGA-PN promotes faster mobility and site-specific activity in MEL in addition to efficiently preserving apigenin photodegradation. The results also showed increased free radical accumulation and antioxidant enzymes depletion inside tumor cells, which exacerbated DNA damage and results in apoptosis through mitochondrial dysfunction [236].

A polymer-based delivery vehicle for cisplatin composed of chitosan and carboxymethylcellulose showed enhanced cytotoxicity (approximately 10-fold greater) in MEL tumor cells compared with nonencapsulated cisplatin. Further, rapid intracellular drug release was observed upon endocytosis of this system by a tumor cell, and only high-density NPs and positively charged-surfaces were capable of releasing cisplatin into MEL. Moreover, it decreased drug loss during blood circulation [237].

Superparamagnetic iron oxide nanoparticles consist of a carboxydextran shell and show increased uptake in human mesenchymal stem cells; the nanoparticle uptake efficiency was related to a higher density of carboxyl groups on the nanoparticle surface [238].

3. ADVANTAGES AND DISADVANTAGES OF NANOCARRIER SYSTEMS

This paper describes umpteen benefits to use nanocarriers system to vehiculate drugs used in melanoma therapy. But, to choose the better system type, it is also necessary to analyze the disadvantage of each system. Table 2 describes main advantages and disadvantages of each system.

Table 2: Main advantages and disadvantages of each system.

Nanocarrier	Advantages	Disadvantages	References
Hydrogels	Cells and fragile drugs, like peptides, proteins, DNA, and oligonucleotides, could be protected by aqueous environment Good transport of nutrients to cells and products from cells Cell adhesion ligands easily modified them Can be injected as a liquid at body temperature; Usually biocompatible	Can be difficult to manufacture Usually mechanically weak Difficulty in encapsulating the drug Problems connecting with the cells Difficult to sterilize	[262]
Liposomes	They can be formed by natural or synthetic lipids Biodegradable Nontoxic Thermosensitive Hydrophilic and lipophilic molecules can be incorporated	High-energy sonication frequently causes oxidation and degradation of phospholipid Low-energy sonication requires long periods of sonication and can also be destructive to phospholipid High-pressure homogenization can confer decreased stability Application of volatile organic solvents	[263–266]
Micelles	Ease to prepare Good stability Many administration routes available	Risk of disintegration after administration	[267]
Cyclodextrins	Potential solubilizing and stabilizing agents Higher order complexes are possible Targeting water-insoluble drugs to the oral route	Some cyclodextrins have been shown to be irritants Safety concerns limited their use for parenteral administration	[268, 269]
Liquid crystals	They are easy to prepare Thermodynamically stable Composed of simple chemicals *In situ* phase transformations	Difficult to prepare and administer due to high viscosity Toxicity related to high surfactant concentration	[267, 270, 271]
Nanoparticles	They can be prepared by different methods Hydrophilic and lipophilic molecules can be incorporated They can change the surface Increased drug stability Possibility of controlled drug release and drug targeting	Toxicological assessment is uncompleted Low drug-loading capacities It is difficult to develop an analytical method for drug delivery Difficult to scale up the production Stability problems during storage	[195, 272]

4. NANOCARRIERS APPLICATION IN ANIMAL MODEL OR CLINICAL STUDIES

Particular nanocarrier structural components were previously described for improving drug targeting to the tumor tissue (Table 1). Now, some animal models studies or clinical efficacy will be portrayed.

A study realized by Shi and coworkers (2014) demonstrated that microRNA-34a and paclitaxel-loaded functional cationic solid lipid nanoparticles presented a synergistic anticancer efficacy. In vivo test was conducted and this system was much more potent inhibitor of B16F10-bearing tumor growth and can eliminate melanoma metastasized to the lungs cells compared to single drug-loaded SLN [273].

C57BL/6 mice were inoculated subcutaneously with B16-F10 melanoma cells (1×10^6 in 100 μL/animal) to verify the effect of free curcumin (CUR) and CUR-loaded nanocapsules on melanoma tumor growth. Results showed that treatments significantly inhibited the tumor growth, 59.6% (1128.4 mm^3 tumor volume) after treatment with free curcumin, 61.4% (1078.2 mm^3 tumor volume) after treatment with Cur-loaded lipid nanocapsule suspensions, and 71.3% (801.4 mm^3 tumor volume) after treatment with Cur-loaded polymeric nanocapsule suspensions, when compared to the control group treated with cell culture medium only (2791.0 mm^3 tumor volume). Cisplatin was used as positive control and decreased 72.4% (770.0 mm^3) of tumor volume [274].

Interleukin-2-loaded polymeric nanoparticle inhibited the tumor growth and can lengthen survival in mice B16F1-bearing melanoma. The antitumor effect was dose-dependent and the system demonstrated low toxicity, representing a new strategy in drug delivery system for melanoma gene therapy [234].

Cai et al., 2012, carried out a study to verify the influence of tumor-targeting nanocarrier in long-circulation effects promoted by PEGylated liposome. Results demonstrated that paclitaxel-loaded targeted PEGylated liposomes (TL-PTX) lengthen the half-life of paclitaxel 2.01-fold of conventional liposome and 3.40-fold of free paclitaxel in plasma. Higher accumulation of TL-PTX in tumor tissue, liver, and spleen was observed compared to conventional liposome and free paclitaxel [275].

Doxil, the first FDA-approved nanodrug [276], is corroborated to development of nanomedicine for melanoma therapy. After that, a lot of clinical trials have been done to verify the efficacy of nanodrugs to improve the survival and quality life in patients with melanoma, especially with poor prognosis [277].

In a clinical phase II study, Ugurel et al. (2004) verified that patients treated with liposomal doxorubicin as monotherapy present survival benefit. Outpatients setting (30 patients) were included on this study. Liposomal doxorubicin is used at 50 mg/m^2 i.v. on days 1, 22, 43 and 64, subsequently at 40 mg/m^2 at day 85 before first staging and in 4-week intervals thereafter. The results demonstrated that 7 patients stay alive more than 300 days and 5 patients more than 400 days [278].

Patients with cancer stage IV melanoma participated in an open-label, phase II study conducted by Hwu and coworkers (2006). Patients received a combination of 75 mg/m^2 per day of temozolomide, during 6 weeks, there was a

2-week break between cycles, and they were continuously subcutaneously administrated PEGylated IFN-2b at 0.5 g/kg/week. Results showed that patient's median survival was 12 months and they were followed for 16 months and brain metastases were developed in any patients. Researchers concluded that a combination therapy promotes antitumor activity in metastatic melanoma [279].

5. NEW APPROACHES AND CHALLENGES

The wide range of compositions, morphologies, and particle sizes exhibited by drug delivery systems makes it difficult to understand their cellular uptake mechanisms. Thus, elucidating fundamental cellular processes that cells used to import and export select extracellular molecules may contribute to understanding the cellular internalization mechanisms of the systems and aid in selecting the appropriate system to transport active compounds [280]. Endocytosis of particles into cells depends not only on particle size, but also on surface coating and cell type [226, 227, 238].

Advances in nanotechnology based drug delivery systems have improved our understanding of the biological effects of nanotechnology-based systems, which will undoubtedly lead to important, clinically relevant improvements in drug delivery. New challenges in developing nanotechnology-based drug delivery systems for MEL antitumoral therapy include the feasibility of upscaling processes to quickly bring the innovative therapeutic techniques to the market and the potential for multifunctional systems that will fulfill several biological and therapeutic requirements, such as the system needing to be able to target tumor cells or tumor environment after systemic delivery. Further challenges include researches on efficiency of targeted anticancer therapies and imaging agents as well as international standards regarding their toxicology and biocompatibility.

So, the possibility of nanocarriers can promote the targeted cancer therapy and potentially early detection of cancer lesions; noninvasive imaging that permits determination of molecular signatures induces the concept of personalized medicine [281]. But not only that, the personalized medicine based on adjusted therapy to individual differences that can be detected by genetic test, guiding the choose of drug and their dosage. So, combining clinical and molecular biomarkers in nanomedicine contribute to improvement of the disease management [282].

6. AUTHOR'S OPINION

Drug delivery systems represent an alternative strategy to carrier antineoplastic agents. Many advantages of drug delivery system have been described in recent studies such as better drug stability, better bioavailability, controlled drug release, long circulation time in blood, selective organs or tissue distribution, a reduction of the total dose required, and minimizing the toxic side effects.

Certain drug delivery characteristic can distinguish their application such as hydrogel that are stimuli-sensitive swelling-controlled release system. Liposome can encapsulate both hydrophobic and hydrophilic compounds. CD can form drug complexes and are biocompatible. LC protects the active ingredients from thermal degradation or photobleaching. SLN and NLC are maintained in the tumor for long period of the time due to low venous return and lymphatic drainage. PN forms reservoir-based drug delivery systems in nanocapsules and matrix organized polymeric chains in nanospheres. The wide range of compositions, morphologies, and particle sizes exhibited by drug delivery systems makes it variable mechanism for successful targeted delivery, while making it difficult to understand their cellular uptake mechanisms.

Another important aspect is identifying pathological systems' metabolism, tumor cell morphology, and microenvironment properties for accumulating the most drug delivery system at the site of action, at which decreased drug toxicity and adverse effects and biomarkers (antibodies, aptamers, peptides, and small molecules) can be identified, and molecules can be attached to the systems surface for successful targeted drug delivery to the site of action.

Thus, elucidating fundamental cellular processes that cells used to import and export select extracellular molecules may contribute to understanding the cellular internalization mechanisms of the systems and aid in selecting the appropriate system to transport active compounds. Endocytosis of particles into cells depends not only on particle size, but also on surface coating and cell type.

Several studies on cancer have been conducted worldwide, but peculiarities of tumor cells that distinguish them from normal cells are not completely elucidated, which made the delineation of targeted drug delivery for cancer therapy difficult.

Another problem is that chemotherapy drug delivery systems remain at an early development stage. Several studies have been executed to physicochemically characterize certain systems. However, the influence of systems to improve drug biological properties is understudied.

Tumor microenvironment plays an important role in tumorigenesis and may also influence the success rate of melanoma therapy. The drug delivery systems need to cross anatomical and physiological barriers of tumor microenvironment. However, many mysteries emphasize the complexity of the task.

In the recent decade, one of the most studied fields is nanotechnology-based drug delivery and various targeting mechanisms were discovered such as cancer-specific ligand for receptor-mediated active targeting (i.e., folate and hyaluronic acid); microenvironment-responsive molecules that respond to changes in pH, temperature, light, chemicals, and electromagnetic fields; PEGylation-induced passive targeting; electrostatics interaction and molecules that prevent the opsonization.

Drug delivery system for melanoma therapy may target the several pathways involved in melanoma development such as three-tiered Ras/Raf/MEK mitogen-activated protein kinase (MAPK); PI(3)K; NF-kappaB; p16INK4a/RB and ARF signalling pathways.

Although breakthrough in melanoma antitumor therapy research has been observed, more studies are necessary to better understand the role of drug delivery system in MEL therapy.

ACKNOWLEDGMENTS

This work was supported by Coordenação de Aperfeiçoamento de Pessoal de Nível Superior (CAPES), Fundação de Amparo à Pesquisa do Estado de São Paulo (FAPESP), Conselho Nacional de Desenvolvimento Científico e Tecnológico (CNPq), and Programa de Apoio ao Desenvolvimento Científico da Faculdade de Ciências Farmacêuticas (PADC-FCF-UNESP).

REFERENCES

1. H. J. Cohen, E. Cox, K. Manton, and M. Woodbury, "Malignant melanoma in the elderly," Journal of Clinical Oncology, vol. 5, no. 1, pp. 100–106, 1987.
2. P. E. LeBoit, Pathology & Genetics: Skin Tumours, edited by: World Health Organization, IARC Press, Lyon, France, 2006.
3. R. N. Saladi and A. N. Persaud, "The causes of skin cancer: a comprehensive review," Drugs of Today, vol. 41, no. 1, pp. 37–53, 2005.
4. E. de Vries, F. I. Bray, J. W. W. Coebergh, and D. M. Parkin, "Changing epidemiology of malignant cutaneous melanoma in Europe 1953–1997: rising trends in incidence and mortality but recent stabilizations in western Europe and decreases in Scandinavia," International Journal of Cancer, vol. 107, no. 1, pp. 119–126, 2003.
5. J. Ferlay, H.-R. Shin, F. Bray, D. Forman, C. Mathers, and D. M. Parkin, "Estimates of worldwide burden of cancer in 2008: GLOBOCAN 2008," International Journal of Cancer, vol. 127, no. 12, pp. 2893–2917, 2010.
6. A. Jemal, F. Bray, M. M. Center, J. Ferlay, E. Ward, and D. Forman, "Global cancer statistics," CA Cancer Journal for Clinicians, vol. 61, no. 2, pp. 69–90, 2011.
7. R. Colombari, F. Bonetti, G. Zamboni et al., "Distribution of melanoma specific antibody (HMB-45) in benign and malignant melanocytic tumours. An immunohistochemical study on paraffin sections,"Virchows Archiv A, vol. 413, no. 1, pp. 17–24, 1988.
8. K. Blessing, D. S. A. Sanders, and J. J. H. Grant, "Comparison of immunohistochemical staining of the novel antibody melan-A with S100 protein and HMB-45 in malignant melanoma and melanoma variants," Histopathology, vol. 32, no. 2, pp. 139–146, 1998.
9. T. Gambichler, P. Regeniter, F. G. Bechara et al., "Characterization of benign and malignant melanocytic skin lesions using optical coherence tomography in vivo," Journal of the American Academy of Dermatology, vol. 57, no. 4, pp. 629–637, 2007.
10. R. Harson and C. Grose, "Egress of varicella-zoster virus from the melanoma cell: a tropism for the melanocyte," Journal of Virology, vol. 69, no. 8, pp. 4994–5010, 1995.

11. A. Ingraffea, "Melanoma," Facial Plastic Surgery Clinics of North America, vol. 21, no. 1, pp. 33–42, 2013.

12. J. Y. Lin and D. E. Fisher, "Melanocyte biology and skin pigmentation," Nature, vol. 445, no. 7130, pp. 843–850, 2007.

13. Y. Yamaguchi and V. J. Hearing, "Physiological factors that regulate skin pigmentation," BioFactors, vol. 35, no. 2, pp. 193–199, 2009.

14. S. L. Winsey, N. A. Haldar, H. P. Marsh et al., "A variant within the DNA repair gene XRCC3 is associated with the development of melanoma skin cancer," Cancer Research, vol. 60, no. 20, pp. 5612–5616, 2000.

15. D. B. McKenna, V. R. Doherty, K. M. Mclaren, and J. A. A. Hunter, "Malignant melanoma and lymphoproliferative malignancy: is there a shared aetiology?" British Journal of Dermatology, vol. 143, no. 1, pp. 171–173, 2000.

16. V. Bataille, "Genetic epidemiology of melanoma," European Journal of Cancer, vol. 39, no. 10, pp. 1341–1347, 2003.

17. M. S. Brose, P. Volpe, M. Feldman et al., "BRAF and RAS mutations in human lung cancer and melanoma," Cancer Research, vol. 62, no. 23, pp. 6997–7000, 2002.

18. K. Omholt, S. Karsberg, A. Platz, L. Kanter, U. Ringborg, and J. Hansson, "Screening of N-ras codon 61 mutations in paired primary and metastatic cutaneous melanomas: mutations occur early and persist throughout tumor progression," Clinical Cancer Research, vol. 8, no. 11, pp. 3468–3474, 2002.

19. M. A. Tucker and A. M. Goldstein, "Melanoma etiology: where are we?" Oncogene, vol. 22, no. 20, pp. 3042–3052, 2003.

20. M.-F. Demierre and V. K. Sondak, "Cutaneous melanoma: pathogenesis and rationale for chemoprevention," Critical Reviews in Oncology/Hematology, vol. 53, no. 3, pp. 225–239, 2005.

21. K. Colston, M. J. Colston, and D. Feldman, "1,25-dihydroxyvitamin D3 and malignant melanoma: the presence of receptors and inhibition of cell growth in culture," Endocrinology, vol. 108, no. 3, pp. 1083–1086, 1981.

22. D. E. Godar, R. J. Landry, and A. D. Lucas, "Increased UVA exposures and decreased cutaneous Vitamin D3 levels may be responsible for the increasing incidence of melanoma," Medical Hypotheses, vol. 72, no. 4, pp. 434–443, 2009.

23. S. V. Madhunapantula and G. P. Robertson, "Chemoprevention of Melanoma," Advances in Pharmacology, vol. 65, pp. 361–398, 2012.

24. D. C. Whiteman, P. Watt, D. M. Purdie, M. C. Hughes, N. K. Hayward, and A. C. Green, "Melanocytic nevi, solar keratoses, and divergent pathways to cutaneous melanoma," Journal of the National Cancer Institute, vol. 95, no. 11, pp. 806–812, 2003.

25. M. Megahed, M. Schön, D. Selimovic, and M. P. Schön, "Reliability of diagnosis of melanoma in situ," The Lancet, vol. 359, no. 9321, pp. 1921–1922, 2002.

26. G. Massi and P. E. Leboit, "Melanoma in situ," in Histological Diagnosis of Nevi and Melanoma, pp. 403–412, Springer, 2004.

27. W. H. Clark Jr. and M. C. Mihm Jr., "Lentigo maligna and lentigo-maligna melanoma," American Journal of Pathology, vol. 55, no. 1, pp. 39–67, 1969.

28. V. Cardile, G. Malaponte, C. Loreto et al., "Raf kinase inhibitor protein (RKIP) and phospho-RKIP expression in melanomas," Acta Histochemica, vol. 115, no. 8, pp. 795–802, 2013.

29. R. T. Prehn, "The paradoxical association of regression with a poor prognosis in melanoma contrasted with a good prognosis in keratoacanthoma," Cancer Research, vol. 56, no. 5, pp. 937–940, 1996.

30. S. M. Swetter, P. M. Ecker, D. L. Johnson, and J. D. Harvell, "Primary dermal melanoma: a distinct subtype of melanoma," Archives of Dermatology, vol. 140, no. 1, pp. 99–103, 2004.

31. K. Hoek, D. L. Rimm, K. R. Williams et al., "Expression profiling reveals novel pathways in the transformation of melanocytes to melanomas," Cancer Research, vol. 64, no. 15, pp. 5270–5282, 2004.

32. D. Grossman and D. C. Altieri, "Drug resistance in melanoma: mechanisms, apoptosis, and new potential therapeutic targets," Cancer and Metastasis Reviews, vol. 20, no. 1-2, pp. 3–11, 2001.

33. M. Karin and A. Lin, "NF-κB at the crossroads of life and death," Nature Immunology, vol. 3, no. 3, pp. 221–227, 2002.

34. L. A. Fecher, S. D. Cummings, M. J. Keefe, and R. M. Alani, "Toward a molecular classification of melanoma," Journal of Clinical Oncology, vol. 25, no. 12, pp. 1606–1620, 2007.

35. L. M. Cohen, "Lentigo maligna and lentigo maligna melanoma," Journal of the American Academy of Dermatology, vol. 33, no. 6, pp. 923–939, 1995.

36. S. Menzies, "Superficial spreading melanoma," in An Atlas of Dermoscopy, J. M. Ashfaq, A. Marghoob, and R. Marghoob, Eds., pp. 203–209, CRC Press, 2004.

37. A. E. Chang, L. H. Karnell, and H. R. Menck, "The national cancer data base report on cutaneous and noncutaneous melanoma: a summary of 84,836 cases from the past decade," Cancer, vol. 83, no. 8, pp. 1664–1678, 1998.

38. E. T. Krementz, R. J. Reed, and W. P. Coleman III, "Acral lentiginous melanoma. A clinicopathologic entity," Annals of Surgery, vol. 195, no. 5, pp. 632–645, 1982.

39. R. A. Scolyer, G. V. Long, and J. F. Thompson, "Evolving concepts in melanoma classification and their relevance to multidisciplinary melanoma patient care," Molecular Oncology, vol. 5, no. 2, pp. 124–136, 2011.

40. N. J. Crowley and H. F. Seigler, "Late recurrence of malignant melanoma: analysis of 168 patients,"Annals of Surgery, vol. 212, no. 2, pp. 173–177, 1990.

41. M.-H. Schmid-Wendtner, J. Baumert, M. Schmidt et al., "Late metastases of cutaneous melanoma: an analysis of 31 patients," Journal of the American Academy of Dermatology, vol. 43, no. 4, pp. 605–609, 2000.

42. A. J. Sober and J. M. Burstein, "Precursors to skin cancer," Cancer, vol. 75, no. 2, supplement, pp. 645–650, 1995.

43. K. Hoffmann, J. Jung, S. El Gammal, and P. Altmeyer, "Malignant melanoma in 20-MHz B scan sonography," Dermatology, vol. 185, no. 1, pp. 49–55, 1992.

44. T. S. Wang, T. M. Johnson, P. N. Cascade, B. G. Redman, V. K. Sondak, and J. L. Schwartz, "Evaluation of staging chest radiographs and serum lactate dehydrogenase for localized melanoma," Journal of the American Academy of Dermatology, vol. 51, no. 3, pp. 399–405, 2004.

45. C. M. Balch, A. C. Buzaid, S.-J. Soong, et al., "Final version of the American Joint Committee on Cancer staging system for cutaneous melanoma," Journal of Clinical Oncology, vol. 19, no. 16, pp. 3635–3648, 2001.

46. S. B. Revin and S. A. John, "Electrochemical marker for metastatic malignant melanoma based on the determination of l-dopa/l-tyrosine ratio," Sensors and Actuators B: Chemical, vol. 188, pp. 1026–1032, 2013.

47. D. S. Tyler, M. Onaitis, A. Kherani et al., "Positron emission tomography scanning in malignant melanoma: clinical utility in patients with Stage III disease," Cancer, vol. 89, no. 5, pp. 1019–1025, 2000.

48. D. A. Sipkins, D. A. Cheresh, M. R. Kazemi, L. M. Nevin, M. D. Bednarski, and K. C. P. Li, "Detection of tumor angiogenesis in vivo by $\alpha_v\beta_3$-targeted magnetic resonance imaging," Nature Medicine, vol. 4, no. 5, pp. 623–626, 1998.

49. D. Rinne, R. P. Baum, G. Hör, and R. Kaufmann, "Primary staging and follow-up of high risk melanoma patients with whole-body 18F-fluorodeoxyglucose positron emission tomography: results of a prospective study of 100 patients," Cancer, vol. 82, no. 9, pp. 1664–1671, 1998.

50. M. González Cao, J. M. Auge, R. Molina et al., "Melanoma inhibiting activity protein (MIA), beta-2 microglobulin and lactate dehydrogenase (LDH) in metastatic melanoma," Anticancer Research, vol. 27, no. 1B, pp. 595–599, 2007.

51. L. L. Yu, T. J. Flotte, K. K. Tanabe et al., "Detection of microscopic melanoma metastases in sentinel lymph nodes," Cancer, vol. 86, no. 4, pp. 617–627, 1999.

52. M. B. Sporn, N. M. Dunlop, D. L. Newton, and J. M. Smith, "Prevention of chemical carcinogenesis by vitamin A and its synthetic analogs (retinoids)," Federation Proceedings, vol. 35, no. 6, pp. 1332–1338, 1976.

53. F. L. Meyskens Jr., P. J. Farmer, and H. Anton-Culver, "Etiologic pathogenesis of melanoma: a unifying hypothesis for the missing attributable risk," Clinical Cancer Research, vol. 10, no. 8, pp. 2581–2583, 2004.

54.   A. V. Anstey, "Systemic photoprotection with α-tocopherol (vitamin E) and β-carotene," Clinical and Experimental Dermatology, vol. 27, no. 3, pp. 170–176, 2002.

55.   W. Stahl and H. Sies, "Carotenoids and protection against solar UV radiation," Skin Pharmacology and Applied Skin Physiology, vol. 15, no. 5, pp. 291–296, 2002.

56.   H. Tsao, M. Feldman, J. E. Fullerton, A. J. Sober, D. Rosenthal, and W. Goggins, "Early detection of asymptomatic pulmonary melanoma metastases by routine chest radiographs is not associated with improved survival," Archives of Dermatology, vol. 140, no. 1, pp. 67–70, 2004.

57.   E. Erdei and S. M. Torres, "A new understanding in the epidemiology of melanoma," Expert Review of Anticancer Therapy, vol. 10, no. 11, pp. 1811–1823, 2010.

58.   D. Kavanagh, A. D. K. Hill, B. Djikstra, R. Kennelly, E. M. W. McDermott, and N. J. O'Higgins, "Adjuvant therapies in the treatment of stage II and III malignant melanoma," Surgeon, vol. 3, no. 4, pp. 245–256, 2005.

59.   S. Schreiber, E. Kämpgen, E. Wagner et al., "Immunotherapy of metastatic malignant melanoma by a vaccine consisting of autologous interleukin 2-transfected cancer cells: outcome of a phase I study,"Human Gene Therapy, vol. 10, no. 6, pp. 983–993, 1999.

60.   R. W. Dubois, S. M. Swetter, M. Atkins et al., "Developing indications for the use of sentinel lymph node biopsy and adjuvant high-dose interferon alfa-2b in melanoma," Archives of Dermatology, vol. 137, no. 9, pp. 1217–1224, 2001.

61.   H. Tsao, M. B. Atkins, and A. J. Sober, "Management of cutaneous melanoma," The New England Journal of Medicine, vol. 351, no. 10, pp. 998–1012, 2004.

62.   S. Hu, Y. Parmet, G. Allen et al., "Disparity in melanoma: a trend analysis of melanoma incidence and stage at diagnosis among whites, Hispanics, and blacks in Florida," Archives of Dermatology, vol. 145, no. 12, pp. 1369–1374, 2009.

63.   V. K. Sondak and G. T. Gibney, "Indications and options for systemic therapy in melanoma," Surgical Clinics of North America, vol. 94, no. 5, pp. 1049–1058, 2014.

64.   S. Stadler, K. Weina, C. Gebhardt, and J. Utikal, "New therapeutic options for advanced non-resectable malignant melanoma," Advances in Medical Sciences, vol. 60, no. 1, pp. 83–88, 2015.

65.   A. Ribas, A. Hauschild, R. Kefford et al., "Phase III, open-label, randomized, comparative study of tremelimumab (CP-675,206) and chemotherapy (temozolomide [TMZ] or dacarbazine [DTIC]) in patients with advanced melanoma," Journal of Clinical Oncology, vol. 26, no. 15, supplement, 2008.

66.   M. A. Postow, J. Chesney, A. C. Pavlick et al., "Nivolumab and ipilimumab versus ipilimumab in untreated melanoma," The New England Journal of Medicine, 2015.

67.   S. Bagcchi, "Pembrolizumab for treatment of refractory melanoma," The Lancet Oncology, vol. 15, no. 10, article e419, 2014.

68. T. K. Burki, "Variation in prostate cancer management," The Lancet Oncology, vol. 15, no. 10, p. e419, 2014.

69. B. V. Bonifácio, P. B. da Silva, M. Aparecido dos Santos Ramos, K. Maria Silveira Negri, T. Maria Bauab, and M. Chorilli, "Nanotechnology-based drug delivery systems and herbal medicines: a review,"International Journal of Nanomedicine, vol. 9, no. 1, pp. 1–15, 2014.

70. R. M. Mainardes, M. C. Cocenza Urban, P. O. Cinto, M. V. Chaud, R. C. Evangelista, and M. P. Daflon Gremião, "Liposomes and micro/nanoparticles as colloidal carriers for nasal drug delivery," Current Drug Delivery, vol. 3, no. 3, pp. 275–285, 2006.

71. R. N. Saha, S. Vasanthakumar, G. Bende, and M. Snehalatha, "Nanoparticulate drug delivery systems for cancer chemotherapy," Molecular Membrane Biology, vol. 27, no. 7, pp. 215–231, 2010.

72. A. E. Grill, N. W. Johnston, T. Sadhukha, and J. Panyam, "A review of select recent patents on novel nanocarriers," Recent Patents on Drug Delivery and Formulation, vol. 3, no. 2, pp. 137–142, 2009.

73. J. Venugopal, M. P. Prabhakaran, S. Low et al., "Continuous nanostructures for the controlled release of drugs," Current Pharmaceutical Design, vol. 15, no. 15, pp. 1799–1808, 2009.

74. V. P. Torchilin, "Multifunctional, stimuli-sensitive nanoparticulate systems for drug delivery," Nature Reviews Drug Discovery, vol. 13, no. 11, pp. 813–827, 2014.

75. A. Kumar, A. Srivastava, I. Y. Galaev, and B. Mattiasson, "Smart polymers: physical forms and bioengineering applications," Progress in Polymer Science, vol. 32, no. 10, pp. 1205–1237, 2007.

76. M. P. Patel, S. T. Churchman, A. T. Cruchley, M. Braden, and D. M. Williams, "Delivery of macromolecules across oral mucosa from polymeric hydrogels is enhanced by electrophoresis (iontophoresis)," Dental Materials, vol. 29, no. 11, pp. e299–e307, 2013.

77. S. Thomas, R. Shanks, and C. Sarathchandran, Nanostructured Polymer Blends, edited by: W. Andrew, Elsevier, Oxford, UK, 1st edition, 2013.

78. T. R. Hoare and D. S. Kohane, "Hydrogels in drug delivery: progress and challenges," Polymer, vol. 49, no. 8, pp. 1993–2007, 2008.

79. M. Casolaro, D. B. Barbara, and M. Emilia, "Hydrogel containing l-valine residues as a platform for cisplatin chemotherapy," Colloids and Surfaces B: Biointerfaces, vol. 88, no. 1, pp. 389–395, 2011.

80. J. L. Drury and D. J. Mooney, "Hydrogels for tissue engineering: scaffold design variables and applications," Biomaterials, vol. 24, no. 24, pp. 4337–4351, 2003.

81. S. W. Kim, Y. H. Bae, and T. Okano, "Hydrogels: swelling, drug loading, and release," Pharmaceutical Research, vol. 9, no. 3, pp. 283–290, 1992.

82. R. Vasita and D. S. Katti, "Nanofibers and their applications in tissue engineering," International Journal of Nanomedicine, vol. 1, no. 1, pp. 15–30, 2006.

83. N. A. Peppas, P. Bures, W. Leobandung, and H. Ichikawa, "Hydrogels in pharmaceutical formulations,"European Journal of Pharmaceutics and Biopharmaceutics, vol. 50, no. 1, pp. 27–46, 2000.

84. D. J. Murphy, M. G. Sankalia, R. G. Loughlin, R. F. Donnelly, M. G. Jenkins, and P. A. Mccarron, "Physical characterisation and component release of poly(vinyl alcohol)-tetrahydroxyborate hydrogels and their applicability as potential topical drug delivery systems," International Journal of Pharmaceutics, vol. 423, no. 2, pp. 326–334, 2012.

85. N. Bhattarai, J. Gunn, and M. Zhang, "Chitosan-based hydrogels for controlled, localized drug delivery,"Advanced Drug Delivery Reviews, vol. 62, no. 1, pp. 83–99, 2010.

86. Y. Gao, J. Xie, H. Chen et al., "Nanotechnology-based intelligent drug design for cancer metastasis treatment," Biotechnology Advances, vol. 32, no. 4, pp. 761–777, 2014.

87. M. J. Alvarez-Figueroa and J. Blanco-Méndez, "Transdermal delivery of methotrexate: iontophoretic delivery from hydrogels and passive delivery from microemulsions," International Journal of Pharmaceutics, vol. 215, no. 1-2, pp. 57–65, 2001.

88. G. Lu and H. W. Jun, "Diffusion studies of methotrexate in Carbopol and Poloxamer gels," International Journal of Pharmaceutics, vol. 160, no. 1, pp. 1–9, 1998.

89. F. C. Carvalho, G. Calixto, I. N. Hatakeyama, G. M. Luz, M. P. D. Gremião, and M. Chorilli, "Rheological, mechanical, and bioadhesive behavior of hydrogels to optimize skin delivery systems,"Drug Development and Industrial Pharmacy, vol. 39, no. 11, pp. 1750–1757, 2013.

90. G. Calixto, A. C. Yoshii, H. Rocha e Silva, B. S. F. Cury, and M. Chorilli, "Polyacrylic acid polymers hydrogels intended to topical drug delivery: preparation and characterization," Pharmaceutical Development and Technology, 2014.

91. M. Redpath, C. M. G. Marques, C. Dibden, A. Waddon, R. Lalla, and S. MacNeil, "Ibuprofen and hydrogel-released ibuprofen in the reduction of inflammation-induced migration in melanoma cells,"British Journal of Dermatology, vol. 161, no. 1, pp. 25–33, 2009.

92. M. J. Moura, M. H. Gil, and M. M. Figueiredo, "Delivery of cisplatin from thermosensitive co-cross-linked chitosan hydrogels," European Polymer Journal, vol. 49, no. 9, pp. 2504–2510, 2013.

93. M. R. Bernsen, J.-W. Tang, L. A. Everse, J. W. Koten, and W. Den Otter, "Interleukin 2 (IL-2) therapy: potential advantages of locoregional versus systemic administration," Cancer Treatment Reviews, vol. 25, no. 2, pp. 73–82, 1999.

94. G. W. Bos, J. J. L. Jacobs, J. W. Koten et al., "In situ crosslinked biodegradable hydrogels loaded with IL-2 are effective tools for local IL-2 therapy," European Journal of Pharmaceutical Sciences, vol. 21, no. 4, pp. 561–567, 2004.

95. T. Takei, K. Sugihara, M. Yoshida, and K. Kawakami, "Injectable and biodegradable sugar beet pectin/gelatin hydrogels for biomedical

applications," Journal of Biomaterials Science, Polymer Edition, vol. 24, no. 11, pp. 1333–1342, 2013.

96. E. Ruel-Gariépy, M. Shive, A. Bichara et al., "A thermosensitive chitosan-based hydrogel for the local delivery of paclitaxel," European Journal of Pharmaceutics and Biopharmaceutics, vol. 57, no. 1, pp. 53–63, 2004.

97. W. S. Shim, J.-H. Kim, K. Kim et al., "pH- and temperature-sensitive, injectable, biodegradable block copolymer hydrogels as carriers for paclitaxel," International Journal of Pharmaceutics, vol. 331, no. 1, pp. 11–18, 2007.

98. C. C. Beh, R. Mammucari, and N. R. Foster, "Lipids-based drug carrier systems by dense gas technology: a review," Chemical Engineering Journal, vol. 188, pp. 1–14, 2012.

99. E. Fahy, S. Subramaniam, H. A. Brown et al., "A comprehensive classification system for lipids," Journal of Lipid Research, vol. 46, no. 5, pp. 839–861, 2005.

100. D.-G. Yu, C. Branford-White, G. R. Williams et al., "Self-assembled liposomes from amphiphilic electrospun nanofibers," Soft Matter, vol. 7, no. 18, pp. 8239–8247, 2011.

101. S. Zhang, H.-J. Sun, A. D. Hughes et al., "'single-single' amphiphilic janus dendrimers self-assemble into uniform dendrimersomes with predictable size," ACS Nano, vol. 8, no. 2, pp. 1554–1565, 2014.

102. J. Muñoz and M. C. Alfaro, "Rheological and phase behaviour of amphiphilic lipids," Grasas y Aceites, vol. 51, no. 1-2, pp. 6–25, 2000.

103. M. Y. Vagin, E. V. Malyh, N. I. Larionova, and A. A. Karyakin, "Spontaneous and facilitated micelles formation at liquid/liquid interface: towards amperometric detection of redox inactive proteins," Electrochemistry Communications, vol. 5, no. 4, pp. 329–333, 2003.

104. M. C. Woodle and D. Papahadjopoulos, "Liposome preparation and size characterization," Methods in Enzymology, vol. 171, pp. 193–217, 1989.

105. W. R. Hargreaves, "Liposomes from ionic, single-chain amphiphiles," Biochemistry, vol. 17, no. 18, pp. 3759–3767, 1978.

106. F. Frézard, D. A. Schettini, O. G. Rocha, and C. Demicheli, "Lipossomas: propriedades físico-químicas e farmacológicas, aplicações na quimioterapia à base de antimônio," Química Nova, vol. 28, no. 3, pp. 511–518, 2005.

107. V. P. Torchilin, "Recent advances with liposomes as pharmaceutical carriers," Nature Reviews Drug Discovery, vol. 4, no. 2, pp. 145–160, 2005.

108. L. C. G. Machado, S. Anne, and M. L. W. Klüppel, "Liposomes applied in pharmacology: a review," Estudos de Biologia, vol. 29, no. 67, pp. 215–224, 2007.

109. A. N. Jătariu, M. Popa, and C. A. Peptu, "Different particulate systems—bypass the biological barriers," Journal of Drug Targeting, vol. 18, no. 4, pp. 243–253, 2010.

110.   N. Berger, A. Sachse, J. Bender, R. Schubert, and M. Brandl, "Filter extrusion of liposomes using different devices: comparison of liposome size, encapsulation efficiency, and process characteristics,"International Journal of Pharmaceutics, vol. 223, no. 1-2, pp. 55–68, 2001.

111.   M. M. Lapinski, A. Castro-Forero, A. J. Greiner, R. Y. Ofoli, and G. J. Blanchard, "Comparison of liposomes formed by sonication and extrusion: rotational and translational diffusion of an embedded chromophore," Langmuir, vol. 23, no. 23, pp. 11677–11683, 2007.

112.   F. Olson, C. A. Hunt, F. C. Szoka, W. J. Vail, and D. Papahadjopoulos, "Preparation of liposomes of defined size distribution by extrusion through polycarbonate membranes," BBA—Biomembranes, vol. 557, no. 1, pp. 9–23, 1979.

113.   E. Feitosa, P. C. A. Barreleiro, and G. Olofsson, "Phase transition in dioctadecyldimethylammonium bromide and chloride vesicles prepared by different methods," Chemistry and Physics of Lipids, vol. 105, no. 2, pp. 201–213, 2000.

114.   D. G. Hunter and B. J. Frisken, "Effect of extrusion pressure and lipid properties on the size and polydispersity of lipid vesicles," Biophysical Journal, vol. 74, no. 6, pp. 2996–3002, 1998.

115.   G. Maulucci, M. De Spirito, G. Arcovito, F. Boffi, A. C. Castellano, and G. Briganti, "Particle distribution in DMPC vesicles solutions undergoing different sonication times," Biophysical Journal, vol. 88, no. 5, pp. 3545–3550, 2005.

116.   M. Owais and C. M. Gupta, "Targeted drug delivery to macrophages in parasitic infections," Current Drug Delivery, vol. 2, no. 4, pp. 311–318, 2005.

117.   R. A. Schwendener, "Liposomes in biology and medicine," Advances in Experimental Medicine and Biology, vol. 620, pp. 117–128, 2007.

118.   S. Bhowmick, R. Ravindran, and N. Ali, "Leishmanial antigens in liposomes promote protective immunity and provide immunotherapy against visceral leishmaniasis via polarized Th1 response,"Vaccine, vol. 25, no. 35, pp. 6544–6556, 2007.

119.   S. E. Treiger Borborema, R. A. Schwendener, J. A. Osso Junior, H. F. de Andrade Junior, and N. do Nascimento, "Uptake and antileishmanial activity of meglumine antimoniate-containing liposomes inLeishmania (Leishmania) major-infected macrophages," International Journal of Antimicrobial Agents, vol. 38, no. 4, pp. 341–347, 2011.

120.   P. Wolf, H. Maier, R. R. Müllegger et al., "Topical treatment with liposomes containing T4 endonuclease V protects human skin in vivo from ultraviolet-induced upregulation of interleukin-10 and tumor necrosis factor-α," Journal of Investigative Dermatology, vol. 114, no. 1, pp. 149–156, 2000.

121.   D. Yarosh, J. Klein, A. O'Connor, J. Hawk, E. Rafal, and P. Wolf, "Effect of topically applied T4 endonuclease V in liposomes on skin cancer in xeroderma pigmentosum: a randomised study," The Lancet, vol. 357, no. 9260, pp. 926–929, 2001.

122.   M. B. Pierre, A. C. Tedesco, J. M. Marchetti, and M. V. Bentley, "Stratum corneum lipids liposomes for the topical delivery of 5-

aminolevulinic acid in photodynamic therapy of skin cancer: preparation and in vitro permeation study," BMC Dermatology, vol. 1, no. 1, p. 5, 2001.

123. Y. Chen, Q. Wu, Z. Zhang, L. Yuan, X. Liu, and L. Zhou, "Preparation of curcumin-loaded liposomes and evaluation of their skin permeation and pharmacodynamics," Molecules, vol. 17, no. 5, pp. 5972–5987, 2012.

124. M. Nobayashi, M. Mizuno, T. Kageshita, K. Matsumoto, T. Saida, and J. Yoshida, "Repeated cationic multilamellar liposome-mediated gene transfer enhanced transduction efficiency against murine melanoma cell lines," Journal of Dermatological Science, vol. 29, no. 3, pp. 206–213, 2002.

125. D. Liu, H. Hu, Z. Lin et al., "Quercetin deformable liposome: preparation and efficacy against ultraviolet B induced skin damages in vitro and in vivo," Journal of Photochemistry and Photobiology B: Biology, vol. 127, pp. 8–17, 2013.

126. E. M. M. Del Valle, "Cyclodextrins and their uses: a review," Process Biochemistry, vol. 39, no. 9, pp. 1033–1046, 2004.

127. M. Singh, R. Sharma, and U. C. Banerjee, "Biotechnological applications of cyclodextrins," Biotechnology Advances, vol. 20, no. 5-6, pp. 341–359, 2002.

128. A. L. Laza-Knoerr, R. Gref, and P. Couvreur, "Cyclodextrins for drug delivery," Journal of Drug Targeting, vol. 18, no. 9, pp. 645–656, 2010.

129. A. Biwer, G. Antranikian, and E. Heinzle, "Enzymatic production of cyclodextrins," Applied Microbiology and Biotechnology, vol. 59, no. 6, pp. 609–617, 2002.

130. T. Loftsson and D. Duchêne, "Cyclodextrins and their pharmaceutical applications," International Journal of Pharmaceutics, vol. 329, no. 1-2, pp. 1–11, 2007.

131. T. Loftsson and M. E. Brewster, "Pharmaceutical applications of cyclodextrins. 1. Drug solubilization and stabilization," Journal of Pharmaceutical Sciences, vol. 85, no. 10, pp. 1017–1025, 1996.

132. T. Loftsson, M. Másson, and M. E. Brewster, "Self-association of cyclodextrins and cyclodextrin complexes," Journal of Pharmaceutical Sciences, vol. 93, no. 5, pp. 1091–1099, 2004.

133. J. Zhang and P. X. Ma, "Host-guest interactions mediated nano-assemblies using cyclodextrin-containing hydrophilic polymers and their biomedical applications," Nano Today, vol. 5, no. 4, pp. 337–350, 2010.

134. K. Uekama, "Design and evaluation of cyclodextrin-based drug formulation," Chemical and Pharmaceutical Bulletin, vol. 52, no. 8, pp. 900–915, 2004.

135. B. Gidwani and A. Vyas, "Synthesis, characterization and application of epichlorohydrin-β-cyclodextrin polymer," Colloids and Surfaces B: Biointerfaces, vol. 114, pp. 130–137, 2014.

136. N. C. Bellocq, S. H. Pun, G. S. Jensen, and M. E. Davis, "Transferrin-containing, cyclodextrin polymer-based particles for tumor-targeted

gene delivery," Bioconjugate Chemistry, vol. 14, no. 6, pp. 1122–1132, 2003.

137.   S. Hu-Lieskovan, J. D. Heidel, D. W. Bartlett, M. E. Davis, and T. J. Triche, "Sequence-specific knockdown of EWS-FLI1 by targeted, nonviral delivery of small interfering RNA inhibits tumor growth in a murine model of metastatic Ewing's sarcoma," Cancer Research, vol. 65, no. 19, pp. 8984–8992, 2005.

138.   S. H. Pun, F. Tack, N. C. Bellocq et al., "Targeted delivery of RNA-cleaving DNA enzyme (DNAzyme) to tumor tissue by transferrin-modified, cyclodextrin-based particles," Cancer Biology and Therapy, vol. 3, no. 7, pp. 641–650, 2004.

139.   D. Michel, J. M. Chitanda, R. Balogh et al., "Design and evaluation of cyclodextrin-based delivery systems to incorporate poorly soluble curcumin analogs for the treatment of melanoma," European Journal of Pharmaceutics and Biopharmaceutics, vol. 81, no. 3, pp. 548–556, 2012.

140.   S. Ganta, H. Devalapally, A. Shahiwala, and M. Amiji, "A review of stimuli-responsive nanocarriers for drug and gene delivery," Journal of Controlled Release, vol. 126, no. 3, pp. 187–204, 2008.

141.   I. F. Tannock and D. Rotin, "Acid pH in tumors and its potential for therapeutic exploitation," Cancer Research, vol. 49, no. 16, pp. 4373–4384, 1989.

142.   Y. Liu, W. Wang, J. Yang, C. Zhou, and J. Sun, "pH-sensitive polymeric micelles triggered drug release for extracellular and intracellular drug targeting delivery," Asian Journal of Pharmaceutical Sciences, vol. 8, no. 3, pp. 159–167, 2013.

143.   Z. Zhang, J. Ding, X. Chen et al., "Intracellular pH-sensitive supramolecular amphiphiles based on host-guest recognition between benzimidazole and β-cyclodextrin as potential drug delivery vehicles,"Polymer Chemistry, vol. 4, no. 11, pp. 3265–3271, 2013.

144.   H. He, S. Chen, J. Zhou et al., "Cyclodextrin-derived pH-responsive nanoparticles for delivery of paclitaxel," Biomaterials, vol. 34, no. 21, pp. 5344–5358, 2013.

145.   F. Huang and H. W. Gibson, "Polypseudorotaxanes and polyrotaxanes," Progress in Polymer Science, vol. 30, no. 10, pp. 982–1018, 2005.

146.   J. Chang, Y. Li, G. Wang, B. He, and Z. Gu, "Fabrication of novel coumarin derivative functionalized polypseudorotaxane micelles for drug delivery," Nanoscale, vol. 5, no. 2, pp. 813–820, 2013.

147.   Z.-H. Chen and E. Niki, "4-Hydroxynonenal (4-HNE) has been widely accepted as an inducer of oxidative stress. Is this the whole truth about it or can 4-HNE also exert protective effects?" IUBMB Life, vol. 58, no. 5-6, pp. 372–373, 2006.

148.   W. Siems and T. Grune, "Intracellular metabolism of 4-hydroxynonenal," Molecular Aspects of Medicine, vol. 24, no. 4-5, pp. 167–175, 2003.

149.   S. Pizzimenti, E. Ciamporcero, P. Pettazzoni et al., "The inclusion complex of 4-hydroxynonenal with a polymeric derivative of β-

cyclodextrin enhances the antitumoral efficacy of the aldehyde in several tumor cell lines and in a three-dimensional human melanoma model," Free Radical Biology and Medicine, vol. 65, pp. 765–777, 2013.

150. S. K. Rodal, G. Skretting, Ø. Garred, F. Vilhardt, B. van Deurs, and K. Sandvig, "Extraction of cholesterol with methyl-β-cyclodextrin perturbs formation of clathrin-coated endocytic vesicles," Molecular Biology of the Cell, vol. 10, no. 4, pp. 961–974, 1999.

151. I. S. Babina, S. Donatello, I. R. Nabi, and A. M. Hopkins, "Lipid rafts as master regulators of breast cancer cell function," in Breast Cancer— Carcinogenesis, Cell Growth and Signalling Pathways, M. Gunduz and E. Gunduz, Eds., chapter 19, InTech, 2011.

152. T. Murai, Y. Maruyama, K. Mio, H. Nishiyama, M. Suga, and C. Sato, "Low cholesterol triggers membrane microdomain-dependent CD44 shedding and suppresses tumor cell migration," Journal of Biological Chemistry, vol. 286, no. 3, pp. 1999–2007, 2011.

153. P. Keller and K. Simons, "Cholesterol is required for surface transport of influenza virus hemagglutinin,"Journal of Cell Biology, vol. 140, no. 6, pp. 1357–1367, 1998.

154. R. Onodera, K. Motoyama, A. Okamatsu et al., "Involvement of cholesterol depletion from lipid rafts in apoptosis induced by methyl-β-cyclodextrin," International Journal of Pharmaceutics, vol. 452, no. 1-2, pp. 116–123, 2013.

155. A. Mazzaglia, M. L. Bondì, A. Scala et al., "Supramolecular assemblies based on complexes of nonionic amphiphilic cyclodextrins and a meso-tetra(4-sulfonatophenyl)porphine tributyltin(IV) derivative: potential nanotherapeutics against melanoma," Biomacromolecules, vol. 14, no. 11, pp. 3820 3829, 2013.

156. G. S. Kienle and H. Kiene, "Complementary cancer therapy: a systematic review of prospective clinical trials on anthroposophic mistletoe extracts," European Journal of Medical Research, vol. 12, no. 3, pp. 103–119, 2007.

157. E. Ernst, "The role of complementary and alternative medicine in cancer," The Lancet Oncology, vol. 1, no. 3, pp. 176–180, 2000.

158. C. M. Strüh, S. Jäger, A. Kersten, C. M. Schempp, A. Scheffler, and S. F. Martin, "Triterpenoids amplify anti-tumoral effects of mistletoe extracts on murine B16.f10 melanoma in vivo," PLoS ONE, vol. 8, no. 4, Article ID e62168, 2013.

159. R. H. Cichewicz and S. A. Kouzi, "Chemistry, biological activity, and chemotherapeutic potential of betulinic acid for the prevention and treatment of cancer and HIV infection," Medicinal Research Reviews, vol. 24, no. 1, pp. 90–114, 2004.

160. C. Şoica, C. Dehelean, C. Danciu et al., "Betulin complex in γ-cyclodextrin derivatives: properties and antineoplasic activities in in vitro and in vivo tumor models," International Journal of Molecular Sciences, vol. 13, no. 11, pp. 14992–15011, 2012.

161. C. Nicholas and G. B. Lesinski, "Immunomodulatory cytokines as therapeutic agents for melanoma,"Immunotherapy, vol. 3, no. 5, pp. 673–690, 2011.

162. R. A. Flavell, S. Sanjabi, S. H. Wrzesinski, and P. Licona-Limón, "The polarization of immune cells in the tumour environment by TGFβ," Nature Reviews Immunology, vol. 10, no. 8, pp. 554–567, 2010.

163. J. Park, S. H. Wrzesinski, E. Stern et al., "Combination delivery of TGF-β inhibitor and IL-2 by nanoscale liposomal polymeric gels enhances tumour immunotherapy," Nature Materials, vol. 11, no. 10, pp. 895–905, 2012.

164. S. B. Brown, E. A. Brown, and I. Walker, "The present and future role of photodynamic therapy in cancer treatment," The Lancet Oncology, vol. 5, no. 8, pp. 497–508, 2004.

165. H. Kolarova, J. Macecek, P. Nevrelova et al., "Photodynamic therapy with zinc-tetra(p-sulfophenyl)porphyrin bound to cyclodextrin induces single strand breaks of cellular DNA in G361 melanoma cells," Toxicology in Vitro, vol. 19, no. 7, pp. 971–974, 2005.

166. S. S. Guterres, M. P. Alves, and A. R. Pohlmann, "Polymeric nanoparticles, nanospheres and nanocapsules, for cutaneous applications," Drug Target Insights, vol. 2, pp. 147–157, 2007.

167. V. Mohanraj and Y. Chen, "Nanoparticles—a review," Tropical Journal of Pharmaceutical Research, vol. 5, no. 1, pp. 561–573, 2006.

168. H. Bhargava, A. Narurkar, and L. Lieb, "Using microemulsions for drug delivery," Pharmaceutical Technology, vol. 11, no. 3, pp. 46–54, 1987.

169. T. P. Formariz, M. C. Urban, A. A. Silva Júnior, M. P. Gremião, and A. G. Oliveira, "Microemulsões e fases líquidas cristalinas como sistemas de liberação de fármacos," Revista Brasileira de Ciências Farmacêuticas, vol. 41, no. 3, pp. 301–313, 2005.

170. M. Chorilli, P. S. Prestes, R. B. Rigon, G. R. Leonardi, L. A. Chiavacci, and M. V. Scarpa, "Desenvolvimento de sistemas líquido-cristalinos empregando silicone fluido de co-polímero glicol e poliéter funcional siloxano," Química Nova, vol. 32, no. 4, pp. 1036–1040, 2009.

171. C. C. Mueller-Goymann and S. G. Frank, "Interaction of lidocaine and lidocaine-HCl with the liquid crystal structure of topical preparations," International Journal of Pharmaceutics, vol. 29, no. 2-3, pp. 147–159, 1986.

172. O. Lehmann, Flüssige Kristalle, Wilhelm Engelmann, Leipzig, Germany, 1904.

173. M. Ferrari, "Obtenção e aplicação de emulsões múltiplas contendo óleo de andiroba e copaíba," inFaculdade de Ciências Farmacêuticas de Ribeirão Preto, p. 147, Sao Paulo University, Ribeirão Preto, Brazil, 1998.

174. M. Urban, "Desenvolvimento de sistemas de liberação micro e nanoestruturados para administração cutânea do acetato de

dexametasona," in Drugs and Medicines, p. 136, São Paulo State University, Araraquara, Brazil, 2004.

175. L. Bitan-Cherbakovsky, A. Aserin, and N. Garti, "Structural characterization of lyotropic liquid crystals containing a dendrimer for solubilization and release of gallic acid," Colloids and Surfaces B, vol. 112, pp. 87–95, 2013.

176. M. Gosenca, M. Bešter-Rogač, and M. Gašperlin, "Lecithin based lamellar liquid crystals as a physiologically acceptable dermal delivery system for ascorbyl palmitate," European Journal of Pharmaceutical Sciences, vol. 50, no. 1, pp. 114–122, 2013.

177. D. Bei, J. Marszalek, and B.-B. C. Youan, "Formulation of dacarbazine-loaded cubosomes—part I: influence of formulation variables," AAPS PharmSciTech, vol. 10, no. 3, pp. 1032–1039, 2009.

178. D. Bei, T. Zhang, J. B. Murowchick, and B.-B. C. Youan, "Formulation of dacarbazine-loaded cubosomes. Part III. physicochemical characterization," AAPS PharmSciTech, vol. 11, no. 3, pp. 1243–1249, 2010.

179. X. Gong, M. J. Moghaddam, S. M. Sagnella et al., "Lyotropic liquid crystalline self-assembly material behavior and nanoparticulate dispersions of a phytanyl pro-drug analogue of capecitabine—a chemotherapy agent," ACS Applied Materials & Interfaces, vol. 3, no. 5, pp. 1552–1561, 2011.

180. K. L. von Eckardstein, S. Patt, C. Kratzel, J. C. W. Kiwit, and R. Reszka, "Local chemotherapy of F98 rat glioblastoma with paclitaxel and carboplatin embedded in liquid crystalline cubic phases," Journal of Neuro-Oncology, vol. 72, no. 3, pp. 209–215, 2005.

181. K. L. von Eckardstein, R. Reszka, and J. C. Kiwit, "Intracavitary chemotherapy (paclitaxel/carboplatin liquid crystalline cubic phases) for recurrent glioblastoma—clinical observations," Journal of Neuro-Oncology, vol. 74, no. 3, pp. 305–309, 2005.

182. S. B. Rizwan, Y.-D. Dong, B. J. Boyd, T. Rades, and S. Hook, "Characterisation of bicontinuous cubic liquid crystalline systems of phytantriol and water using cryo field emission scanning electron microscopy (cryo FESEM)," Micron, vol. 38, no. 5, pp. 478–485, 2007.

183. N. Zeng, Q. Hu, Z. Liu et al., "Preparation and characterization of paclitaxel-loaded DSPE-PEG-liquid crystalline nanoparticles (LCNPs) for improved bioavailability," International Journal of Pharmaceutics, vol. 424, no. 1-2, pp. 58–66, 2012.

184. D. Bei, J. Marszalek, and B.-B. C. Youan, "Formulation of dacarbazine-loaded cubosomes—part II: influence of process parameters," AAPS PharmSciTech, vol. 10, no. 3, pp. 1040–1047, 2009.

185. L. Mu and S. S. Feng, "A novel controlled release formulation for the anticancer drug paclitaxel (Taxol): PLGA nanoparticles containing vitamin E TPGS," Journal of Controlled Release, vol. 86, no. 1, pp. 33–48, 2003.

186. A. Pampel, D. Michel, and R. Reszka, "Pulsed field gradient MAS-NMR studies of the mobility of carboplatin in cubic liquid-crystalline

phases," Chemical Physics Letters, vol. 357, no. 1-2, pp. 131–136, 2002.

187. A. Dowling, R. Clift, N. Grobert et al., "Nanoscience and nanotechnologies: opportunities and uncertainties," The Royal Society & The Royal Academy of Engineering Report, Royal Academy of Engineering, London, UK, 2004.

188. P. K. Jain, K. S. Lee, I. H. El-Sayed, and M. A. El-Sayed, "Calculated absorption and scattering properties of gold nanoparticles of different size, shape, and composition: applications in biological imaging and biomedicine," The Journal of Physical Chemistry B, vol. 110, no. 14, pp. 7238–7248, 2006.

189. K. A. Howard, U. L. Rahbek, X. Liu et al., "RNA interference in vitro and in vivo using a novel chitosan/siRNA nanoparticle system," Molecular Therapy, vol. 14, no. 4, pp. 476–484, 2006.

190. P. L. Apopa, Y. Qian, R. Shao et al., "Iron oxide nanoparticles induce human microvascular endothelial cell permeability through reactive oxygen species production and microtubule remodeling," Particle and Fibre Toxicology, vol. 6, no. 1, article 1, 2009.

191. M. Auffan, J. Rose, J.-Y. Bottero, G. V. Lowry, J.-P. Jolivet, and M. R. Wiesner, "Towards a definition of inorganic nanoparticles from an environmental, health and safety perspective," Nature Nanotechnology, vol. 4, no. 10, pp. 634–641, 2009.

192. Food and Drug Administration (FDA), Guidance for Industry: Assessing the Effects of Significant Manufacturing Process Changes, Including Emerging Technologies, on the Safety and Regulatory Status of Food Ingredients and Food Contact Substances, Including Food Ingredients That are Color Additives, Food and Drug Administration (FDA), Silver Spring, Md, USA, 2012.

193. R. Müller and J. Lucks, "Arzneistoffträger aus festen lipidteilchen, feste lipidnanosphären (sln)," European Patent, 1996.

194. R. H. Müller, R. D. Petersen, A. Hommoss, and J. Pardeike, "Nanostructured lipid carriers (NLC) in cosmetic dermal products," Advanced Drug Delivery Reviews, vol. 59, no. 6, pp. 522–530, 2007.

195. W. Mehnert and K. Mäder, "Solid lipid nanoparticles: production, characterization and applications," Advanced Drug Delivery Reviews, vol. 47, no. 2-3, pp. 165–196, 2001.

196. E. B. Souto, A. J. Almeida, and R. H. Müller, "Lipid nanoparticles (SLN, NLC) for cutaneous drug delivery: structure, protection and skin effects," Journal of Biomedical Nanotechnology, vol. 3, no. 4, pp. 317–331, 2007.

197. Y.-C. Kuo and C.-C. Wang, "Cationic solid lipid nanoparticles with primary and quaternary amines for release of saquinavir and biocompatibility with endothelia," Colloids and Surfaces B: Biointerfaces, vol. 101, pp. 101–105, 2013.

198. Y.-C. Kuo and C.-T. Liang, "Catanionic solid lipid nanoparticles carrying doxorubicin for inhibiting the growth of U87MG

cells," Colloids and Surfaces B: Biointerfaces, vol. 85, no. 2, pp. 131–137, 2011.

199. F. Canfarotta, M. J. Whitcombe, and S. A. Piletsky, "Polymeric nanoparticles for optical sensing,"Biotechnology Advances, vol. 31, no. 8, pp. 1585–1599, 2013.

200. C.-M. J. Hu and L. Zhang, "Nanoparticle-based combination therapy toward overcoming drug resistance in cancer," Biochemical Pharmacology, vol. 83, no. 8, pp. 1104–1111, 2012.

201. H. Fessi, F. Piusieux, J. P. Devissaguet, N. Ammoury, and S. Benita, "Nanocapsule formation by interfacial polymer deposition following solvent displacement," International Journal of Pharmaceutics, vol. 55, no. 1, pp. R1–R4, 1989.

202. N. Al Khouri Fallouh, L. Roblot-Treupel, and H. Fessi, "Development of a new process for the manufacture of polyisobutylcyanoacrylate nanocapsules," International Journal of Pharmaceutics, vol. 28, no. 2-3, pp. 125–132, 1986.

203. B. Magenheim and S. Benita, "Nanoparticle characterization: a comprehensive physicochemical approach," STP Pharma Sciences, vol. 1, no. 4, pp. 221–241, 1991.

204. K. S. Soppimath, T. M. Aminabhavi, A. R. Kulkarni, and W. E. Rudzinski, "Biodegradable polymeric nanoparticles as drug delivery devices," Journal of Controlled Release, vol. 70, no. 1-2, pp. 1–20, 2001.

205. G. Orive, E. Anitua, J. L. Pedraz, and D. F. Emerich, "Biomaterials for promoting brain protection, repair and regeneration," Nature Reviews Neuroscience, vol. 10, no. 9, pp. 682–692, 2009.

206. K. Letchford and H. Burt, "A review of the formation and classification of amphiphilic block copolymer nanoparticulate structures: micelles, nanospheres, nanocapsules and polymersomes," European Journal of Pharmaceutics and Biopharmaceutics, vol. 65, no. 3, pp. 259–269, 2007.

207. D. Hanahan and R. A. Weinberg, "Hallmarks of cancer: the next generation," Cell, vol. 144, no. 5, pp. 646–674, 2011.

208. A. Jain, K. Jain, P. Kesharwani, and N. K. Jain, "Low density lipoproteins mediated nanoplatforms for cancer targeting," Journal of Nanoparticle Research, vol. 15, no. 9, article 1888, 38 pages, 2013.

209. H. Hillaireau and P. Couvreur, "Nanocarriers' entry into the cell: relevance to drug delivery," Cellular and Molecular Life Sciences, vol. 66, no. 17, pp. 2873–2896, 2009.

210. J. H. Sakamoto, A. L. van de Ven, B. Godin et al., "Enabling individualized therapy through nanotechnology," Pharmacological Research, vol. 62, no. 2, pp. 57–89, 2010.

211. Y.-C. Kuo and H.-H. Chen, "Effect of electromagnetic field on endocytosis of cationic solid lipid nanoparticles by human brain-microvascular endothelial cells," Journal of Drug Targeting, vol. 18, no. 6, pp. 447–456, 2010.

212.   D. E. Owens III and N. A. Peppas, "Opsonization, biodistribution, and pharmacokinetics of polymeric nanoparticles," International Journal of Pharmaceutics, vol. 307, no. 1, pp. 93–102, 2006.

213.   A. Gabizon and F. Martin, "Polyethylene glycol-coated (pegylated) liposomal doxorubicin. Rationale for use in solid tumours," Drugs, vol. 54, no. 4, pp. 15–21, 1997.

214.   T. Ameller, V. Marsaud, P. Legrand, R. Gref, G. Barratt, and J.-M. Renoir, "Polyester-poly(ethylene glycol) nanoparticles loaded with the pure antiestrogen RU 58668: physicochemical and opsonization properties," Pharmaceutical Research, vol. 20, no. 7, pp. 1063–1070, 2003.

215.   K. Na, E. S. Lee, and Y. H. Bae, "Adriamycin loaded pullulan acetate/sulfonamide conjugate nanoparticles responding to tumor pH: pH-dependent cell interaction, internalization and cytotoxicity in vitro," Journal of Controlled Release, vol. 87, no. 1–3, pp. 3–13, 2003.

216.   F. Sonvico, S. Mornet, S. Vasseur et al., "Folate-conjugated iron oxide nanoparticles for solid tumor targeting as potential specific magnetic hyperthermia mediators: synthesis, physicochemical characterization, and in vitro experiments," Bioconjugate Chemistry, vol. 16, no. 5, pp. 1181–1188, 2005.

217.   H. L. Wong, R. Bendayan, A. M. Rauth, H. Y. Xue, K. Babakhanian, and X. Y. Wu, "A mechanistic study of enhanced doxorubicin uptake and retention in multidrug resistant breast cancer cells using a polymer-lipid hybrid nanoparticle system," Journal of Pharmacology and Experimental Therapeutics, vol. 317, no. 3, pp. 1372–1381, 2006.

218.   E. Ristorcelli, E. Beraud, P. Verrando et al., "Human tumor nanoparticles induce apoptosis of pancreatic cancer cells," The FASEB Journal, vol. 22, no. 9, pp. 3358–3369, 2008.

219.   R. A. Parlo and P. S. Coleman, "Enhanced rate of citrate export from cholesterol-rich hepatoma mitochondria. The truncated Krebs cycle and other metabolic ramifications of mitochondrial membrane cholesterol," Journal of Biological Chemistry, vol. 259, no. 16, pp. 9997–10003, 1984.

220.   H. L. Wong, R. Bendayan, A. M. Rauth, Y. Li, and X. Y. Wu, "Chemotherapy with anticancer drugs encapsulated in solid lipid nanoparticles," Advanced Drug Delivery Reviews, vol. 59, no. 6, pp. 491–504, 2007.

221.   Y. Noguchi, J. Wu, R. Duncan et al., "Early phase tumor accumulation of Macromolecules: a great difference in clearance rate between tumor and normal tissues," Japanese Journal of Cancer Research, vol. 89, no. 3, pp. 307–314, 1998.

222.   A. K. Iyer, G. Khaled, J. Fang, and H. Maeda, "Exploiting the enhanced permeability and retention effect for tumor targeting," Drug Discovery Today, vol. 11, no. 17-18, pp. 812–818, 2006.

223.   H. Maeda, J. Wu, T. Sawa, Y. Matsumura, and K. Hori, "Tumor vascular permeability and the EPR effect in macromolecular therapeutics: a review," Journal of Controlled Release, vol. 65, no. 1-2, pp. 271–284, 2000.

224. Z. Xu, L. Chen, W. Gu et al., "The performance of docetaxel-loaded solid lipid nanoparticles targeted to hepatocellular carcinoma," Biomaterials, vol. 30, no. 2, pp. 226–232, 2009.

225. J. Dausend, A. Musyanovych, M. Dass et al., "Uptake mechanism of oppositely charged fluorescent nanoparticles in HeLa cells," Macromolecular Bioscience, vol. 8, no. 12, pp. 1135–1143, 2008.

226. E. Chang, N. Thekkek, W. W. Yu, V. L. Colvin, and R. Drezek, "Evaluation of quantum dot cytotoxicity based on intracellular uptake," Small, vol. 2, no. 12, pp. 1412–1417, 2006.

227. T. Xia, M. Kovochich, M. Liong, J. I. Zink, and A. E. Nel, "Cationic polystyrene nanosphere toxicity depends on cell-specific endocytic and mitochondrial injury pathways," ACS Nano, vol. 2, no. 1, pp. 85–96, 2008.

228. L. Guo, Y. Peng, J. Yao, L. Sui, A. Gu, and J. Wang, "Anticancer activity and molecular mechanism of resveratrol-bovine serum albumin nanoparticles on subcutaneously implanted human primary ovarian carcinoma cells in nude mice," Cancer Biotherapy and Radiopharmaceuticals, vol. 25, no. 4, pp. 471–477, 2010.

229. K. Teskač and J. Kristl, "The evidence for solid lipid nanoparticles mediated cell uptake of resveratrol,"International Journal of Pharmaceutics, vol. 390, no. 1, pp. 61–69, 2010.

230. Z.-R. Huang, S.-C. Hua, Y.-L. Yang, and J.-Y. Fang, "Development and evaluation of lipid nanoparticles for camptothecin delivery: a comparison of solid lipid nanoparticles, nanostructured lipid carriers, and lipid emulsion," Acta Pharmacologica Sinica, vol. 29, no. 9, pp. 1094–1102, 2008.

231. D. Liu, Z. Liu, L. Wang, C. Zhang, and N. Zhang, "Nanostructured lipid carriers as novel carrier for parenteral delivery of docetaxel," Colloids and Surfaces B: Biointerfaces, vol. 85, no. 2, pp. 262–269, 2011.

232. S.-H. Hsu, C.-J. Wen, S. A. Al-Suwayeh, Y.-J. Huang, and J.-Y. Fang, "Formulation design and evaluation of quantum dot-loaded nanostructured lipid carriers for integrating bioimaging and anticancer therapy," Nanomedicine, vol. 8, no. 8, pp. 1253–1269, 2013.

233. K. S. Ho and M. S. Shoichet, "Design considerations of polymeric nanoparticle micelles for chemotherapeutic delivery," Current Opinion in Chemical Engineering, vol. 2, no. 1, pp. 53–59, 2013.

234. H. Yao, S. S. Ng, L.-F. Huo et al., "Effective melanoma immunotherapy with interleukin-2 delivered by a novel polymeric nanoparticle," Molecular Cancer Therapeutics, vol. 10, no. 6, pp. 1082–1092, 2011.

235. A. R. Khuda-Bukhsh, S. S. Bhattacharyya, S. Paul, and N. Boujedaini, "Polymeric nanoparticle encapsulation of a naturally occurring plant scopoletin and its effects on human melanoma cell A375,"Journal of Chinese Integrative Medicine, vol. 8, no. 9, pp. 853–862, 2010.

236. S. Das, J. Das, A. Samadder, A. Paul, and A. R. Khuda-Bukhsh, "Strategic formulation of apigenin-loaded PLGA nanoparticles for

intracellular trafficking, DNA targeting and improved therapeutic effects in skin melanoma in vitro," Toxicology Letters, vol. 223, no. 2, pp. 124–138, 2013.

237. D. Vieira, V. Kim, D. Petri, C. Menck, and A. Carmona-Ribeiro, "Polymer-based delivery vehicle for cisplatin," Nanotechnology, vol. 3, pp. 382–385, 2011.

238. V. Mailänder, M. R. Lorenz, V. Holzapfel et al., "Carboxylated superparamagnetic iron oxide particles label cells intracellularly without transfection agents," Molecular Imaging and Biology, vol. 10, no. 3, pp. 138–146, 2008.

239. A. Ades, J. P. Carvalho, S. R. Graziani et al., "Uptake of a cholesterol-rich emulsion by neoplastic ovarian tissues," Gynecologic Oncology, vol. 82, no. 1, pp. 84–87, 2001.

240. D. Gal, M. Ohashi, P. C. MacDonald, H. J. Buchsbaum, and E. R. Simpson, "Low-density lipoprotein as a potential vehicle for chemotherapeutic agents and radionucleotides in the management of gynecologic neoplasms," American Journal of Obstetrics and Gynecology, vol. 139, no. 8, pp. 877–885, 1981.

241. M. J. Rudling, V. P. Collins, and C. O. Peterson, "Delivery of aclacinomycin A to human glioma cells in vitro by the low-density lipoprotein pathway," Cancer Research, vol. 43, no. 10, pp. 4600–4605, 1983.

242. M. Masquelier, S. Vitols, and C. Peterson, "Low-density lipoprotein as a carrier of antitumoral drugs: in vivo fate of drug-human-low-density lipoprotein complexes in mice," Cancer Research, vol. 46, no. 8, pp. 3842–3847, 1986.

243. B. Lundberg, "Preparation of drug low density lipoprotein complexes for delivery of antitumoral drugs via the low density lipoprotein pathway," Cancer Research, vol. 47, no. 15, pp. 4105–4108, 1987.

244. H. Jin, J. F. Lovell, J. Chen et al., "Cytosolic delivery of LDL nanoparticle cargo using photochemical internalization," Photochemical and Photobiological Sciences, vol. 10, no. 5, pp. 810–816, 2011.

245. R. R. Allison and K. Moghissi, "Oncologic photodynamic therapy: clinical strategies that modulate mechanisms of action," Photodiagnosis and Photodynamic Therapy, vol. 10, no. 4, pp. 331–341, 2013.

246. C. J. Field and P. D. Schley, "Evidence for potential mechanisms for the effect of conjugated linoleic acid on tumor metabolism and immune function: lessons from n-3 fatty acids," The American Journal of Clinical Nutrition, vol. 79, no. 6, pp. 1190S–1198S, 2004.

247. J. S. Falconer, J. A. Ross, K. C. H. Fearon, R. A. Hawkins, M. G. O'Riordain, and D. C. Carter, "Effect of eicosapentaenoic acid and other fatty acids on the growth in vitro of human pancreatic cancer cell lines," British Journal of Cancer, vol. 69, no. 5, pp. 826–832, 1994.

248. S. Jaracz, J. Chen, L. V. Kuznetsova, and I. Ojima, "Recent advances in tumor-targeting anticancer drug conjugates," Bioorganic and Medicinal Chemistry, vol. 13, no. 17, pp. 5043–5054, 2005.

249. X.-Y. Yang, Y.-X. Li, M. Li, L. Zhang, L.-X. Feng, and N. Zhang, "Hyaluronic acid-coated nanostructured lipid carriers for targeting paclitaxel to cancer," Cancer Letters, vol. 334, no. 2, pp. 338–345, 2013.

250. K. Y. Choi, H. Chung, K. H. Min et al., "Self-assembled hyaluronic acid nanoparticles for active tumor targeting," Biomaterials, vol. 31, no. 1, pp. 106–114, 2010.

251. K. Y. Choi, K. H. Min, H. Y. Yoon et al., "PEGylation of hyaluronic acid nanoparticles improves tumor targetability in vivo," Biomaterials, vol. 32, no. 7, pp. 1880–1889, 2011.

252. J. D. Byrne, T. Betancourt, and L. Brannon-Peppas, "Active targeting schemes for nanoparticle systems in cancer therapeutics," Advanced Drug Delivery Reviews, vol. 60, no. 15, pp. 1615–1626, 2008.

253. J. F. Kukowska-Latallo, K. A. Candido, Z. Cao et al., "Nanoparticle targeting of anticancer drug improves therapeutic response in animal model of human epithelial cancer," Cancer Research, vol. 65, no. 12, pp. 5317–5324, 2005.

254. C. P. Leamon and J. A. Reddy, "Folate-targeted chemotherapy," Advanced Drug Delivery Reviews, vol. 56, no. 8, pp. 1127–1141, 2004.

255. Y. Malam, M. Loizidou, and A. M. Seifalian, "Liposomes and nanoparticles: nanosized vehicles for drug delivery in cancer," Trends in Pharmacological Sciences, vol. 30, no. 11, pp. 592–599, 2009.

256. D. Labarre, C. Vauthier, C. Chauvierre, B. Petri, R. Müller, and M. M. Chehimi, "Interactions of blood proteins with poly(isobutylcyanoacrylate) nanoparticles decorated with a polysaccharidic brush," Biomaterials, vol. 26, no. 24, pp. 5075–5084, 2005.

257. R.-L. Hong, C.-J. Huang, Y.-L. Tseng et al., "Direct comparison of liposomal doxorubicin with or without polyethylene glycol coating in C-26 tumor-bearing mice: is surface coating with polyethylene glycol beneficial?" Clinical Cancer Research, vol. 5, no. 11, pp. 3645–3652, 1999.

258. Y. Sheng, C. Liu, Y. Yuan et al., "Long-circulating polymeric nanoparticles bearing a combinatorial coating of PEG and water-soluble chitosan," Biomaterials, vol. 30, no. 12, pp. 2340–2348, 2009.

259. S. Krasnici, A. Werner, M. E. Eichhorn et al., "Effect of the surface charge of liposomes on their uptake by angiogenic tumor vessels," International Journal of Cancer, vol. 105, no. 4, pp. 561–567, 2003.

260. R. Kunstfeld, G. Wickenhauser, U. Michaelis et al., "Paclitaxel encapsulated in cationic liposomes diminishes tumor angiogenesis and melanoma growth in a 'humanized' SCID mouse model," Journal of Investigative Dermatology, vol. 120, no. 3, pp. 476–482, 2003.

261. S. Ran, A. Downes, and P. E. Thorpe, "Increased exposure of anionic phospholipids on the surface of tumor blood vessels," Cancer Research, vol. 62, no. 21, pp. 6132–6140, 2002.

262.   A. S. Hoffman, "Hydrogels for biomedical applications," Advanced Drug Delivery Reviews, vol. 64, pp. 18–23, 2012.

263.   T. P. Chelvi and R. Ralhan, "Designing of thermosensitive liposomes from natural lipids for multimodality cancer therapy," International Journal of Hyperthermia, vol. 11, no. 5, pp. 685–695, 1995.

264.   S. Batzri and E. D. Korn, "Single bilayer liposomes prepared without sonication," Biochimica et Biophysica Acta—Biomembranes, vol. 298, no. 4, pp. 1015–1019, 1973.

265.   J. J. Escobar-Chávez, "Nanocarriers for transdermal drug delivery," Skin, vol. 19, p. 22, 2012.

266.   M. R. Mozafari, "Liposomes: an overview of manufacturing techniques," Cellular and Molecular Biology Letters, vol. 10, no. 4, pp. 711–719, 2005.

267.   M. Malmsten, "Soft drug delivery systems," Soft Matter, vol. 2, no. 9, pp. 760–769, 2006.

268.   R. A. Rajewski and V. J. Stella, "Pharmaceutical applications of cyclodextrins. 2. In vivo drug delivery," Journal of Pharmaceutical Sciences, vol. 85, no. 11, pp. 1142–1169, 1996.

269.   A. F. Soares, R. D. A. Carvalho, and F. Veiga, "Oral administration of peptides and proteins: nanoparticles and cyclodextrins as biocompatible delivery systems," Nanomedicine, vol. 2, no. 2, pp. 183–202, 2007.

270.   P. S. Prestes, M. Chorilli, L. A. Chiavacci, M. V. Scarpa, and G. R. Leonardi, "Physicochemical characterization and rheological behavior evaluation of the liquid crystalline mesophases developed with different silicones," Journal of Dispersion Science and Technology, vol. 31, no. 1, pp. 117–123, 2009.

271.   M. Rückert and G. Otting, "Alignment of biological macromolecules in novel nonionic liquid crystalline media for NMR experiments," Journal of the American Chemical Society, vol. 122, no. 32, pp. 7793–7797, 2000.

272.   E. B. Souto, P. Severino, M. H. A. Santana, and S. C. Pinho, "Solid lipid nanoparticles: classical methods of lab production," Quimica Nova, vol. 34, no. 10, pp. 1762–1769, 2011.

273.   S. Shi, L. Han, L. Deng et al., "Dual drugs (microRNA-34a and paclitaxel)-loaded functional solid lipid nanoparticles for synergistic cancer cell suppression," Journal of Controlled Release, vol. 194, pp. 228–237, 2014.

274.   L. Mazzarino, L. F. C. Silva, J. C. Curta et al., "Curcumin-loaded lipid and polymeric nanocapsules stabilized by nonionic surfactants: an in vitro and in vivo antitumor activity on B16-F10 melanoma and macrophage uptake comparative study," Journal of Biomedical Nanotechnology, vol. 7, no. 3, pp. 406–414, 2011.

275.   L. Cai, X. Wang, W. Wang et al., "Peptide ligand and PEG-mediated long-circulating liposome targeted to FGFR overexpressing tumor in vivo," International Journal of Nanomedicine, vol. 7, pp. 4499–4510, 2012.

276. Y. Barenholz, "Doxil—The first FDA-approved nano-drug: lessons learned," Journal of Controlled Release, vol. 160, no. 2, pp. 117–134, 2012.

277. A. S. Yang and P. B. Chapman, "The history and future of chemotherapy for melanoma,"Hematology/Oncology Clinics of North America, vol. 23, no. 3, pp. 583–597, 2009.

278. M. Ugurel, D. Schadendorf, W. Fink et al., "Clinical phase II study of pegylated liposomal doxorubicin as second-line treatment in disseminated melanoma," Onkologie, vol. 27, no. 6, pp. 540–544, 2004.

279. W.-J. Hwu, K. S. Panageas, J. H. Menell et al., "Phase II study of temozolomide plus pegylated interferon-α-2b for metastatic melanoma," Cancer, vol. 106, no. 11, pp. 2445–2451, 2006.

280. P. Decuzzi and M. Ferrari, "The receptor-mediated endocytosis of nonspherical particles," Biophysical Journal, vol. 94, no. 10, pp. 3790–3797, 2008.

281. H. K. Sajja, M. P. East, H. Mao, Y. A. Wang, S. Nie, and L. Yang, "Development of multifunctional nanoparticles for targeted drug delivery and noninvasive imaging of therapeutic effect," Current Drug Discovery Technologies, vol. 6, no. 1, pp. 43–51, 2009.

282. D. Rosenblum and D. Peer, "Omics-based nanomedicine: the future of personalized oncology," Cancer Letters, vol. 352, no. 1, pp. 126–136, 2013.

CHAPTER 4

Modular Hydrogels for Drug Delivery

Susana Simões[1,2], Ana Figueiras[1,2,3], Francisco Veiga[1,2]

[1]Laboratory of Pharmaceutical Technology, University of Coimbra, Coimbra, Portugal
[2]Pharmaceutical Studies Center (CEF), University of Coimbra, Coimbra, Portugal;
[3]Health Sciences Center (CICS), Faculty of Health Sciences, University of Beira Interior, Covilhã, Portugal.

ABSTRACT

The development of novel drug delivery systems is an essential step toward controlled site-specific administration of therapeutics within the body. It is desirable for delivery vehicles to be introduced into the body through minimally invasive means and, these vehicles should be capable of releasing drug to their intended location at a controlled rate. Furthermore, it is desirable to develop drug delivery systems that are capable of in vivo to suffer degradation and to deliver the drug completely, avoiding the need to surgically remove the vehicle at the end of its useful lifetime. Hydrogels are of particular interest for drug delivery applications due to their ability to address these needs in addition to their good biocompatibility, tunable network structure to control the diffusion of drugs and, tunable affinity for drugs. However, hydrogels are also limited for drug delivery applications due to the often quick elution of drug from their highly swollen polymer matrices as well as the difficulty inherent in the injection of macroscopic hydrogels into the body. This paper presents an overview to the advances in hydrogels based drug delivery. Different types of hydrogels can be used for drug delivery to specific sites in the gastrointestinal tract ranging from the oral cavity to the colon. These novel systems exhibit a range of several peculiar properties which make them attractive as controlled drug release formulations. Moreover, such materials are biocompatible and can be formulated to give controlled, pulsed, and triggered drug release profiles in a variety of tissues.

Keywords: Hydrogels; Swelling; Stimuli-Environmental; Controlled Drug Deliver

1. INTRODUCTION

Natural polymers are derived from renewable resources widely distributed in nature [1]. These materials exhibit a large diversity of structures, different physiological functions and, may offer a variety of potential applications in the field of tissue engineering due to their various properties, such as pseudoplastic behavior, gelation ability, water binding capacity and biodegradability. Many of these polymers form hydrogels that can respond to external stimuli. Hydrogels resemble natural living tissue more than any other class of synthetic biomaterials due to their high water content and soft consistency which is similar to natural tissue [2]. Furthermore, the high water content of these materials contribute to their biocompatibility and can be used as contact lenses, linings for artificial hearts, materials for artificial skin, membranes for biosensors and drug delivery devices [2-10].

Hydrogels are polymeric materials that do not dissolve in water at physiological conditions. However, they swell considerably in aqueous medium [11] and demonstrate extraordinary capacity (>20%) for imbibe water into their network structure. Gels that exhibit a phase transition in response to change in external conditions such as pH, ionic strength, temperature and, electric currents are known as "stimuli-responsive" or "smart" gels [12]. Being insoluble, these three-dimensional hydrophilic networks can retain a large amount of water that contributes to their good blood compatibility and maintains a certain degree of structural integrity and elasticity [13]. This phenomenon can be explained by the presence of hydrophilic functional groups in their structure, such as -OH, -COOH, -CONH$_2$, and -SO$_3$H, capable of absorbing water without undergoing dissolution. Despite these many advantageous properties, hydrogels also have several limitations. The low tensile strength of many hydrogels limits their use in load-bearing applications and can result in the premature dissolution or flow away of the hydrogel from a targeted local site [14]. However, this limitation may not be important in some typical drug delivery applications (e.g. subcutaneous injection). In the case of hydrophobic drugs, the quantity and homogeneity of drug loading into the hydrogel may be limited. On the other hand, the high water content and large pore sizes of the most hydrogels often result in relatively rapid drug release, over a few hours to a few days.

The present review addresses recent important research developments, which have focused in hydrogel systems. Actually, the literature is exhaustive on general aspects about hydrogels. However, the article intends to concentrate on the hydrogel applications that are responsive to pH and temperature with particular emphasis on the drug delivery. Each of the next issues intends to summarize the practical use of hydrogel-based drug delivery therapies for clinical use.

2. CHARACTERIZATION AND NETWORK STRUCTURE

Hydrogels can be designed to have some specifications, such as swelling and mechanical characteristics, justifying their variety of biomedical applications, from contact lenses to controlled-release drug delivery and tissue engineering [15,16]. They can be prepared from natural and synthetic polymer materials [17] and classified using various criteria depending on their preparation method and physicochemical properties (**Table 1**). Natural polymers, such as proteins [18], polysaccharides [19], and deoxyribonucleic acids (DNAs) are cross-linked by either physical or chemical bonds, and synthetic hydrogels can be easily prepared by cross-linking polymerization of synthetic monomers [20]. In addition, natural polymers can be combined with synthetic polymers to obtain different properties in the same hydrogel [21]. For example, the biodegradable property of natural polymers has been combined with several functionalities of synthetic polymers to give new functional hydrogels [22,23]. Numerous monomers and cross-linking agents have been used for the synthesis of hydrogels with wide range of chemical compositions [24].

Hydrogels, particularly, those intended for application in drug delivery and biomedical purposes are required to have acceptable biodegradability and biocompatibility, relevant requisites on the development of novel synthesis and cross-linking methods to design the desired products. Thus, a great variety of cross-linking approaches have been developed to prepare hydrogels for each particular application [41]. In the polymeric network, hydrophilic domains are present, which are hydrated in an aqueous environment, creating the hydrogel structure. As the term "network" implies cross-link, this kind of linkage have to be present to avoid dissolution of the hydrophilic polymer chains into the aqueous phase. A great variety of chemical and physical methods to establish cross-linking have been used to prepare hydrogels [41]. In chemically cross-linked gels, covalent bonds are present between different polymer chains. In physically cross-linked gels, dissolution is prevented by physical interactions, which exist between different polymer chains. The network structure of a hydrogel will determine its properties as a drug delivery device.

Generally, hydrogels are characterized for their morphology, swelling properties and elasticity [39]. Morphology is indicative of their porous structure, swelling determines the release mechanism of the drug from the swollen polymeric mass while elasticity affects the mechanical strength of the network and determines the stability of these drug carriers [37]. **Table 2** highlights some of the important methods to prepare and measure hydrogels and some crucial characterization parameters.

3. RESPONSIVE HYDROGELS

Over the last thirty years, researches have dedicated much attention to the so called "stimuli-responsive" or "environment-sensitive" polymers. This kind of polymers are the ability to answer concerning to small physiccal or chemical stimuli. Hydrogels can exhibit dramatic changes in their swelling behavior, network structure, permeability or mechanical strength in response to different internal or external stimuli [42]. **Figure 1** shows various stimuli that have been explored for modulating drug delivery. External stimuli are produced with the help of different stimuli-generating devices, whereas internal stimuli are produced within the body to control the structural changes in the polymer network and to exhibit the desired drug release [43]. Most of the time drug release is observed during the swelling of the hydrogel. However, a few instances drug release was observed during syneresis of the hydrogel as a result of a squeezeing mechanism [44].

Table 1. Hydrogels classification.

Classification	Contents	Ref.
Origin	• Natural • Synthetic	[17]
Ionic charge (based on the nature of the pendent groups)	• Neutral • Anionic • Cationic • Ampholytic	[25-27]
Water content or degree of swelling	• Low swelling • Medium swelling • High swelling • Superabsorbent	[28]
Network Structure (Porosity)	• Nonporous • Microporous • Macroporous • Superporous	[29,30]
Network morphology	• Amorphous • Semicrystaline • Hydrogen bonded structures • Super molecular structures • Hydrocolloidal agregates	[31,32]
Cross-linking method	• Chemical (or covalent) • Physical (or noncovalent)	[33]
Component (based on the method of preparation)	• Homopolymer • Copolymer • Multipolymer • Interpenetrating	[34]
Function (based on the organization of the monomers)	• Biodegradable or Non-biodegradable • Stimuli responsive • Superabsorbent	[33,35,36]
Mechanism controlling the drug release	• Diffusion controlled release systems • Swelling controlled release systems • Chemically controlled release systems • Environment responsive systems.	[14,37-40]

Table 2. Hydrogels preparation and characterization parameters

Preparation	Characterization Parameter	Techniques of measurement
Isostatic ultra high pressure (IUHP)	Morphology/ Network pore size	• Quasi-elastic laser-light scattering; • Electron microscopy; • Mercury morosimetry; • Rubber elasticity measurements; • Equilibrium swelling experiments.
Use of cross linkers		• Dimensional changes with time;
Use of water and critical conditions of drying	Degree of swelling	• Aqueous medium or medium specific pH; • Volume or mass degree of swelling; • Equilibrium water content.
Use of gelling agents	Cross-linking and mechanical strength	• Ultimate compressive strength, change in polymer solubility with time.
Use of nucleophilic substitutio reaction		• Membrane permeability. • Controlled strength experiments. • Nuclear magnetic resonance (NMR).
	Drug diffusion	• Fourier transform infrared (FTIR) spectroscopy. • Scanning electron microscopy (SEM). • Quasi-elastic laser light scattering.
Use of irradiation and freeze thawing		
	Drug distribution	• FTIR microscopy. • SEM

Such systems are evidently attractive for biotechnology and medicine studies [24,45-49]. Versatile stimuli-sensitive controlled release systems can be fabricated, provided that the hydrogels are well designed to change their configuration in response to these stimuli based on almost infinitely available mechanisms [50]. Typical examples of environmental-sensitive hydrogels are listed in **Table 3**.

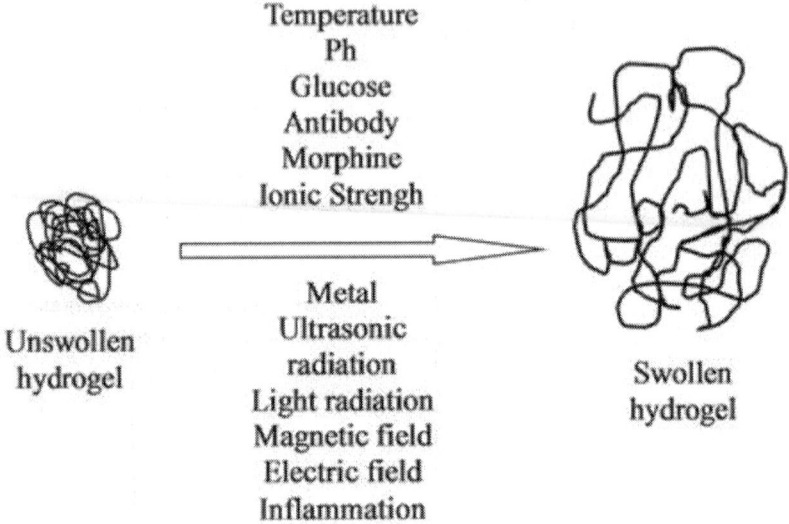

Temperature
Ph
Glucose
Antibody
Morphine
Ionic Strengh

Unswollen hydrogel

Metal
Ultrasonic radiation
Light radiation
Magnetic field
Electric field
Inflammation

Swollen hydrogel

Figure 1. Representation of hydrogels stimuli responsive swelling.

For a polymer system to become "sensitive", (i.e. capable of responding strongly to slight changes provided from the external medium), a first-order phase transition, accompanied by a sharp decrease in the specific volume of their macromolecules must occur. The theoretical foundations of such processes were stated by Flory and collaborators [51].

One of the main conditions for the manifestation of the critical phenomena in the swollen polymer networks or linear macromolecules is the presence of "poor" solvent [52]. In such solvent, the forces of attraction between the segments of the polymer chain may overcome the repulsion forces associated with the excluded volume, which leads to the collapse of the polymer chain [45].

The responsive hydrogels are highly sensitive to changes in the environment and have been used in several applications, such as biosensors [46,53-55], superabsorbent polymers [56-58], site-specific drug delivery systems [46,53,54,59-69], emerging nanoscale technologies [70-77] and tissue engineering [16,78-89]. other important area for the use of polyelectrolytic hydrogels is bioand mucoadhesive drug delivery systems [90-92]. Responsive hydrogels are unique concerning many different mechanisms for drug release and a lot of many different types of release systems based on these materials are formulated. For instance, in most cases drug release occurs when the gel is highly swollen or swelling and is typically controlled by the rate of swelling, the drug diffusion, or a coupling of swelling and diffusion.

Other interesting characteristic of many responsive hydrogels is that the mechanism causing the network structural changes can be entirely reversible. This behavior is depicted in **Figure 2** for a pHor temperature-responsive hydrogels. The ability of these systems to exhibit rapid changes in their swelling behavior and pore structure in response to changes in environmental conditions lend to these materials favorable characteristics as carriers for delivery of bioactive agents, including peptides and proteins. This type of behavior may allow these materials to serve as self-regulated and pulsatile drug delivery systems.

Table 3. Various environmental stimuli used for triggering drug release from responsive hydrogels.

	Environmental stimuli	Mechanism	Applications	Polymers	References
Physical	Temperature	Competition between hydrophobic interaction and hydrogen bonding.	On/off drug release, squeezing device.	PNIPAAm; PDEAAm	[93-100]
	Electrical signal	Reversible swelling or deswelling in the presence of electrical field.	Actuator, artificial muscle, on/off drug release.	Polyelectrolytes PHEMA	[101-104]
	Light	Temperature change via the incorporated photosensitive molecules; dissociation into ion pairs by UV irradiation.	Optical switches, ophthalmic drug delivery.	Copolymer of PNIPAAm	[55,105-108]
	Magnetic fields	Applied magnetic field causes pore in gel and swelling followed by drug release.	Controlled drug delivery while the magnetic particles, used for medical therapy.	EVAc, Copolymer of PNIPAAm	[109-114]
	Ultrasonic irradiation	Temperature increase causes release of drug.	Drug delivery.	EVAh	[115-118]
Chemical	Ionic strength	Change in concentration of ions inside the gel causes swelling and release of drug	Biosensor for glucose, used for medical therapy.	Nonionic PNIPAAm	[119-121]
	pH	Ionization of polymer chain upon pH change; pH change causes swelling and release of drug.	pH-dependent oral drug delivery	PAA, PDEAEM	[52,121-125]
	Chemical agents	Formation of charge-transfer complex causes swelling and release of drug.	Controlled drug delivery.	Chitosan -PEO, polyelectrolytes	[62,119,126,127]
Bio-chemical	Glucose	pH change causes by glucose oxidase; reversible interaction between glucose-containing polymers and Concanavalin A; reversible sol-gel transformation.	Self-regulated insulin delivery.	EVAc; pH-sensitive hydrogels; Concanavalin A-grafted polymers.	[27,36,128-133]
	Antigen	Competition between polymer-grafted antigen and free antigen.	Modulated drug release in the presence of a specific antigen; sensor for immunoassay and antigen.	Semi-IPN with grafted antibodies or antigens.	[48,134,135]

Note: PNIPAAm = poly(N-isopropylacrylamide); PDEAAm = poly(N,N'-diethylacrylamide); PHEMA = poly(2-hydroxyethyl methacrylate); EVAc = ethylene-co-vinyl acetate; EVAh = ethylene-co-vinyl alcohol; PEO = polyethylene oxide; IPN = interpenetrating network.

4. APPLICATIONS OF HYDROGELS IN DRUG DELIVERY

Hydrogels have been extremely useful in biomedical and pharmaceutical applications due to their unique swelling properties and their structure. Based on their functionalities, these biomaterials can be an excellent candidate for controlled drug release systems, bioadhesive or targetable devices and, self-regulated release formulations. According to the delivery administration, hydrogel-based devices can be used for oral, nasal, ocular, rectal, vaginal, epidermal and subcutaneous applications [50,137].

Controlled-release or controlled-delivery systems are intended to provide the drug at a predetermined temporal and/or spatial way within the body to fulfill the specific therapeutic needs. Hydrogels, among the different controlled-release systems exploited so far, have particular properties which make them to be potentially considered as one of the ideal future controlled-release systems. There are two major categories of hydrogel based delivery systems: 1) time-controlled systems and 2) stimuliinduced release systems [40,138]. Sensitive hydrogel systems are developed to deliver their content(s) in response to a fluctuating condition in a way that desirably coincides with the physiological requirements at the right time and proper place [40]. Despite the huge attraction centered towards the novel drug delivery systems based on the environment-sensitive hydrogels in the past and current times, these systems have disadvantages of their own. The most considerable drawback of stimuli-sensitive hydrogels is their significantly slow response time, with the easiest way to achieve fast-acting responsiveness being to develop thinner and smaller hydrogels which, in turn, bring about fragility and loss of mechancal strength in the polymer network and the hydrogel device itself [48].

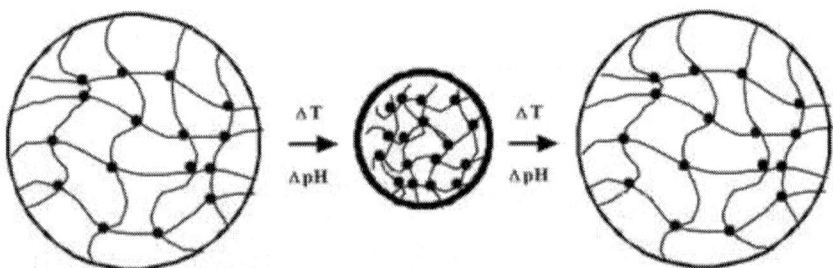

Figure 2. Swollen temperatureand pH-sensitive hydrogels may exhibit an abrupt change from the expanded (left) to the collapsed (syneresed) state (center) and then back to the expanded state (right), adapted from [136].

Controlled drug delivery can be used to achieve some objectives. That is:
- Sustained constant concentration of therapeutically active compounds in the blood with minimum fluctuations;
- Predictable and reproducible release rates over a long period of time;
- Protection of bioactive compounds having a very short half-time;
- Elimination of side-effects, waste of drug and frequent dosing;
- Optimized therapy and better patient compliance;
- Solution for drug stability problems.

Hydrogels have a unique characteristics combination that makes them useful in drug delivery applications. Due to their hydrophilicity, hydrogels can imbibe large amounts of water (>90%, w/v). Therefore, the molecular release mechanisms from hydrogels are very different from hydrophobic polymers. Both simple and sophisticated models have been previously developed to predict the release of a drug from a hydrogel device as a function of time. These models are based on the rate limiting step for controlled release and are therefore categorized as follows [136]:

1) Diffusion-controlled systems:

Matrix (monolithic systems)

Reservoir (membrane systems)

2) Swelling-controlled systems:

Solvent-activated systems Osmotically controlled systems

3) Chemically controlled systems:

Bioerodible and biodegradable systems Pendent chain systems Diffusion-controlled is the most widely applicable mechanism for describing drug release from hydrogels. This property can be described by Fick's law, according with equations (1) and (2) [139],

$$J = -D\left(\frac{dc}{dx}\right)$$

(1)

$$\frac{\partial c}{\partial t} = -D\left(\frac{\partial^2 c}{\partial x^2}\right)$$

(2)

Assuming a binary system, t, c, J, and D represent the time, the solute concentration, mass flux and the diffusion mutual differential coefficient, respectively.

In reservoir devices the drug is contained in a core which is surrounded by a rate-controlling polymeric membrane. Drug transport from the core through the external polymer membrane occurs by dissolution at one interface of the membrane and diffusion driven by a gradient in thermodynamic activity. Drug transport can be described by Fick's first law. If the activity of the drug in the reservoir remains constant and infinite sink conditions are maintained, the drug release rate may be continued to be constant and can be predicted since it depends on the membrane permeability and device configuration. Then, drug release will be independent of time and, zero-order kinetics can be achieved. Drug diffusivities is generally determined empirically or estimated a priori using free volume, hydrodynamic or obstructtion-based theories [140].

Swelling-controlled release occurs when diffusion of drug is faster than hydrogel swelling. The modeling of this mechanism usually involves moving boundary conditions where molecules are released at the interface of rubbery and glassy phases of swollen hydrogels [141]. The release of many small molecule drugs from hydroxypropyl methylcellulose (HPMC) hydrogel tablets is commonly modeled using this mechanism [141,142]. In swelling-controlled devices, the drug can be released following two different mechanisms: diffusion and relaxation of polymer chains, which occur at the glassyrubbery interface. Ritger and Peppas [143,144] proposed a simple equation to determine the relative importance of diffusion and macromolecular relaxation on the overall drug delivery process,

$$\frac{M_t}{M_\infty} = kt^n$$

(3)

M_t and M_∞ are the amounts of drug released at time t and at the equilibrium, respectively. k is a proportionality constant and n is the diffusional exponent.

Ritger and Peppas [143,144] introduced this exponenttial equation to describe the drug release behavior from polymeric matrixes and, analysis of the Fickian and nonFickian diffusional behavior relative to the value of the exponent n was performed. Diffusional exponent values for planar, cylindrical and spherical drug release systems were related with the mechanism of delivery.

Chemically-controlled release is used to describe drug release determined by reactions occurring within a delivery matrix. The most common reactions that occur within hydrogel delivery systems are cleavage of polymer chains via hydrolytic or enzymatic degradation or reversible/ irreversible reactions occurring between the polymer network and the drug released [145]. Under certain conditions, the surface or bulk erosion of hydrogels will control the rate of drug release. Alternatively, if drug-binding moieties are incorporated in the hydrogels, the binding equilibrium can determine the drug release rate. Chemically-controlled release can be further categorized according to the type of chemical reaction occurring during drug release. Generally, the delivery of encapsulated or tethered drugs can occur through the degradation of pendant chains or during surface-erosion or bulk-degradation of the polymer backbone [40]. In pendant chain systems, the drug is covalently attached to a polymer backbone. The bond between the drug and the polymer is labile and can be broken by hydrolysis or enzymatic degradation.

In erodible drug delivery systems the release of the drug is controlled by the dissolution or degradation of the polymer. Contrary to pendant chain systems, the drug diffuses from erodible systems. Depending on whether diffusion or polymer degradation controls the release rate, the drug is released following different mechanisms. If erosion of the polymer is much slower than the diffusion of the drug through the polymer, then drug release can be treated as a diffusion controlled process. If the diffusion of the drug from the polymer matrix is very slow, then polymer degradation or erosion is the rate-controlling step. Two different types of erodible polymers can be found: hydrophilic and hydrophobic [136]. Hydrophilic erodible polymers are completely permeated by water and they undergo a bulk erosion process. Erosion takes place throughout the polymer matrix. Hydrophobic erodible polymers can experience bulk or surface erosion. In bulk erosion, degradation occurs throughout the bulk of the polymer and generally the analysis of the drug release kinetics is complex since it comprises erosion and diffusion [136].

In all routes of drug administration, oral administration has been considered to be most convenient, and hence the majority of dosage forms are designed for oral drug delivery. Different types of hydrogels can be used for delivery of drugs to certain areas in the gastrointestinal tract ranging from the oral cavity to the colon, as shown in **Figure 3**.

The next section summarizes the applications of hydrogels in different routes of administration, including its challenges and current status of development.

4.1. Oral Administration

Oral drug delivery is the most desirable and preferred method of therapeutic agents administration. In addition, the oral therapy is generally considered as the first strategy investigated in the discovery and development of new drug entities and pharmaceutical formulations, mainly because the patient acceptance, the convenience in administration, and the cost-effective manufacturing process. Oral drug delivery using controllable hydrogels has attracted considerable attention in the past 20 years due its enormous market potential [146].

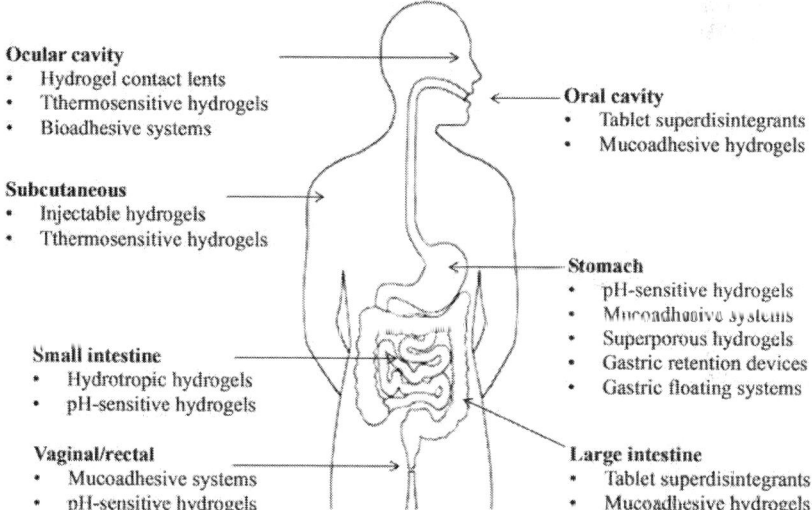

Figure 3. Tissue localization of hydrogel-based drug delivery systems, adapted from [20].

In oral administration, hydrogels can deliver drugs to four major specific sites: mouth, stomach, small intestine and colon. By controlling their swelling properties or mucoadhesive characteristics in the presence of a biological fluid, hydrogels can be a useful carrier to release drugs in a controlled manner at these desired sites. Furthermore, the mucoadhesive hydrogels offer an attractive property for drug targeting at certain specific regions, leading to a locally increased drug concentration and, consequently, enhancing the drug absorption at the release site [20,50].

The application of mucoadhesive hydrogels in buccal drug delivery seems to have some advantages like a rapid drug action, the absence of hepatic first-pass metabolism and the absence of drug degradation in the gastrointestinal tract due the low pH values [147,148]. For maximum benefit from drug delivery buccal bioadhesive hydrogels were developed. They can be prepared using different polymers namely hydroxyethylcellulose (HEC), hydroxypropylcellulose (HPC), polyvinylpyrrolidone (PVP), and poly(vinyl alcohol) [149], or mixtures of different polymers, such as HPC, hydroxypropylmethylcellulose (HPMC), karaya gum and polyethylene glycol (PEG) 400 [146], or HPC and Carbopol 934 [150].

4.2. Ocular Route

The ocular route is mainly used for the local treatment of eye pathologies. Many physiological constraints prevent a desired drug delivery to the eye due to its protective mechanisms, such as effective tear drainage, blinking and low permeability of the cornea. Therefore, conventional eyedrops containing a drug solution tend to be eliminated rapidly from the eye, and drugs administered exhibit limited absorption, leading to poor ophthalmic bioavailability (2% - 10%). Additionally, their short retention time in the local of action often results in a frequent dosing regimen to achieve the therapeutic efficacy for a sufficiently long duration. These challenges have motivated researchers to develop drug delivery systems to provide a prolonged ocular residence time of drugs [151].

The following types of mucoadhesive formulations have been evaluated for ocular drug delivery: viscous liquids (suspensions and ointments), hydrogels and, solids (inserts). Certain dosage forms, such as suspensions and ointments, can be retained in the eye, although these formulations sometimes give to the patients an unpleasant feeling because of the solid and semi-solid characteristics. Due to their elastic properties, hydrogels can also represent an ocular drainage-resistant device. In particular, in-situ hydrogels are attractive as an ocular drug delivery system because of their facility in dosing as a liquid and, their long-term retention property as a gel after dosing. Hui and Robinson [152] introduced hydrogels consisting of cross-linked PAA for ocular delivery of progesterone in rabbits. These preparations increased progesterone concentration in the aqueous humor four times over aqueous suspensions. Cohen and collaborators [153] developed an in situ gel system of alginate with high guluronic acid content for ophthalmic delivery of pilocarpine. This system significantly extended the duration of the pressure-reducing effect of pilocarpine. Carlfors [154] investigated the rheological properties of the deacetylated gellan gum. These gels upon instillation in the eye due to the presence of cations and, the high rate of the sol/gel transition results in a long precorneal contact time. An approach for ocular inserts was presented by Chetoni and collaborators [155]. In this study, cylindrical devices for oxytetracycline were developed using mixtures of silicone clastomer and grafted on the surface of the inserts with an interpenetrating mucoadhesive polymeric network of PAA or PMAA. The ocular retention of IPN-grafted inserts was significantly higher than the ungrafted ones. An in vivo study using rabbits showed a prolonged release of oxytetracycline from the inserts for several days.

4.3. Nasal Route

The nasal route of drug administration is the most suitable alternative of drug delivery for poorly absorbable compounds such as peptide and proteins. The nasal epithelium exhibits relatively high permeability and, only two cell layers separate the nasal lumen from the dense blood-vessel network in the lamina propria. The respiretory epithelium covered by a mucus layer is the major lining of the human nasal cavity and is essential in the clearance of mucus by the mucociliary system [156].

Several structural mucoadhesive polymers were tested for their ability to retard the nasal mucociliary clearance in rats [157]. The clearance was measured using microspheres labeled with a fluorescent marker incorporated into the formulation. The clearance rate of each polymer gel was found to be lower than the control microsphere suspension, resulting in an increased residence time of the gel formulations in the nasal cavity. Ilium and collaborators [158] evaluated chitosan solutions as delivery platforms for nasal administration of insulin to rats and sheep. They reported a concentration-dependent absorption-enhancing effect with minimal histological changes of the nasal mucosa. Nakamura and collaborators [159] described a microparticulated dosage form of budesonide, consisting of bioadhesive and pH-dependent graft copolymers of PMAA and PEG, resulting in an increase and constant plasma levels of budesonide for 8 h after nasal administration in rabbits.

4.4. Transdermal Route

The transdermal route has been considered as a possible site for the systemic delivery of drugs. The benefits of transdermal drug delivery systems include ease of application and delivery, sustained and steady drug release, reduced systemic side effects, avoidance of drug degradation in the GI tract and first-pass hepatic metabolism. Furthermore, swollen hydrogels with high water content can provide a better feeling for the skin compared with conventional ointments and patches. Versatile hydrogelbased devices for transdermal delivery have been proposed. Sun and collaborators [160] prepared composite membranes comprising of cross-linked poly (hydroxyethylmethacrylate) (PHEMA) with a non-woven polyester support. Depending on the preparation conditions, the composite membranes could be tailored to give a permeation flux ranging from 4 to 68 mg/cm^2 per h for nitroglycerin. Gayet and Fortier [161] reported the use of the (bovine serum albumin) BSA-PEG hydrogels containing high water content over 96% as controlled release devices in the field of wound dressing. However, skin acts as a barrier to foreign substances, preventing the entrance of the majority of drugs. Therefore, researchers have been developed several electrically assisted methods to enhance the drug permeation across the skin, including electroporation, ionophoresis, sonophoresis and, laser irradiation [162-164].

5. CONCLUSIONS AND FUTURE TRENDS

Actually, there are enough scientific evidences for the potentiality of hydrogels in the delivery of drug molecules to a desired site by triggering the release through an external stimulus such as temperature, pH, glucose or light. These systems being biocompatible and biodegradable in nature have been used in the development of nanobiotechnological products and have excellent applications in the field of controlled drug delivery as well. This is the reason because these turn-able biomedical devices are gaining attention as intelligent drug carrier systems.

In fact, the design and synthesis of environment-sensitive hydrogels has significant potential in future biomedical and nanotechnology applications. The success of these materials relies on the development of novel materials that can address specific biological and medical challenges. This development will occur through the synthesis of new polymers or by modification of natural polymers. Hydrogels being used for cartilage or tissue engineering should be capable to providing mechanical properties as well as the molecular signals that are present in the native or regenerative organ. Finally, advanceing the knowledge and the use of hydrogels and sensitive polymers for nanotechnology is an important area with significant potential that remains to be fully investigation.

Finally, it is possible to conclude that many significant recent advances in biomaterials occur at the interface of clinical medicine, materials science and engineering. This aspect creates opportunities and training programs for individual cross-disciplinary research and the engaged in these areas can significant accelerate the advance of biomaterials and create new applications for these materials in medicine.

ACKNOWLEDGEMENTS

This work was financially supported by a Grant (Praxis SFRH/BD/48324/2008) from FCT (Fundação para a Ciência e a Tecnologia, Portugal).

REFERENCES

1.   S. M. Gomes, H. Azevedo, P. Malafaya, S. Silva, J. Oliveira, G. Silva, R. Sousa, J. Mano and R. Reis, "Natural Polymers in Tissue Engineering Applications," In: C. V. Blitterswijk, Ed., Tissue Engineering, Academic Press, Waltham, 2008, pp. 146-191.
2.   B. D. Ratner and A. S. Hoffman, "Hydrogels for Medical and Related Applications," ACS Publications, Washington DC, 1976.
3.   N. A. Peppas, "Hydrogels in medicine and pharmacy," CRC Press, Boca Raton, 1986.
4.   N. A. Peppas and R. Langer, "New challenges in biomaterials," Science, Vol. 263, No. 5154, 1994, pp. 1715- 1720.

5.   A. S. Hoffman, "Hydrogels for biomedical applications," Advanced Drug Delivery Reviews, Vol. 54, No. 1, 2002, pp. 3-12.

6.   J. Jagur-Grodzinski, "Polymeric gels and hydrogels for biomedical and Pharmaceutical Applications," Polymers for Advanced Technologies, Vol. 21, No. 1, 2009, pp. 27- 47.

7.   B. K. Nanjawade, F. V. Manvi and A.S. Manjappa, "In situ-forming hydrogels for sustained ophthalmic drug delivery," Journal of Controlled Release, Vol. 122, No. 2, 2007, pp. 119-134.

8.   L. Xinming, C. Yingde, A. W. Lloyd, S. V. Mikhalovsky, S. R. Sandeman, C. A. Howel and L. Liewen, "Polymeric hydrogels for Novel Contact Lens-Based Ophthalmic Drug Delivery Systems: A Review," Contact Lens & Anterior Eye, Vol. 31, No. 2, 2008, pp. 57-64.

9.   S. Brahim, D. Narinesingh and A. Guiseppi-Elie, "Polypyrrole-hydrogel composites for the construction of Clinically Important Biosensors," Biosensors and Bioelectronics, Vol. 17, No. 1-2, 2002, pp. 53-59.

10.   N. B. Graham and M. E. McNeill, "Hydrogels for controlled Drug Delivery," Biomaterials, Vol. 5, No. 1, 1984, pp. 27-36.

11.   N. A. Peppas, Y. Huang, M. Torres-Lugo, J. H. Ward and J. Zhang, "Physicochemical, foundations and Structural Design of hydrogels in medicine and biology," Annual Review of Biomedical Engineering, Vol. 2, 2000, pp. 9-29.

12.   L. Chen, Z. Tian and Y. Du, "Synthesis and pH sensitivity of Carboxymethyl Chitosan-Based Polyampholyte Hydrogels for Protein Carrier Matrices," Biomaterials, Vol. 25, No. 17, 2004, pp. 3725-3732.

13.   Q. Li, J. Wang, S. Shahani, D. D. N. Sun, B. Sharma, J. H. Elisseeff and K. W. Leong, "Biodegradable and photocrosslinkable polyphosphoester hydrogel," Biomaterials, Vol. 27, No. 17, 2006, pp. 1027-1034.

14.   T. R. Hoare and D. S. Kohane, "Hydrogels in Drug Delivery: Progress and Challenges," Polymer, Vol. 49, No. 8, 2008, pp. 1993-2007.

15.   R. Langer and J. P. Vacanti, "Tissue engineering," Science, Vol. 260, No. 5110, 1993, pp. 920-926.

16.   J. L. Drury and D. J. Mooney, "Hydrogels for Tissue Engineering: Scaffold Design Variables and Applications," Biomaterials, Vol. 24, No. 24, 2003, pp. 4337- 4351.

17.   K. A. Davis and K. S. Anseth, "Controlled Release from Crosslinked Degradable Networks," Critical Reviews in Therapeutic Drug Carrier Systems, Vol. 19, No. 4-5, 2002, pp. 385-423.

18.   K. Y. Lee and S. H. Yuk, "Polymeric Protein Delivery Systems," Progress in Polymer Science, Vol. 32, No. 7, 2007, pp. 669-697.

19.   T. Coviello, P. Matricardi, C. Marianecci and F. Alhaique, "Polysaccharide hydrogels for Modified Release Formulations," Journal of Controlled Release, Vol. 119, No. 1, 2007, pp. 5-24.

20.   S. H. Jeong, K. M. Huh and K. Park, "Hydrogel Drug Delivery Systems, in Polymers in Drug Delivery," CRC press, Boca Raton, 2006.

21.   K. R. Kamath and K. Park, "Biodegradable hydrogels in Drug Delivery," Advanced Drug Delivery Reviews, Vol. 11, No. 1-2, 1993, pp. 59-84.

22.   P. M. de la Torre, S. Torrado and S. Torrado, "Interpolymer complexes of poly(acrylic acid) and chitosan: influence of the Ionic Hydrogel-Forming Medium," Biomaterials, Vol. 24, No. 8, 2003, pp. 1459-1468.

23.   Y. Kumashiro, K. M. Huh, T. Ooya and N. Yui, "Modulatory Factors on Temperature-Synchronized Degradation of Dextran Grafted with Thermoresponsive Polymers and Their Hydrogels," Biomacromolecules, Vol. 2, No. 3, 2001, pp. 874-879.

24.   J. Kopecek and J. Yang, "Hydrogels as Smart Biomaterials," Polymer International, Vol. 56, No. 9, 2007, pp. 1078-1098.

25.   S. A. Lapidot, J. Kost, K. H. J. Buschow, W. C. Robert, C. F. Merton, I. Bernard, J. K. Edward, M. Subhash and V. Patrick, "Hydrogels," in: K. H. J. Buschow, et al., Eds., Encyclopedia of Materials: Science and Technology, Elsevier, Oxford, 2001, pp. 3878- 3882.

26.   T. Tamar, K. Joseph and A. L. Smadar, "Modeling ionic Hydrogels Swelling: Characterization of the Non-Steady State," Biotechnology and Bioengineering, Vol. 84, No. 1, 2003, pp. 20-28.

27.   T. Traitel, Y. Cohen and J. Kost, "Characterization of Glucose-Sensitive Insulin Release Systems in simulated in Vivo Conditions," Biomaterials, Vol. 21, No. 16, 2000, pp. 1679-1687.

28.   K. Wang, J. Burban and E. Cussler, "Hydrogels as separation agents," Responsive Gels: Volume Transitions II, Vol. 110, 1993, pp. 67-79.

29.   L. Brannon-Peppas and N. A. Peppas, "Dynamic and equilibrium Swelling Behaviour of pH-Sensitive Hydrogels Containing 2-Hydroxyethyl Methacrylate," Biomaterials, Vol. 11, No. 9, 1990, pp. 635-644.

30.   C. Jun, P. Haesun and P. Kinam, "Synthesis of superporous hydrogels: Hydrogels with Fast Swelling and superabsorbent properties," Journal of Biomedical Materials Research A, Vol. 44, No. 1, 1999, pp. 53-62.

31.   P. Kofinas and R. E. Cohen, "Development of methods for Quantitative Characterization of Network Morphology in Pharmaceutical Hydrogels," Biomaterials, Vol. 18, No. 20, 1997, pp. 1361-1369.

32.   K. Pathmanathan and G. P. Johari, "Relaxation and crystallization of water in a hydrogel," Journal of Chemical Society Faraday Transactions, Vol. 90, No. 8, 1994, pp. 1143-1148.

33.   H. Park, K. Park and W. S. W. Shalaby, "Biodegradable Hydrogels for Drug Delivery," CRC Press, Boca Raton, 1993.

34.   A. M. Lowman and N. A. Peppas, "Analysis of the Complexation/Decomplexation Phenomena in Graft Copolymer Networks," Macromolecules, Vol. 30, No.17, 1997, pp. 4959-4965.

35.   B. Jeong, Y.K. Choi, Y. H. Bae, G. Zentner and S. W. Kim, "New biodegradable polymers for Injectable Drug Delivery Systems," Journal of Controlled Release, Vol. 62, No, 1-2, 1999, pp. 109-114.

36.   N. A. Peppas, "Physiologically Responsive Hydrogels," Journal of Bioactive and Compatible Polymers, Vol. 6, No. 3, 1991, pp. 241-246.

37. A. R. Khare and N. A. Peppas, "Swelling/deswelling of Anionic Copolymer Gels," Biomaterials, Vol. 16, No. 7, 1995, pp. 559-567.

38. N. A. Peppas, A. R. Khare, "Preparation, structure and Diffusional Behavior of hydrogels in Controlled Release," Advanced Drug Delivery Reviews, Vol. 11, No. 1- 2, 1993, pp. 1-35.

39. S. Amin, S. Rajabnezhad and K. Kohli, "Hydrogels as Potential Drug Delivery Systems," Scientific Research and Essays, Vol. 4, No. 11, 2009, pp. 1175-1183.

40. C. C. Lin and A. T. Metters, "Hydrogels in Controlled Release Formulations: Network Design and Mathematical Modeling," Advanced Drug Delivery Reviews, Vol. 58, No. 12-13, 2006, pp. 1379-1408.

41. W. E. Hennink and C. F. van Nostrum, "Novel crosslinking methods to Design Hydrogels," Advanced Drug Delivery Reviews, Vol. 54, No. 1, 2002, pp. 13-36.

42. A. M. Lowman and N. A. Peppas, "Encyclopedia of Controlled Drug Delivery," John Wiley & Sons, Hoboken, 1999.

43. S. W. Kim, Y. H. Bae, "Stimuli-Modulated Delivery Systems," in: G. L.Amidon, P. I. Lee and E. M. Topp, Eds., Transport Processes in Pharmaceutical Systems, Marcel Dekker, New York, 2000, pp. 547-573.

44. A. Gutowska, J. S. Bark, I. Chan Kwon, Y. Han Bae, Y. Cha and S. Wan Kim, "Squeezing hydrogels for controlled Oral Drug Delivery," Journal of Controlled Release, Vol. 48, No. 2-3, 1997, pp. 141-148.

45. I. Y. Galaev, "Smart polymers in biotechnology and medicine," Russian Chemical Reviews, Vol. 64, No. 5, 1995, pp. 471-489.

46. I. Y. Galaev and B. Mattiasson, "Smart' polymers and What They Could Do in Biotechnology and Medicine," Trends in Biotechnology, Vol. 17, No. 8, 1999, pp. 335- 340.

47. P. S. Stayton, M. E. H. El-Sayed, N. Murthy, V. Bulmus, C. Lackey, C. Cheung and A. S. Hoffman, "Smart delivery systems for biomolecular therapeutics," Orthodontics and Craniofacial Research, Vol. 8, No. 3, 2005, pp. 219-225.

48. Y. Qiu and K. Park, "Environment-sensitive hydrogels for Drug Delivery," Advanced Drug Delivery Reviews, Vol. 53, No. 3, 2001, pp. 321-339.

49. J. C. Ruiz, C. Alvarez-Lorenzo, P. Taboada, G. Burillo, E. Bucio, K. de Prijck, H. J. Nelis, T. Coenye and A. Concheiro, "Polypropylene grafted with Smart Polymers (PNIPAAm/PAAc) for loading and Controlled Release of vancomycin," European Journal of Pharmaceutics and Biopharmaceutics, Vol. 70, No. 2, 2008, pp. 467-477.

50. N. A. Peppas, P. Bures, W. Leobandung and H. Ichikawa, "Hydrogels in Pharmaceutical Formulations," European Journal of Pharmaceutics and Biopharmaceutics, Vol. 50, No. 1, 2000, pp. 27-46.

51. B. Erman and P. J. Flory, "Critical phenomena and transitions in Swollen Polymer Networks and in Linear Macromolecules," Macromolecules, Vol. 19, No. 9, 1986, pp. 2342-2353.

52.   J. Ma, X. Liu, Z. Yang and Z. Tong, "A pH-Sensitive Hydrogel with Hydrophobic Association for Controlled Release of Poorly Water-Soluble Drugs," Journal of Macromolecular Science: Pure and Applied Chemistry, Vol. 46, No. 8, 2009, pp. 816-820.

53.   C. H. Alarcon, S. Pennadam and C. Alexander, "Stimuli Responsive Polymers for Biomedical Applications," Chemical Society Reviews, Vol. 34, No. 3, 2005, pp. 276- 285.

54.   A. S. Hoffman and P. S. Stayton, "Conjugates of stimuliresponsive polymers and proteins," Progress in Polymer Science, Vol. 32, No. 8-9, 2007, pp. 922-932.

55.   R. J. Mart, R. D. Osborne, M. M. Stevens and R. V. Ulijn, "Peptide-Based Stimuli-Responsive Biomaterials," Soft Matter, Vol. 2, No. 10, 2006, pp. 822-835.

56.   P. Ali and B. Shahram, "Synthesis and Evaluation of pH and Thermosensitive Pectin-Based Superabsorbent Hydrogel for Oral Drug Delivery Systems," Starch-Stärke, Vol. 61, No. 3-4, 2009, pp. 161-172.

57.   J. Chen and K. Park, "Synthesis and characterization of Superporous Hydrogel Composites," Journal of Controlled Release, Vol. 65, No. 1-2, 2000, pp. 73-82.

58.   J. Zhang, K. Yuan, Y. P. Wang, S. J. Gu and S. T. Zhang, "Preparation and properties of Polyacrylate/Bentonite Superabsorbent Hybrid via Intercalated Polymerization," Materials Letters, Vol. 61, No. 2, 2007, pp. 316-320.

59.   B. G. Geest, C. Déjugnat, G. B. Sukhorukov, K. Braeckmans, S. C. De Smedt and J. Demeester, "Self-Rupturing Microcapsules," Advanced Materials, Vol. 17, No. 19, 2005, pp. 2357-2361.

60.   D. Yonghui, W. Changchun, S. Xizhong, Y. Wuli, J. Lan, G. Hong and F. Shoukuan, "Preparation, Characterization, and Application of Multistimuli-Responsive Microspheres with Fluorescence-Labeled Magnetic Cores and Thermoresponsive Shells," Chemistry European Journal, Vol. 11, No. 20, 2005, pp. 6006-6013.

61.   J. D. Ehrick, S. K. Deo, T. W. Browning, L. G. Bachas, M. J. Madou and S. Daunert, "Genetically Engineered Protein in Hydrogels Tailors Stimuli-Responsive Characteristics," Nature Materials, Vol. 4, No. 4, 2005, pp. 298- 302.

62.   J. Gu, F. Xia, Y. Wu, X. Qu, Z. Yang and L. Jiang, "Programmable delivery of Hydrophilic Drug Using Dually Responsive Hydrogel Cages," Journal of Controlled Release, Vol. 117, No. 3, 2007, pp. 396-402.

63.   K. S. V. K. Rao, B. V. K. Naidu, M. C. S. Subha, M. Sairam and T. M. Aminabhavi, "Novel Chitosan-Based pH-sensitive Interpenetrating Network Microgels for the Controlled Release of Cefadroxil," Carbohydrate Polymers, Vol. 66, No. 3, 2006, pp. 333-344.

64.   K. S. Soppimath, A. R. Kulkarni and T. M. Aminabhavi, "Chemically Modified Polyacrylamide-g-Guar Gum-Based Crosslinked Anionic Microgels as pH-Sensitive Drug Delivery Systems: Preparation and

Characterization," Journal of Controlled Release, Vol. 75, No. 3, 2001, pp. 331- 345.

65.   G. Tae, M. Scatena, P. S. Stayton and A. S. Hoffman, "PEG-Cross-Linked Heparin Is an Affinity Hydrogel for Sustained Release of Vascular Endothelial Growth Factor," Journal of Biomaterials Science Polymer Edition, Vol. 17, No. 1-2, 2006, pp. 187-197.

66.   P. D. Thornton, R. J. Mart and R. V. Ulijn, "EnzymeResponsive Polymer Hydrogel Particles for Controlled Release," Advanced Materials, Vol. 19, No. 9, 2007, pp. 1252-1256.

67.   I. R. Wheeldon, S. Calabrese Barton and S. Banta, "Bioactive Proteinaceous Hydrogels from Designed Bifunctional Building Blocks," Biomacromolecules, Vol. 8, No. 10, 2007, pp. 2990-2994.

68.   K. C. Wood, H. F. Chuang, R. D. Batten, D. M. Lynn and P. T. Hammond, "Controlling Interlayer Diffusion to Achieve Sustained, Multiagent Delivery from Layer-ByLayer Thin Films," Proceedings of the National Academy of Sciences, Vol. 103, No. 27, 2006, pp. 10207-10212.

69.   J. Zhou, G. Wang, L. Zou, L. Tang, M. Marquez and Z. Hu, "Viscoelastic Behavior and in Vivo Release Study of Microgel Dispersions with Inverse Thermoreversible Gelation," Biomacromolecules, Vol. 9, No. 1, 2008, pp. 142- 148.

70.   N. Singh, A. W. Bridges, A. J. Garcia and L. A. Lyon, "Covalent Tethering of Functional Microgel Films onto Poly(ethylene terephthalate) Surfaces," Biomacromolecules, Vol. 8, No. 10, 2007, pp. 3271-3275.

71.   C. M. Nolan, C. D. Reyes, J. D. Debord, A. J. Garcia and L. A. Lyon, "Phase Transition Behavior, Protein Adsorption, and Cell Adhesion Resistance of Poly(ethylene glycol) Cross-Linked Microgel Particles," Biomacromolecules, Vol. 6, No. 4, 2005, pp. 2032-2039.

72.   D. Gan and L. A. Lyon, "Synthesis and Protein Adsorption Resistance of PEG-Modified Poly(N-isopropylacrylamide) Core/Shell Microgels," Macromolecules, Vol. 35, No. 26, 2002, pp. 9634-9639.

73.   P. Kim, D. Kim, B. Kim, S. Choi, S. Lee, A. Khademhosseini, R. Langer and K. Y. Suh, "Fabrication of nanostructures of Polyethylene Glycol for applications to protein adsorption and Cell Adhesion," Nanotechnology, Vol. 16, No. 10, 2005, pp. 2420-2426.

74.   H. Cong, A. Revzin and T. Pan, "Non-adhesive PEG Hydrogel Nanostructures for Self-Assembly of Highly Ordered Colloids," Nanotechnology, Vol. 20, 2009, Article ID: 075307.

75.   K. S. Jeong, P. S. Jun, L. S. Min, L. Y. Moo, K. H. Chan and I. K. Sun, "Electroactive characteristics of Interpenetrating Polymer Network Hydrogels Composed of Poly(Vinyl Alcohol) and Poly(N-Isopropylacrylamide)," Journal of Applied Polymer Science, Vol. 89, No. 4, 2003, pp. 890-894.

76.   J. Raula, J. Shan, M. Nuopponen, A. Niskanen, H. Jiang, E. I. Kauppinen and H. Tenhu, "Synthesis of Gold Nanoparticles Grafted with a Thermoresponsive Polymer by Surface-Induced Reversible-

Addition-Fragmentation ChainTransfer Polymerization," Langmuir, Vol. 19, No. 8, 2003, pp. 3499-3504.

77. M. Yamato, M. Utsumi, A. Kushida, C. Konno, A. Kikuchi and T. Okano, "Thermo-Responsive Culture Dishes Allow the Intact Harvest of Multilayered Keratinocyte Sheets without Dispase by Reducing Temperature," Tissue Engineering, Vol. 7, No. 4, 2001, pp. 473-480.

78. K. S. Anseth, A. T. Metters, S. J. Bryant, P. J. Martens, J. H. Elisseeff, C. N. Bowman, "In situ Forming Degradable Networks and Their Application in Tissue Engineering and Drug Delivery," Journal of Controlled Release, Vol. 78, No. 1-3, 2002, pp. 199-209.

79. C. Chung and J. A. Burdick, "Engineering Cartilage Tissue," Advanced Drug Delivery Reviews, Vol. 60, No. 2, 2008, pp. 243-262.

80. D. Y. Fozdar, W. Zhang, M. Palard, C. W. Patrick Jr. and S. Chen, "Flash Imprint Lithography Using a Mask Aligner: A Method for Printing Nanostructures in Photosensitive Hydrogels," Nanotechnology, Vol. 19, No. 21, 2008, pp. 215-303.

81. K. Y. Suh, M. C. Park and P. Kim, "Capillary Force Lithography: A Versatile Tool for Structured Biomaterials Interface towards Cell and Tissue Engineering", Advanced Functional Materials, Vol. 19, No. 17, 2009, pp. 2699-2712.

82. M. P. Lutolf and J. A. Hubbell, "Synthetic biomaterials as Instructive Extracellular Microenvironments for morphogenesis in Tissue Engineering," Nature Biotechnology, Vol. 23, No. 1, 2005, pp. 47-55.

83. B. K. Mann, "Biologic gels in Tissue Engineering," Clinics in Plastic Surgery, Vol. 30, No. 4, 2003, pp. 601-609.

84. C. Schwall and I. Banerjee, "Microand Nanoscale Hydrogel Systems for Drug Delivery and Tissue Engineering," Materials, Vol. 2, No. 2, 2009, pp. 577-612.

85. M. Sokolsky-Papkov, K. Agashi, A. Olaye, K. Shakesheff and A. J. Domb, "Polymer carriers for Drug Delivery in Tissue Engineering," Advanced Drug Delivery Reviews, Vol. 59, No. 4-5, 2007, pp. 187-206.

86. H. J. Wang, L. Di, Q. S. Ren and J. Y. Wang, "Applications and Degradation of Proteins Used as Tissue Engineering Materials," Materials, Vol. 2, No. 2, 2009, pp. 613-635.

87. C. Weinand, I. Pomerantseva, C. M. Neville, R. Gupta, E. Weinberg, I. Madisch, F. Shapiro, H. Abukawa, M. J. Troulis and J. P. Vacanti, "Hydrogel-β-TCP scaffolds and Stem Cells for Tissue Engineering Bone," Bone, Vol. 38, No. 4, 2006, pp. 555-563.

88. S. X. Zheng, A. Shama, L. Yanchun and D. P. Glenn, "Synthesis and evaluation of injectable, in situ Crosslinkable Synthetic Extracellular Matrices for tissue engineering," Journal of Biomedical Materials Research A, Vol. 79A, No. 4, 2006, pp. 902-912.

89. X. q. Jia and L. K. Kristi, "Hybrid Multicomponent Hydrogels for Tissue Engineering," Macromolecular Bioscience, Vol. 9, No. 2, 2009, pp. 140-156.

90. L. Serra, J. Doménech and N. A. Peppas, "Engineering design and Molecular Dynamics of Mucoadhesive Drug Delivery Systems as

Targeting Agents," European Journal of Pharmaceutics and Biopharmaceutics , Vol. 71, No. 3, 2009, pp. 519-528.

91. G. P. Andrews, T. P. Laverty and D. S. Jones, "Mucoadhesive polymeric platforms for controlled Drug Delivery," European Journal of Pharmaceutics and Biopharmaceutics, Vol. 71, No. 3, 2009, pp. 505-518.

92. N. A. Peppas and J. J. Sahlin, "Hydrogels as mucoadhesive and Bioadhesive Materials: A Review," Biomaterials, Vol. 17, No. 16, 1996, pp. 1553-1561.

93. Y. H. Bae, T. Okano and S. W. Kim, "'On-Off' Thermocontrol of Solute Transport. I. Temperature Dependence of Swelling of N-Isopropylacrylamide Networks Modified with Hydrophobic Components in Water," Pharmaceutical Research, Vol. 8, No. 4, 1991, pp. 531-537.

94. Y. H. Bae, T. Okano and S. W. Kirn, "'On-Off' Thermocontrol of Solute Transport. II. Solute Release from Thermosensitive Hydrogels," Pharmaceutical Research, Vol. 8, No. 5, 1991, pp. 624-628.

95. B. Jeong, S. W. Kim and Y. H. Bae, "Thermosensitive Sol-Gel Reversible Hydrogels," Advanced Drug Delivery Reviews, Vol. 54, No. 1, 2002, pp. 37-51.

96. E. Ruel-Gariépy and J. C. Leroux, "In Situ-Forming Hydrogels—Review of Temperature-Sensitive Systems," European Journal of Pharmaceutics and Biopharmaceutics, Vol. 58, No. 2, 2004, pp. 409-426.

97. X. Zhang and R. Zhuo, "Synthesis of Temperature-Sensitive Poly(N-isopropylacrylamide) Hydrogel with Improved Surface Property," Journal of Colloid and Interface Science, Vol. 223, No. 2, 2000, pp. 311-313.

98. C. Alvarez-Lorenzo, A. Concheiro, A. S. Dubovik, N. V. Grinberg, T. V. Burova and V. Y. Grinberg, "Temperature-Sensitive Chitosan Poly(N-isopropylacrylamide) inTerpenetrated Networks with Enhanced Loading Capacity and Controlled Release Properties," Journal of Controlled Release, Vol. 102, No. 3, 2005, pp. 629-641.

99. M. Yamato, Y. Akiyama, J. Kobayashi, J. Yang, A. Kikuchi and T. Okano, "Temperature-Responsive Cell Culture Surfaces for Regenerative Medicine with Cell Sheet Engineering," Progress in Polymer Science, Vol. 32, No. 8-9, 2007, pp. 1123-1133.

100. L. Pérez-Alvarez, V. S. Martínez, E. Hernáez and I. Katime, "Novel pHand Temperature-Responsive Methacrylamide Microgels," Macromolecular Chemistry and Physics, Vol. 210, No. 13-14, 2009, pp. 1120-1126. I. C. Kwon, Y. H. Bae, T. Okano and S. W. Kim, "Drug release from Electric Current Sensitive Polymers," Journal of Controlled Release, Vol. 17, No. 2, 1991, pp. 149- 153.

101. K. Sawahata, M. Hara, H. Yasunaga and Y. Osada, "Electrically Controlled Drug Delivery System Using Polyelectrolyte Gels," Journal of Controlled Release, Vol. 14, No. 3, 1990, pp. 253-262.

102.   H. Li, R. Luo and K. Y. Lam, "Modeling of Ionic Transport in Electric-Stimulus-Responsive Hydrogels," Journal of Membrane Science, Vol. 289, No. 1-2, 2007, pp. 284- 296.

103.   H. Li, "Kinetics of Smart Hydrogels Responding to Electric Field: A Transient Deformation Analysis," International Journal of Solids and Structures, Vol. 46, No. 6, 2009, pp. 1326-1333.

104.   A. Mamada, T. Tanaka, D. Kungwatchakun and M. Irie, "Photoinduced phase transition of gels," Macromolecules, Vol. 23, No. 5, 1990, pp. 1517-1519.

105.   F. M. Andreopoulos, E. J. Beckman and A. J. Russell, "Light-Induced Tailoring of PEG-hydrogel properties," Biomaterials, Vol. 19, No. 15, 1998, pp. 1343-1352.

106.   C. Alvarez-Lorenzo, S. Deshmukh, L. Bromberg, T. A. Hatton, I. Sandez-Macho and A. Concheiro, "Temperatureand Light-Responsive Blends of Pluronic F127 and poly(N,N-dimethylacrylamide-co-methacryloyloxyazobenzene)," Langmuir, Vol. 23, No. 23, 2007, pp. 11475- 11481.

107.   C. Alvarez-Lorenzo, L. Bromberg and A. Concheiro, "Light-sensitive Intelligent Drug Delivery Systems," Photochemistry and Photobiology, Vol. 85, No. 4, 2009, pp. 848-860.

108.   J. Kost, J. Wolfrum and R. Langer, "Magnetically enhanced Insulin Release in Diabetic Rats," Journal of Biomedical Materials Research A, Vol. 21, No. 12, 1987, pp. 1367-1373.

109.   L. L. Lao and R. V. Ramanujan, "Magnetic and Hydrogel Composite Materials for Hyperthermia Applications," Journal of Materials Science: Materials in Medicine, Vol. 15, No. 10, 2004, pp. 1061-1064.

110.   K. L. Ang, S. Venkatraman and R.V. Ramanujan, "Magnetic PNIPA hydrogels for Hyperthermia Applications in Cancer Therapy," Materials Science and Engineering C, Vol. 27, No. 3, 2007, pp. 347-351.

111.   R. Ramanujan, K. Ang and S. Venkatraman, "MagnetPNIPA hydrogels for bioengineering applications," Journal of Materials Science, Vol. 44, No. 5, 2009, pp. 1381- 1387.

112.   N. S. Satarkar and J. Z. Hilt, "Magnetic Hydrogel Nanocomposites for Remote Controlled Pulsatile Drug Release," Journal of Controlled Release, Vol. 130, No. 3, 2008, pp. 246-251.

113.   M. Namdeo, S. K. Bajpai and S. Kakkar, "Preparation of a Magnetic-Field-Sensitive Hydrogel and Preliminary Study of Its Drug Release Behavior," Journal of Biomaterials Science Polymer Edition, Vol. 20, No. 12, 2009, pp. 1747-1761.

114.   L. C. Sederel, L. Does, B. J. Euverman, A. Bantjes, C. Kluft and H. J. M. Kempen, "Hydrogels by irradiation of a Synthetic Heparinoid Polyelectrolyte," Biomaterials, Vol. 4, No. 1, 1983, pp. 3-8.

115.   I. Lavon and J. Kost, "Mass Transport Enhancement by ultrasound in Non-Degradable Polymeric Controlled Release Systems," Journal of Controlled Release, Vol. 54, No. 1, 1998, pp. 1-7.

116.   M. H. Casimiro, J. P. Leal and M. H. Gil, "Characterisation of Gamma Irradiated Chitosan/pHEMA membranes for biomedical purposes,"

Nuclear Instruments and Methods in Physics Research B, Vol. 236, No. 1-4, 2005, pp. 482-487.

117. H. Zhang, H. Xia, J. Wang and Y. Li, "High Intensity Focused Ultrasound-Responsive Release Behavior of PLA-b-PEG Copolymer Micelles," Journal of Controlled Release, Vol. 139, No. 1, 2009, pp. 31-39.

118. J. Berger, M. Reist, J. M. Mayer, O. Felt, N. A. Peppas and R. Gurny, "Structure and interactions in covalently and Ionically Crosslinked Chitosan Hydrogels for biomedical applications," European Journal of Pharmaceutics and Biopharmaceutics, Vol. 57, No. 1, 2004, pp. 19- 34.

119. M. Karbarz, W. Hyka and Z. Stojek, "Swelling Ratio Driven Changes of Probe Concentration in pHand Ionic Strength-Sensitive Poly(Acrylic Acid) Hydrogels," Electrochemistry Communications, Vol. 11, No. 6, 2009, pp. 1217-1220.

120. H. Li and Y. K. Yew, "Simulation of Soft Smart Hydrogels Responsive to pH stimulus: Ionic Strength Effect and Case studies," Materials Science and Engineering C, Vol. 29, No. 7, 2009, pp. 2261-2269.

121. L. Brannon-Peppas and N. A. Peppas, "Equilibrium swelling behavior of pH-Sensitive Hydrogels," Chemical Engineering Science, Vol. 46, No. 3, 1991, pp. 715-722.

122. E. O. Akala, P. Kopecková and J. Kopecek, "Novel pHSensitive Hydrogels with Adjustable Swelling Kinetics," Biomaterials, Vol. 19, No. 11-12, 1998, pp. 1037-1047.

123. M. Torres-Lugo, M. García, R. Record and N. A. Peppas, "pH-Sensitive Hydrogels as Gastrointestinal Tract Absorption Enhancers: Transport Mechanisms of Salmon Calcitonin and Other Model Molecules Using the Caco-2 Cell Model," Biotechnology Progress, Vol. 18, No. 3, 2002, pp. 612-616.

124. A. Richter, G. Paschew, S. Klatt, J. Lienig, K.-F. Arndt and H. J. Adler, "Review on Hydrogel based pH Sensors and Microsensors," Sensors, Vol. 8, No. 1, 2008, pp. 561- 581.

125. H. He, X. Cao and L. J. Lee, "Design of a Novel Hydrogel-Based Intelligent System for Controlled Drug Release," Journal of Controlled Release, Vol. 95, No. 3, 2004, pp. 391-402.

126. J. M. Varghese, Y. A. Ismail, C. K. Lee, K. M. Shin, M. K. Shin, S. I. Kim, I. So and S. J. Kim, "Thermoresponsive Hydrogels Based on Poly(N-Isopropylacrylamide) /Chondroitin Sulfate," Sensors and Actuators B Chemical, Vol. 135, No. 1, 2008, pp. 336-341.

127. T. Miyata, T. Uragami and K. Nakamae, "Biomoleculesensitive hydrogels," Advanced Drug Delivery Reviews, Vol. 54, No. 1, 2002, pp. 79-98.

128. R. Zhang, M. Tang, A. Bowyer, R. Eisenthal and J. Hubble, "Synthesis and characterization of a D-Glucose Sensitive Hydrogel Based on CM-dextran and concanavalin A," Reactive and Functional Polymers, Vol. 66, No. 7, 2006, pp. 757-767.

129. M. Goldraich and J. Kost, "Glucose-Sensitive Polymeric Matrices for Controlled Drug Delivery," Clinical Materials, Vol. 13, No. 1-4, 1993, pp. 135-142.

130. J. J. Kim and K. Park, "Modulated Insulin Delivery from Glucose-Sensitive Hydrogel Dosage Forms," Journal of Controlled Release, Vol. 77, No. 1-2, 2001, pp. 39-47.

131. R. Luo and H. Li, "Simulation analysis of effect of ionic strength on physiochemical and mechanical characteristics of glucose-sensitive hydrogels," Journal of Electroanalytical Chemistry, Vol. 635, No. 2, 2009, pp. 83-92.

132. Y. J. Kim, S. Choi, J. J. Koh, M. Lee, K. S. Ko and S. W. Kim, "Controlled Release of Insulin from Injectable Biodegradable Triblock Copolymer," Pharmaceutical Research, Vol. 18, No. 4, 2001, pp. 548-550.

133. T. Miyata, N. Asami and T. Uragami, "A reversibly antigen-Responsive Hydrogel," Nature, Vol. 399, No. 6738, 1999, pp. 766-769.

134. I. Schöll, G. Boltz-Nitulescu and E. Jensen-Jarolim, "Review of Novel Particulate Antigen Delivery Systems with Special Focus on treatment of type I allergy," Journal of Controlled Release, Vol. 104, No. 1, 2005, pp. 1-27.

135. A. K. Bajpai, S. K. Shukla, S. Bhanu and S. Kankane, "Responsive Polymers in Controlled Drug Delivery," Progress in Polymer Science, Vol. 33, No. 11, 2008, pp. 1088-1118.

136. N. Kashyap, N. Kumar and M. N. V. Kumar, "Hydrogels for pharmaceutical and Biomedical Applications," Critical Reviews in Therapeutic Drug Carrier Systems, Vol. 22, No. 2, 2005, pp. 107-150.

137. A. Kikuchi and T. Okano, "Pulsatile drug Release Control Using Hydrogels," Advanced Drug Delivery Reviews, Vol. 54, No. 1, 2002, pp. 53-77.

138. R. R. Burnette, "Theory of Mass Transfer", in: J. R. Robinson and V. H. L. Lee, Eds., Controlled Drug Delivery: Fundamentals and Applications, Marcel Dekker, New York, 1987, pp. 95-138.

139. B. Amsden, "Solute Diffusion within Hydrogels. Mechannisms and Models," Macromolecules, Vol. 31, No. 23, 1998, pp. 8382-8395.

140. J. Siepmann and N. A. Peppas, "Modeling of Drug Release from Delivery Systems Based on Hydroxypropyl Methylcellulose (HPMC)," Advanced Drug Delivery Reviews, Vol. 48, No. 2-3, 2001, pp. 139-157.

141. R. Bettini, P. Colombo, G. Massimo, P. L. Catellani and T. Vitali, "Swelling and Drug Release in Hydrogel Matrices: Polymer Viscosity and Matrix Porosity Effects," European Journal of Pharmaceutical Sciences, Vol. 2, No. 3, 1994, pp. 213-219.

142. P. L. Ritger and N. A. Peppas, "A simple equation for description of Solute Release I. Fickian and Non-Fickian Release from Non-Swellable Devices in the form of slabs, Spheres, Cylinders or discs," Journal of Controlled Release, Vol. 5, No. 1 1987, pp. 23-36.

143. N. A. Peppas, "Analysis of Fickian and non-Fickian drug release from polymers," Pharmaceutica Acta Helvetiae, Vol. 60, No. 4, 1985, pp. 110-111.

144. D. G. Kanjickal and S. T. Lopina, "Modeling of drug release from polymeric delivery systems—A Review," Critical Reviews in Therapeutic Drug Carrier Systems, Vol. 21, No. 5, 2004, pp. 345-386.

145. T. Nagai and Y. Machida, "Buccal Delivery Systems Using Hydrogels," Advanced Drug Delivery Reviews, Vol. 11, No. 1-2, 1993, pp. 179-191.

146. M. M. Veillard, M. A. Longer, T. W. Martens and J. R. Robinson, "Preliminary studies of oral Mucosal Delivery of Peptide Drugs," Journal of Controlled Release, Vol. 6, No. 1, 1987, pp. 123-131.

147. M. J. Rathbone and I. G. Tucker, "Mechanisms, barriers and pathways of oral Mucosal Drug Permeation," Advanced Drug Delivery Reviews, Vol. 12, No. 1-2, 1993, pp. 41-60.

148. R. Anders and H. P. Merkle, "Evaluation of laminated Muco-Adhesive Patches for Buccal Drug Delivery," International Journal of Pharmaceutics, Vol. 49, No. 3, 1989, pp. 231-240.

149. J. D. Smart, "Drug Delivery Using Buccal-Adhesive Systems," Advanced Drug Delivery Reviews, Vol. 11, No. 3, 1993, pp. 253-270.

150. C. A. L. Bourlais, L. Treupel-Acar, C. T. Rhodes, P. A. Sado and R. Leverge, "New Ophthalmic Drug Delivery Systems," Drug Development and Industrial Pharmacy, Vol. 21, No. 1, 1995, pp. 19-59.

151. H. W. Hui and J. R. Robinson, "Ocular delivery of progesterone using a Bioadhesive Polymer," International Journal of Pharmaceutics, Vol. 26, No. 3, 1985, pp. 203-213.

152. S. Cohen, E. Lobel, A. Trevgoda and Y. Peled, "A novel in Situ-Forming Ophthalmic Drug Delivery System from Alginates Undergoing Gelation in the eye," Journal of Controlled Release, Vol. 44, No. 2-3, 1997, pp. 201-208.

153. J. Carlfors, K. Edsman, R. Petersson and K. Jörnving, "Rheological evaluation of Gelrite® in situ gels for ophthalmic use," European Journal of Pharmaceutical Sciences, Vol. 6, No. 2, 1998, pp. 113-119.

154. P. Chetoni, G. Di Colo, M. Grandi, M. Morelli, M. F. Saettone and S. Darougar, "Silicone rubber/hydrogel Composite Ophthalmic Inserts: Preparation and preliminary in vitro/in vivo evaluation," European Journal of Pharmaceutical Sciences, Vol. 46, No. 1, 1998, pp. 125-132.

155. S. Türker, E. Onur and Y. Özer, "Nasal route and drug delivery systems," Pharmacy World & Science, Vol. 26, No. 3, 2004, pp. 137-142.

156. M. Zhou and M. D. Donovan, "Intranasal mucociliary clearance of Putative Bioadhesive Polymer gels," International Journal of Pharmaceutics, Vol. 135, No. 1-2, 1996, pp. 115-125.

157. L. Illum, N. F. Farraj and S. S. Davis, "Chitosan as a Novel Nasal Delivery System for Peptide Drugs," Pharmaceutical Research, Vol. 11, No. 8, 1994, pp. 1186-1189.

158. K. Nakamura, Y. Maitani, A. M. Lowman, K. Takayama, N. A. Peppas and T. Nagai, "Uptake and release of budesonide from mucoadhesive, pH-Sensitive Copolymers and Their Application to Nasal Delivery," Journal of Controlled Release, Vol. 61, No. 3, 1999, pp. 329-335.

159.   Y. M. Sun, J. J. Huang, F. C. Lin and J. Y. Lai, "Composite poly(2-Hydroxyethyl Methacrylate) Membranes as Rate-Controlling Barriers for Transdermal Applications," Biomaterials, Vol. 18, No. 7, 1997, pp. 527-533.

160.   J. C. Gayet and G. Fortier, "High water content BSAPEG hydrogel for Controlled Release Device: Evaluation of the Drug Release Properties," Journal of Controlled Release, Vol. 38, No. 2-3, 1996, pp. 177-184.

161.   B. J. Bellhouse and M. A. F. Kendall, "Dermal Powder Ject Device," in: M. J. Rathbone, J. Hadgraft and M. S. Roberts, Eds., Modified-Release Drug Delivery Technology, Marcel Dekker, New York, 2003.

162.   M. R. Prausnitz, S. Mitragotri and R. Langer, "Current status and Future Potential of Transdermal Drug Delivery," Nature Reviews Drug Discovery, Vol. 3, No. 2, 2004, pp. 111-124.

163.   S. Mehier-Humbert, R. H. Guy, "Physical methods for Gene Transfer: Improving the kinetics of Gene Delivery into cells," Advanced Drug Delivery Reviews, Vol. 57, No. 5, 2005, pp. 733-753.

CHAPTER 5

Application of Nanotechnology in Drug Delivery

Joana Silva[1], Alexandra R. Fernandes[1, 2] and Pedro V. Baptista[1, 3]

[1]Department of Life Sciences, Faculdade de Ciências e Tecnologia, Universidade Nova de Lisboa, Portugal
[2]Centro de Química Estrutural, Instituto Superior Técnico, Lisboa, Portugal
[3]CIGMH, Departamento de Ciências da Vida, Faculdade de Ciências e Tecnologia, Universidade Nova de Lisboa, Campus de Caparica, Portugal

1. INTRODUCTION

1.1. Nanomedicine for Cancer

Cancer is one of the leading causes of death worldwide, occupying the second place in developing countries, and showing a growing incidence over time [1]. Current cancer therapy strategies are based in surgery, radiotherapy and chemotherapy, being the chemotherapy the one that shows the greater efficiency for cancer treatment, mainly in more advanced stages [2, 3]. Despite of this great response, anticancer agents are administrated at higher amounts in order to provide a final suitable concentration to the target tissues or organs, and this procedure is repeated in each cycle of chemotherapy [4]. Introduction of new agents to cancer therapy has greatly improved patient survival but still there are several biological barriers that antagonize drug delivery to target cells and tissues, namely unfavorable blood half-life and physiologic behavior with high off-target effects and effective clearance from the human organism [2, 5, 6]. Moreover, in cancer, there is a small subset of cancer cells-cancer stem cells (CSC)-that, like normal stem cells, can self-renew, give rise to heterogeneous populations of daughter cells, and proliferate extensively [7, 8]. Standard chemotherapy is directed against rapidly dividing cells, the bulk of non-stem cells of a tumor, and thus CSC often appear relatively refractory to those agents [7-9]. The development of side effects in normal tissues (e.g. nephrotoxicity, neurotoxicity, cardiotoxicity, etc) and multidrug resistance (MDR) mechanisms by cancer cells leads to a reduction in drug concentration at target location, a poor accumulation in the tumor with consequent reduction of efficacy that may associate to patient relapse [9-13]. To overcome these issues and still improve the efficiency of chemotherapeutic agents there is a demand for less toxic and more target specific therapies towards cancer cells, i.e. novel drugs, drug delivery systems (DDSs) and also gene delivery systems [3, 4, 14-17].

Nanotechnology is the manipulation of matter on an atomic, molecular, and supramolecular scale involving the design, production, characterization and application of different nanoscale materials in several key areas providing novel technological advances mainly in the field of medicine (so called Nanomedicine) [6, 18-20]. The development and optimization of drug delivery approaches based in nanoparticles concerns the early detection of cancer cells and/or specific tumor biomarkers, and the enhancement of the efficacy of the treatments applied [21]. The most important biomedical applications of nanoscale materials can be organized as shown in Figure 1.

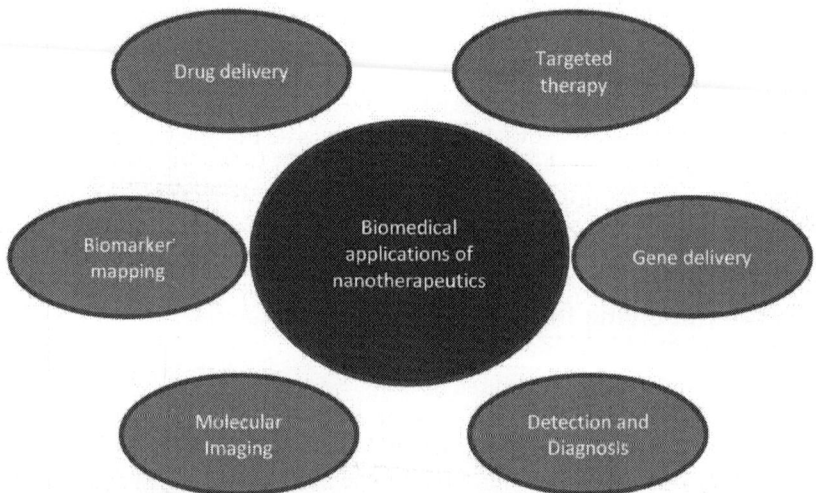

Figure 1. Biomedical application of nanotherapeutics (adapted from [6]).

These nanotherapeutics' potential in cancer relies on i) passive targeting due to the enhance of the permeability and retention (EPR) effect promoted by angiogenic vessels with defective vasculature and improper lymphatic flow surrounding the tumor [18] that can be reinforced by ii) specific targeting based on multifunctional nanomaterials that bypass the biological barriers and reach cancer cells [4]. Nanotechnology for drug vectorization provides for new and more specific drug targeting and delivery platforms that can reduce toxicity and other side effects and also maintain or improve the therapeutic index [9, 22, 23]. In fact, the development of targeting delivery systems is the ultimate goal in cancer therapy, which has been taking the lead in what concerns overcoming the MDR problem [9, 13, 24,25].

Here, we will discuss recent applications on AuNPs as platforms for anticancer therapy, emphasizing strategies for targeted delivery for gene silencing focusing on the optimal pathways to test these therapeutics *in vitro* and *in vivo*. Also, an overview of the toxicological aspects of these materials will be provided.

2. NANOPARTICLES AS DELIVERY SYSTEMS

Nanoparticles have been developed as effective target specific strategies for cancer treatment, acting as nanocarriers and also as active agents [4, 6, 5, 26]. Over the last decades, different types of nanoparticles have been developed based on various components, including carbon, silica oxides, metal oxides, nanocrystals, lipids, polymers, dendrimers, and quantum dots, together with increasing variety of newly developed materials [4, 27-34]. These nanomaterials are capable to provide a high degree of biocompatibility before and after conjugation to biomolecules for specific function so as to translate into nanomedicines and clinical practice. Nanomaterials provide for a favorable blood half-life and physiologic behavior with minimal off-target effects, effective clearance from the human organism, and minimal or no toxicity to healthy tissues in living organisms [35, 36].

In fact, the protection from adsorption to plasma proteins and/or degradation by circulating nucleases allows for an increased availability of effector molecule at site of interest. This is further enhanced by the considerable decrease to clearance from the organism that conjugation to nanoparticles confers. The modulation of pharmacokinetic and pharmacodynamics parameter constitutes a key factor when modifying the mode of administration (and vehicle and route of administration associated) that is usually neglected when compared to the ability of therapeutic nanoconjugates to offer the possibility of enhanced targeting (active and/or passive) and cell uptake. When considering nanoparticles for therapeutics one should also evaluate the effect on cellular metabolism and fate that can be attained via optimal conjugation with (bio)molecules of interest.

DDSs can improve the properties of free drugs by increase their *in vivo* stability and biodistribution, solubility and even by modulation of pharmacokinetics, promoting the transport and even more important the release of higher doses of the drug in the target site in order to be efficient [18, 22, 37,38].

DDSs can be constructed by direct conjugation with the drugs and further surface modifications can lead to a better delivery for such systems, promoting a targeted delivery to specific types of cells and reaching cell compartments such as nucleus and mitochondria [15, 39]. As far as drug delivery is concerned, the most important nanoparticle platforms are liposomes, polymer conjugates, metallic nanoparticles (for example AuNPs), polymeric micelles, dendrimers, nanoshells, and protein and nucleic acid-based nanoparticles (for a more complete review see [40-42].

Among a wide variety of nanosystems, only a few nanomedicines, such as Doxil® (Janssen Biotech Inc., Horsham, PA, USA), DaunoXome® (Galen US Inc., Souderton, PA, USA), Depocyt® (Pacira Pharmaceuticals Inc., San Diego, CA, USA), Genexol-PM® (Samyang Biopharmaceuticals Corporation, Jongno-gu, Seoul, Korea), Abraxane® (Celgene Corporation, Inc., Berkeley Heights, NJ, USA), Myocet® (Sopherion Therapeutics Inc., Princeton, NJ, USA) and Oncaspar® (Enzon Pharmaceuticals Inc., Bridgewater, NJ, USA), are approved for use in the treatment of cancer (for a review see [6]).

The implementation of nanoparticles towards cancer treatment can be based in certain characteristics as their size, surface properties and the possibility of a variety of specific ligands in their surface [18]. The high surface properties and other physicochemical features of nanoparticles can be modulated for the development of valuable systems that detect tumor cells either qualitatively or quantitatively [10,19].

Targeting the cancer cells occurs via two different strategies: passive targeting and active targeting [4,43, 44]. The passive targeting of tumor cells by nanoparticles depends upon an EPR effect promoted by angiogenic vessels with defective vasculature and improper lymphatic flow, reaching a higher accumulation in tumor cells compared to normal cells [15]. The increased accumulation of a drug in the tumor *interstitium* achieved by nanoparticles can be more than ten times higher compared to the drug alone [4]. This type of deliver is based in nanoparticle's half-time of circulation on the bloodstream, size and surface properties, and even depends on the degree of angiogenesis [45]. Despite the increased drug accumulation inside the tumor, this strategy rise some concerns about the targeting specificity of such mechanism based in the controversial influence of the EPR effect on drug externalization, which promotes a widespread distribution all over the tumor [4, 46]. The lack of specificity of such targeting led to further innovation with the implementation of an active targeting, which is achieved by the functionalization of nanoparticle's surface with a plethora of functional moieties such as antibodies and other biomolecules that recognized the specific surface antigens or specific biomarker of tumor cells [4, 44]. The targets choice depends on its high abundance in cell surface and its unique expression, and consequently the capacity of internalization of the nanoconjugate [4, 47, 48]. Although it is considered that active targeting does not have a direct association to the total nanoparticles accumulated within the tumor, it will influence the uptake of nanoparticles via receptor-mediated internalization and improve the efficiency of anti-tumor agents that have intracellular targets [49, 50]. Active targeting can be the potential way of polymeric nanoparticles to deliver chemotherapeutic drugs to cancer cells and is, therefore, one of the main vectors of DDS development at present involving tailoring of nanoparticles to deliver the effective cargo without compromising the selective targeting.

3. GOLD NANOPARTICLES (AUNPS)

Metallic nanostructures are more flexible particles compared to other nanomaterials owed to the possibility of controlling the size, shape, structure, composition, assembly, encapsulation and tunable optical properties [51, 52]. Between the metallic nanostructures possible applied, AuNPs appears of great interest in the medical field, 3showing great efficiency towards cancer therapy [51-54]. The continuous interest in AuNPs is based in their tunable optical properties that can be controlled and modulated for the treatment and diagnosis of diseases [9, 54, 52].

3.1. Synthesis, Functionalization, Characterization and Properties of Aunps

The synthesis of nanoparticles follows some aspects relying in a high homogeneity of the materials in physical properties that greatly influence the size, shape and surface characteristics. The main process for nanoparticles development requires chemical administration of capping agents that adsorb in the surface of nanoparticles ([55] and references therein). AuNPs can be synthesized with different sizes through the reduction of gold with different agents such molecules bearing a thiol group, an aliphatic chain and a charged end group, and that can avoid particle aggregation [37]. Furthermore, this dense layer of stabilizing agent promotes a general change in the surface charge of AuNPs allowing ligand exchange with several molecules, promoting AuNPs functionalisation and then an increase in particle stability in physiological environments [55, 56]. AuNPs deliver systems can be formulated based in their capacity to bearing different functional groups, once it can be involved in covalent and non-covalent bindings by a thiol-linker [37, 55]. In fact, robust AuNPs appear by the stabilization with thiolates once the bond between Au and the thiol (S) is very strong [57]. This process enhances the affinity of the AuNPs surface for several types of ligands such as polyethylene glycol (PEG) molecules, nucleic acids (DNA and RNA), peptides, antibodies, and also small drug molecules (Figure 2) [9, 13, 37, 47, 52, 56, 57].

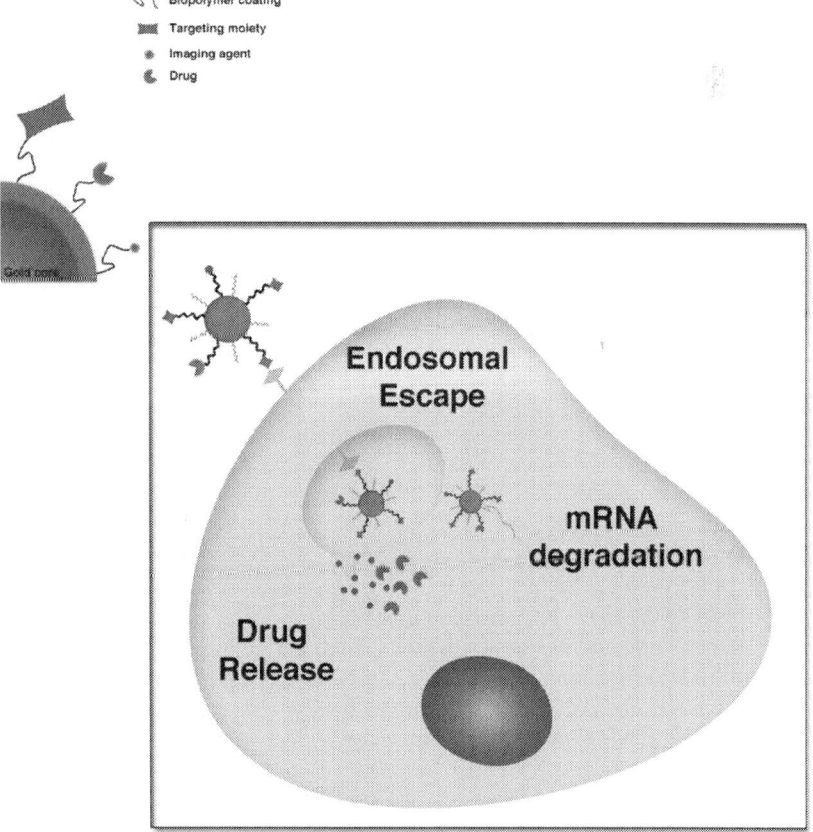

Figure 2. Multifunctional NP-based systems for tumor targeting, delivery and imaging. These innovative NPs comprise a targeting moiety, a silencing moiety and anticancer drug molecules for delivery to the target tissue. Depending on the targeting mechanism, they can be on the surface or inside the NPs. Multifunctional systems can carry reporter molecules tethered to the particle surface and employed as tracking and/or contrast agents.

Most passive targeting AuNPs have a surface coated with PEG for biocompatibility and "stealth" purposes [58]. Importantly, it should be noted that increased hydrophilicity on the AuNPs surface can impede its uptake by cancer cells, thereby hampering efficient drug delivery to tumors by passive targeting nanoparticles [58, 59].

As far as the targeting approach is concerned, one key issue relies on the choice of optimal targeting ligands, possibly by balancing their stoichiometry in comparison with the antibiofouling surface of AuNPs. More specifically, two important ligand properties, ie, affinity and density, can have a key role in effective targeting of nanoparticles to the cell surface membrane. Again, the ligand binding affinity is the result of the equilibrium between enthalpic advantages (for ligand-receptor interaction) and entropic losses (stretching, flexibility, or compressibility of the nanosystem). For example, greater ligand density does not necessarily lead to a higher intracellular concentration, given the decrease in "stealth" surface characteristics. Moreover, although the uptake of AuNPs usually increases with an increasing+/− charge ratio of nanoparticles (in terms of zeta potential values), an excess positive charge can induce toxicity and promote an immunologic reaction. Therefore, the optimal ligand density and charge on the AuNPs surface should be investigated on a case-by-case basis. AuNPs can be incorporated into larger structures such as polymeric nanoparticles or liposomes that deliver large payloads for enhanced diagnostic applications, efficiently encapsulate drugs for concurrent therapy or add additional imaging labels. This array of features has led to their application in biomedical fields, but more recently in approaches where multifunctional gold nanoparticles are used for multiple methods, such as concurrent diagnosis and therapy, so-called theranostics [53, 60-63].

AuNPs characterization is based on UV-Vis spectroscopy for the determination of the surface plasmon resonance (SPR) of the metallic gold, Transmission Electron Microscopy (TEM) for the determination of the average size of the particles, Scanning Electron Microscopy (SEM) for the characterization of the morphological features and Atomic Absorption Spectrometry that quantify the amount of gold [64]. AuNPs biodistribution can be monitored before the delivery of its payload which allows the establishment of treatment plan [65].

The application of AuNPs for *in vitro* diagnosis, *in vivo* imaging, therapy and also as DDSs relies in their chemical stability, high solubility in water, suitable morphology and limited dispersiiity, high surface-to-volume ratio, non-toxicity in biologic systems and an easy synthesis and functionalisation with a plethora of biomolecules (targeting and also silencing moieties) and drugs (Figure 2) [19, 21,55, 56, 66-68].

3.2. Aunps in Cancer Therapy

3.2.1. Photothermal Therapy

AuNPs formulations gain a major impact in cancer therapy in different contexts based in their properties that gain particular interest given some cancer specificities. AuNPs presents tunable optical properties that allow the absorption of light at near UV to near infrared, being the last one a characteristic that allows nanoparticles to enter cells, constituting a major breakthrough for its application in photothermal therapy or hyperthermia [57, 69]. This is thought due to the fact that increasing temperature of the cells above 42°C lead to a loss of cell viability [5]. Thus, nanoparticles heat up after irradiation of the body or local area with a magnetic field or another source of energy and consequently induce an increase in cancer cells temperature until cell death [5]. Several gold nanostructures are being referred as successful candidates as photothermal agents, such as the case described by Sirotkina and coworkers where AuNPs reach a high concentration in the skin tumor tissue and lead to an apoptotic response [70]. AuNPs compared to the simple irradiation method, the laser hyperthermia (a methodology though to circumvent the side effects associated to the current cancer therapies), has an advantageous of needing less irradiation energy to promote tumor ablation [57].

3.2.2. Radiotherapy
AuNPs have been review in radiotherapy experiments in order to overcome the problems associated to the healthy tissue damage imposed by radiotherapy [5, 57]. This strategy is based in the well-known accumulation of AuNPs in the tumor that will be acting as a decoy to focus the radiation in the tumor and limit its action in normal tumor vicinity, being able to decrease the initial quantity of radiation administrated [5, 71, 72]. A long term study using AuNPs and irradiation in mices bearing implanted tumors in order to eliminate the possibility of tumor regression, results in a reduction of the tumor size until not be detected and 86% long term cure, i.e. for more than a year, which was much higher than the 20% survival for the implementation of just radiotherapy [65].

3.2.3. Angiogenesis Inhibition
The inhibition of angiogenesis, i.e. the formation process of new blood vessels, is also a potent mechanism by which AuNPs can operate for cancer therapy [57, 73]. AuNPs have the ability to prevent phosphorylation of the proteins involved in this process of angiogenesis, by their binding to the cysteine residues in heparin-binding growth factors [73]. Complementary, the AuNPs intravenously administrated can be irradiated which leads to endothelium damage and then a break in the oxygen and nutrient supply to the tumors involved, another way of angiogenic therapy [65]. Radiotherapy, once have a major impact in rapidly divided cells, presents reduced activity towards the niche of central cells that become independent of blood supply and then hypoxic and with reduced proliferation capacity, leading to a continuous survival of such cells, which is the main cause of tumor relapse [65]. So the complete abolish of angiogenesis is a potential strategy to eradicate these cancer cells and then eradicate cancer [65].

3.3. Aunps As Delivery Systems
The well-known application of AuNPs in cancer therapy described above, lead to further investigation of new potential therapeutic strategies and was verified that AuNPs can be used in the design of delivery systems [74, 75]. The motivation for the implementation of AuNPs as drug delivery platforms is built in their easy to synthesis, functionalisation and also great biocompatibility, demonstrating that functionalisation with specific payloads have a great

potential to destroy cancer cells [15, 76]. As described above, AuNPs as a potential nanocarrier have the possibility to carry different payloads, such as small drug molecules for drug delivery or biomolecules like DNA, proteins and RNA (siRNAs), being recognized as an attractive gene delivery system (Figure 2) [37]. The conjugation of the two types of therapeutic functions in nanoparticles, i.e. a cytotoxic drug and a specific cancer cell target moiety, act as a single platform in a synergetic way to promote a higher affinity to cancer cells in order to signaling them to the efficiently release of the anticancer agent, circumventing the biological and biophysical barriers [23, 5, 77].

3.3.1. Specific Targeting

Nanoconjugates for drug delivery can accommodate a myriad of anti-cancer molecules that will be release and have its therapeutic effects in cancer cells, however healthy tissues can also be affected and to avoid this problem a targeting strategy is an important feature for combined therapy [78]. Taking advantage of tumor molecular markers as docking sites to concentrate the therapeutic effect at tumors, it is possible to increase therapeutic efficacy while reducing systemic exposure and off-target effects. Several of these tumor molecular markers are surface proteins/receptors present in cancer cells and in tumor vasculature that are not expressed or are expressed at much lower levels in normal cells, thereby distinguishing tumor masses from the surrounding normal tissues [79]. For a selective and with great potential delivery systems based in nanoparticles it is needed an efficient targeting to uniquely overexpressed receptors in cancer cells [76]. The potency of such systems is achieved by the enhancement of cellular accumulation of AuNPs by an active targeting to cancer cells compared to a free drug that passively enters the cells, which simultaneously avoid the biological response and biophysical barriers *in vivo* [5, 80]. In fact, nanoparticles need to continue in the bloodstream for enough time and cannot be eliminated in order to target the tumor site in the body, and surface modifications can be a useful property to avoid the mononuclear phagocytic system [4].

Based in this specificity, this platform can be a potential methodology for cancer therapy once it can differentiate with high certainty between cancer cells and non-cancer cells, one major concern related to the current cancer therapies [5, 81]. The increase of the surface area of AuNPs associated with other features such as distance-and refractive index-spectroscopic properties appears important for the construction of relevant biodetection molecules [55]. This capacity of AuNPs can be exploited for the improving of the therapeutic capacity of such systems, once the application of diagnostic and therapeutic strategies at the same time, Theranostics, can lead to a greater release at a specific tumor site, the targeting moiety or the drug, and it can be tracking in the whole body [82].

There are several types of tumor-targeting biomolecules such as peptides like RGD [47], proteins (transferrin, epidermal growth factor (EGF)) and carbohydrates [5], oligonucleotides such as aptamers [83], and monoclonal antibodies [85-87]. For example, the anti-epithelial growth factor receptor (EGFR) monoclonal antibody has been used as an active targeting agent, since EGFR and its ligands are commonly overexpressed in a variety of solid tumors [86-88].

Functionalization of AuNPs with specific ssDNA molecules (Au-nanoprobes) concerns a great detection system for specific DNA targets, being a rapid, sensitive, specific and inexpensive system [19]. For example, cancer cells overexpressing the folate receptors can be specifically targeted by AuNPs functionalized with folate ligands, and then the chemotherapeutic agent Doxorubicin (DOX) can be release into them and induce a higher toxicity compared to the one in healthy cells (do not express these receptors) and when compared to the DOX alone [89]. Additionally, the cell-surface specific markers that characterize the CSC pool constitute a way for targeting those cells in the whole tumor [7, 8]. Therapeutic strategies need to be focused in this specific CSC niche instead of the rapidly divided in order to prevent tumor regression [7, 8].

The targeting of tumor angiogenic vessels is also gaining increased attention, as it may improve therapeutic efficiency in cases where tumor cells are less accessible [79]. In combination with passive targeting strategies, active targeting may enhance receptor-mediated endocytosis of the AuNP conjugates in target cells [80, 90]. The intracellular localization of effector molecules can be further directed with the use of cell-penetrating peptides, which are short peptides that facilitate the delivery of various cargoes to cells, or with nuclear localization sequences, which direct cargos to the nucleus [91,92]. Decorating AuNPs with proton sponge groups or using photothermal heating can further assist escape from endosomal sequestration/degradation [93]. Active targeting by AuNPs [94-96] has been shown to result in greater tumor accumulation than passive targeting, when AuNPs are administered systemically in vivo (6-13% versus 2-5%) [94, 95].

3.3.2. Aunps for Drug/Cargo Delivery

The construction of DDSs depends on size, charge and surface functionalities of the AuNPs, once they dictate the uptake capacity of such nanovectorization systems as well as its intracellular fate [5, 75]. The possibility to functionalize AuNPs with a plethora of different cargos allows the development of several distinct approaches for drug delivery [76]. Moreover, stable nanovectorization systems in the blood-stream, drug release rate and clearance of the vector are two other important properties for the use of nanoparticles as DDSs [5, 57]. The use of vectorisation systems based in nanoparticles reveal the capacity to transpose the biological barriers imposed, with the release of low molecular-weight molecules that rapidly diffuse into the body, promoting a selective distribution towards cancer cells [5].

The active release of the drug into cells depends on the interaction between the drug and AuNPs (covalent or non-covalent binding) and even on methods of release after reaching the cells [76]. Non-covalent binding, such as the one used for hydrophobic drugs, does not need further alterations to the drug in order to be released [97] while for covalent interactions, establish for prodrugs, it implies the application of an internal or external mechanisms [98, 99]. The tunable optical properties of AuNPs surface is a prominent feature for the release of a drug either by internal or external stimuli [5]. External stimuli can be administrated by a photo-regulated release, which depends on the administration of light to photo-cleavage of nanoparticles-drug interaction to activate the drug once free [100-102], like was demonstrated with AuNPs functionalized with 5-fluorouracil [103]. You and collaborators [104] reported that up to 60% of a doxorubicin payload could be loaded onto hollow gold nanospheres because the drug molecules were adsorbed to both the inner and the outer surfaces of the hollow gold nanospheres via electrostatic attraction. Owing to the strong SPR absorption of novel gold nanostructures, drug release can be activated by NIR light [104, 105]. Glutathione can be a good internal stimulus of release for covalent interactions, by exchange reactions between disulfide of AuNPs and the intracellular glutathione [75]. AuNPs loaded with doxorubicin have been shown to be able to reverse cancer cells' resistance to the drug [106].

AuNPs with a high surface area attractive to establish interactions to a plethora of platinum drugs have been described [107]. Brown and coworkers designed a platinum-tethered AuNP system bearing the active compound of oxalipatin as its platinum molecule, and tested its platform in lung and colorectal cancer cell lines, demonstrating a more cytotoxic effect compared to oxaliplatin alone and also a higher accumulation of the active compound in those cancer cells reaching the nucleus for possible DNA interaction, what constitute a good delivery system [107].

3.3.3. Aunps for Gene Therapy

Gene therapy is though as a hopeful strategy in cancer therapy being considered as a powerful treatment like chemotherapy and radiotherapy, however the implementation of such systems is based in viral vectors that raise cytotoxic and immune response problems [57, 108]. When conjugated to AuNPs, siRNAs have been shown to exhibit increased stability, cellular uptake and efficacy in physiological conditions, retaining the ability to act through the RNAi pathway [109, 110]. The first demonstration that DNA-AuNP conjugates could be easily internalized into cells, without the need for transfection agents, and induced gene silencing by an antisense mechanism was reported by Rosi and co-workers in 2006 [111]. AuNPs as gene-delivery vectors emerged initially with cationic ligands that appears a good gene delivery system once protects the DNA molecule from degradation by DNAse I [112]. Han and coworkers have identified that cationic AuNPs can trigger DNA release into cells by glutathione intracellular concentrations [113].These remarkable studies prompted others to use AuNPs as siRNA delivery systems and contributed to the development of many strategies to improve intracellular siRNA delivery in vitro and in vivo. These strategies can be grouped into two major categories that are currently used for tethering siRNAs to AuNPs, namely (1) the gold-thiol bond and (2) electrostatic interactions. Both categories involve, in some way, the use of poly(ethylene glycol (PEG) or other passivating agents for stabilization and to promote endosomal escape of the AuNP conjugates into the cell cytoplasm [114].

The silencing of cancer-related molecules can be addressed by this delivery platform, being of major concern the oncogenes that have specific involvement in cell survival and proliferation [115]. siRNAs can be pointed as potential therapeutic molecules once its function relies in the suppression of gene expression [78]. siRNA molecules present limitations when administrated alone: do not cross the cell membrane, are rapidly degradable by endo-and exonucleases, have low stability in the blood and induces systemic toxicity [116]. Nanoparticles functionalized with siRNAs that have been tested for targeting reporter genes in *in vitro* cell cultures and recently AuNPs functionalized with siRNAs were investigated for *in vitro* and *in vivo* targeting genes [67, 116]. AuNPs can in fact be a good system for antisense and siRNA delivery since they can protect these molecules from degradation [109]. Thus, in order to improve cytoplasmatic translocation of siRNAs and promoting a complete gene silencing, is of utmost important the formulation of these nanoconjugates with siRNAs once its smaller size can potentiate and improve the interactions between biomolecules in the surface or into a cell [4, 116].

In vivo studies using this system are still scarce, alerting us for the need to overcome remaining barriers that prevent its translation into the clinics. Some recent studies are highlighted. Zhang and co-workers have developed an anti-metastasis therapy consisting of gold nanorods (AuNRs) conjugated electrostatically with siRNAs, which targeted the protease-activated receptor 1 (PAR-1) [117]. These conjugates were then delivered to highly metastatic human breast cancer cells. The authors observed efficient downregulation of PAR-1 mRNA and protein levels and decreased metastatic ability of the cancer cells [117]. By allowing any short nucleic acid to be hybridized to the cargo DNA covalently linked to the AuNP, the former can be designed for a specific purpose, such as gene knockdown, redirection of alternative splicing, and modulation of signal transduction pathways, Ryou and collaborators delivered shRNAs targeting the Mcl-1L mRNA to a xenografted tumor in a mouse model, and showed a ~5% reduction in protein expression which was sufficient to induce apoptosis of the xenograft tumor cells [118]. These studies did not include a targeting strategy because they were performed either in vitro or in vivo by directly injecting the conjugates into tumors. However, for systemic delivery, an additional targeting moiety is generally required to improve treatment efficacy and reduce off-target effects. Lu and co-workers [93] used Au nanocages targeted to folate receptors (overexpressed in many types of cancer) and carrying a siRNA against the NFkBp65, which encodes a transcription factor highly involved in tumor formation and progression. They injected these constructs intravenously into in nude mice bearing HeLa cervical cancer xenografts and observed a significantly higher tumor uptake of the targeted conjugates compared to the non-targeted ones. They additionally took advantage of the photothermal properties of the Au nanocages to achieve a controlled cytoplasmic delivery of siRNA upon NIR light irradiation and observed efficient NF-kappaB p65 downregulation only when tumors were irradiated with NIR light [93].

AuNPs functionalized with c-myc siRNAs were studied in a cervix adenocarcinoma cell line and demonstrate a great accumulation in the cytoplasm of the tumor cells and an evident ability to silencing the *c-myc* oncogene [67]. They functionalized AuNPs with PEG, cell penetration (TAT) and cell adhesion peptides (RGD, which binds to the integrin $\alpha V\beta3$ receptor family) [119], and c-myc targeting-siRNAs [5]. They have also shown in a more recent study that these same nanoconjugates are capable of targeting tumor cells in a lung cancer murine model and of inducing significant downregulation of the c-myc oncogene, followed by tumor growth inhibition and prolonged survival of lung tumor bearing mice [116].

Furthermore, miRNAs can appear with aberrant patterns of expression in tumors and then be related to its development, progression and tumor differentiation [47]. miRNAs can act as oncogenes or tumor suppressor genes accounting to its deregulation in cells, then it's down-regulation or up-regulation respectively, can be a major breakthrough for cancer therapy [120]. Conde and coworkers revealed a platform based in Au-nanobeacons to targeting and efficiently silencing miR-21, an oncogenic miRNA commonly up-regulated in almost all types of cancers [47, 121].

This targeting approached can also become a potential strategy to overcome the problem of multidrug resistance of cancer cells to the application of several drugs (Fernandes and Baptista, 2014). One of the major mechanisms of multidrug resistance in cancer is associated to ATP-binding cassette (ABC) membrane transporters, such as P-glycoprotein (P-gp), and others efflux pumps such as BCRP, which imply these as potential targets of silencing for cancer therapy [122, 123]. Cancer stem cells (CSCs) can also express these membrane proteins which confer to this subpopulation resistance to the current chemotherapeutic agents [123, 124]. Thus, the implication of a silencing strategy towards these cancer related genes evolve in order to minimize cancer resistance barriers to the actual therapy and then obtain an efficient response towards the chemotherapeutic agents applied [115, 125]. It was demonstrated by a system of lipid-modified dextran nanoparticles bearing siRNAs towards *ABCB1* gene (P-gp), that this approach can efficiently deliver the siRNA molecule and reduces the expression of P-gp although at the same order of greatness as the siRNA alone [126]. This reveals the necessity to continuously develop nanoparticles systems that can target and silencing these genes and proteins.

Another multidrug resistance mechanism is associated to the capacity of cancer cells to evade apoptotic response, when resistance induced by efflux pumps is not seen [127]. Apoptosis is the major cellular process induced by chemotherapeutic agents, so cancers bearing apoptosis defects cannot be efficiently treated by those agents, then discovery of the molecular basis of such system can formulate novel therapeutic approaches [127, 128]. For example, the anti-apoptotic protein Bcl2 is considered a proto-oncogene, and nano-based vector delivery systems has been establish with great efficacy towards this molecule [127, 129].

4. TOXICITY OF AUNPS

One major concern regarding AuNPs application in medical field relies in its toxicity in the biological systems, i.e. the production of a general toxicity response not only in cancer cells but also reaching healthy cells at the vicinity [78]. Taken into account the size, surface modifications and solubility in promoting biocompatibility of the nanovectorization systems, they can be safer to apply in the medical field to the treatment of cancer [130]. In fact, nanoparticles size is an important feature because it turns possible to circumvent the immune response and renal clearance, which maintains the therapeutic capacity of such systems [5].

Toxicity of AuNPs is generally accepted to be dependent on particle size, shape, and surface charge and chemistry [131-134]. However, it is thought that once AuNPs have a smaller size, approximately the size of biomolecules, it can be taken like one and then evade cellular barriers, with access to different tissues, and in the end can lead to the disruption of cell biological processes [75, 135]. A control of the size dependent cytotoxicity of AuNPs, revealed that AuNPs with a 1-2 nm size represents more toxicity towards four cancer cells lines compared to AuNPs with 15 nm that do not display any toxicity (Pan *et al.*, 2007). Additionally, the main organs affected by AuNPs are the liver and the spleen (Sun *et al.*, 2011). Also, very small particles (1.4 and 5 nm in diameter) seem to be capable to enter the nucleus, where they can interact with DNA and cause molecular disturbance [136,137]. Larger particles (16 nm and 33 nm) are retained in endosomes and accumulate in the periphery of the nuclear region [138, 139]. At least three different studies reported that cellular uptake of AuNPs reach maximum levels for a particle size of about 50 nm [140-142]. Also, surface functionalisation seem to be capable of inducing higher level of apoptotic cell death, probably related to increased cell uptake when compared to unmodified 40 nm AuNPs [141]. According to data from in vitro studies, AuNPs' toxicity is believed to result mainly from the induction of oxidative stress [143-145]. Indeed, up-regulation of stress related genes was found to result from cell exposure to AuNPs, which also promoted the down-regulation of cell cycle related genes [145-147]. Nevertheless, most of these studies paid little attention to genome damage, such as DNA strand breaks and nuclear abnormalities, or characterization of protein markers for toxicity. An integrated toxicology evaluation encompassing DNA damage, stress related enzymes and a proteome profiling approach showed no significant cytotoxicity of PEGylated AuNPs and no up-regulation of proteins related to oxidative damage [148]. Nevertheless, previous studies using metallic nanoparticles showed acute toxicity, mainly by the introduction of damages to the DNA molecule and also by oxidative damage [146, 149, 150].

AuNPs are however generally considered a system that do not cause acute or adverse toxicity, and then are been taken as safer systems for therapeutic use [135]. AuNPs demonstrate to be a safe system due to their easy of functionalization [151]. This ideal is based in the assumption that gold nanoparticles do not lead to any effect in the cell, and instead, the function moiety in its surface promote the cytotoxic effect expected [139]. In the other way, expression studies revealed an overexpression of stress and inflammation related genes after AuNPs treatment, being associated to the action of AuNPs in oxidative stress induction [75]. A decrease in cell cycle genes expression was simultaneously observed, which symbolizes an irreversible damage that leads to cell death by necrosis [75].

Nanoparticles surface composition is another relevant point when talking about toxicity of nanoparticles systems [5]. The ligands and surface capping agents of AuNPs as the first line of contact with the different actors in the cell pathways can promote toxicity that in the end represents the overall toxicity associated to these nanoconjugates [5]

Also, both positive and negatively charged AuNPs were found to be similarly more cytotoxic against human keratinocytes (HaCaT cells) when compared to neutral AuNPs, with LD50 values of roughly half of those determined for the latter [152]. Despite the disruption in cell morphology and the dose-dependent toxicity observed for all three types of AuNPs, both anionic and cationic AuNPs induce mitochondrial stress and apoptosis in opposition to the necrotic cell death caused by neutral particles [152]. Another in vitro study comparing positive and negatively charged AuNPs reported that cationic NPs were far more toxic to Cos-1 cells, human red blood cells and E. coli than anionic NPs, possibly as a result of cell lysis, as shown by a dye leakage technique [133]. However, Alkilany and co-workers clearly showed that serum proteins become readily adsorbed to the surface of charged NPs, inducing an inversion of surface charge in particles that were originally cationic [153]. This would reduce electrostatic interaction between the original positive NPs and the negative cell membrane, the first step towards cell lysis mediated toxicity of cationic NPs [133].

Regarding in vivo experiments, several studies have demonstrated that AuNPs of 50 nm and larger were non-toxic to mice, conversely to what has been observed for AuNPs <40 nm [54, 55]. In fact, there are concordant data from different studies on the biodistribution and accumulation of AuNPs in mice showing that most of the intravenously injected nanoparticles are retained in the liver, regardless of their size [156-158]. There is also an agreement in that AuNPs have the ability to transpose the blood-brain barrier and thus reach the brain, with a cut-off limit in diameter of around 20 nm [159], and that smaller particles have the most widespread organ distribution [156-158]. Organ distribution seems to be ruled by a more or less complex relationship with nanoparticle size. For instance, it is known that renal excretion of AuNPs is maximized for a narrow size range of 6-8 nm, resulting in an accelerated clearance rate [160]. Despite the valuable use of animal models, the effect of size on the toxicity of AuNPs in humans is difficult to predict since the size of endothelial cells' fenestrae is highly variable between individuals, thereby affecting nanoparticle clearance [75]. Therefore, more consistent data on the toxicological profile of AuNPs in vivo is necessary. For a more complete review on biodistribution, encompassing earlier studies and administration routes other than intravenous injection, see Khlebtsov and Dykman [159]. Furthermore, core size, charge and surface chemistry of AuNPs seems to correlate to toxicity on the development of zebrafish embryos, with positive and negatively charged AuNPs causing mortality and malfunctions to the embryos, respectively [161]. Adverse effects were also found in the model system Drosophila melanogaster after exposure to citrate-capped AuNPs, which were shown to reduce fertility in a dose-dependent manner and also the life span [144, 162].

Nonetheless, long-term studies in higher organisms are necessary to further characterise the safety of AuNPs as therapeutic agents, so they can be safely administered to humans without concerns about late toxicity symptoms.

5. CONCLUSIONS & FUTURE

Cancer is a complex disease with a plethora of cell types and differentiation stages that trigger standard molecular mechanisms towards recruitment of cells and nutrients to enhance survival and proliferation. Cancer complexity is also dependent from the specific and multifaceted umor microenvironment. All these different molecular pathways, mechanism and markers can be used as potential targets for therapeutics. However, current therapeutics (drugs and molecules) show serious cell toxicity that is not merely directed at the cancer cells but instead promote off-target cellular disarray and cell death, usually reported ad undesirable side effects and systemic toxicity.

Nanomedicine has been putting forward several therapeutic concepts that disrupt the way we have been dealing with cancer therapy, i.e. nanoparticles as drug delivery agents, minimising side effects and toxicity of the drugs. Furthermore, these nanoparticle platforms allow for selective targeting of cancer cells or tumor vessels either by incorporating novel or standard anticancer drugs and/or the delivery of therapeutic genetic modulators. These approaches, often based on the robustness and chemical properties of AuNPs, have shown great promise in preclinical models. Some recent advances in ligand-targeted NPs have begun to demonstrate improvement in cancer therapy.

What is more, many tumors become resistant to drugs, requiring that novel strategies involving drug targeting vehicles that deliver high concentrations of combinatorial therapeutics to the selected targets. For this to happen, it is crucial that these nanoconjugates are capable to withstand the body's clearance and reaction to non-self particulates. The robustness of AuNPs as target delivery platforms will be achieved when reticuloendothelial system clearance is avoid and occur an enhance of the endothelial penetration, once the first one can load to a longer time in circulation and the second leads to an increase of targeting and drug accumulation (Kumar et al., 2013).

The use of multiple nanoparticles that can be used together may overcome current limitations of each individual nanoformulation alone. For example, AuNPs have proven to be outstanding vectorisation systems for gene delivery and can be used to target molecular pathways, including those involved in drug resistance and in survival of cancer cells. These NPs may be used in combination with any other polymeric and/or metallic nanoparticles in therapeutic approaches that include drug and thermal ablation, selective delivery via out of the boy triggering (light source).

All of these applications of AuNPs in therapeutics still lack enough toxicology and pharmacology studies and data that can support the effective translation into the clinics. However, the efficacy in fighting cancer cells shows that the effort to push forward with the needed regulatory requirements and compliance is worth pursuing since the enhanced properties allow for outstanding improvements to biocompatibility, circulation and therapeutic response.

ACKNOWLEDGEMENTS

The authors acknowledge Fundação para a Ciência e Tecnologia (FCT/MEC) for funding: CIGMH (PEst-OE/SAU/UI0009/2011); PTDC/BBB-NAN/1812/2012.

REFERENCE

1. Jemal A, Bray F, Center MM, Ferlay J, Ward E, Forman D. Global cancer statistics. CA: A Cancer Journal for Clinicians 2011;61(2): 69-90.

2. Blanco E, Hsiao A, Mann AP, Landry MG, Meric-Bernstam F, Ferrari M. Nanomedicine in cancer therapy: Innovative trends and prospects. Cancer Science 2011;102(7): 1247-1252.

3. Peer D, Karp JM, Hong S, Farokhzad OC, Margalit R, Langer R. Nanocarriers as an emerging platform for cancer therapy. Nature Nanotechnology 2007;2(12): 751-760.

4. Baptista PV. Cancer nanotechnology—prospects for cancer diagnostics and therapy. Current Cancer Therapy Reviews 2009;5(2): 80-88.

5. Conde J, Doria G, Baptista P. Noble Metal Nanoparticles Applications in Cancer. Journal of Drug Delivery 2012;2012(2012): 1-12.

6. Sanna A, Pala N, Sechi M. Targeted therapy using nanotechnology: focus on cancer. International Journal of Nanomedicine 2014;9: 467-483.

7. Clarke MF, Dick JE, Dirks PB, Eaves CJ, Jamieson CHM, Jones DL, Visvader J, Weissman IL, Wahl GM. Cancer Stem Cells-Perspectives on Current Status and Future Directions: AACR Workshop on Cancer Stem Cells. Cancer Research 2006;66(19): 9339-9344.

8. Jordan CT, Guzman ML, Noble M. Mechanisms of Disease: Cancer Stem Cells. The New England Journal of Medicine 2006;355(12): 1253-1261.

9. Fernandes AR, Baptista PV. Nanotechnology for Cancer Diagnostics and Therapy – An Update on Novel Molecular Players. Current Cancer Therapy Reviews 2014;9(3): 1-9.

10. Brigger I, Dubernet C, Couvreur P. Nanoparticles in cancer therapy and diagnosis. Advanced Drug Delivery Reviews 2012;64: 24-36.

11. Oerlemans C, Bult W, Bos M, Storm G, Nijsen JFW, Hennink WE. Polymeric Micelles in Anticancer Therapy: Targeting, Imaging and Triggered Release. Pharmaceutical Research 2010;27(12): 2569-2589.

12. Larsen AK, Escargueil AE, Skladanowski A. Resistance mechanisms associated with altered intracellular distribution of anticancer agents. Pharmacology & Therapeutics 2000;85(3): 217-229.

13. Conde J, de la Fuente JM, Baptista PV. Nanomaterials for reversion of multidrug resistance in cancer: a new hope for an old idea?. Frontiers and Pharmacology 2013;4(134): 1-5.

14. Brannon-Peppas L, Blanchette JO. Nanoparticle and targeted systems for cancer therapy. Advanced Drug Delivery Reviews 2012;64: 206-212.

15. Tiwari G, Tiwari R, Sriwastawa B, Bhati L, Pandey S, Pandey P, Barnnerjee SK. Drug delivery systems: An updated review. International Journal of Pharmaceutical Investigation 2012;2(1): 2-11.

16. Ganta S, Devalapally H, Shahiwala A, Amiji M. A review of stimuli-responsive nanocarriers for drug and gene delivery. Journal of Controlled Release 2008;126(3): 187-204.

17. Koo OM, Rubinstein I, Onyuksel H. Role of nanotechnology in targeted drug delivery and imaging: a concise review. Nanomedicine 2005;1(3): 193-212.

18. Drbohlavova J, Chomoucka J, Adam V, Ryvolova M, Eckschlager T, Hubalek J, Kizek R. Nanocarriers for Anticancer Drugs-New Trends in Nanomedicine. Current Drug Metabolism 2013;14(5): 547-564.

19. Baptista PV, Pereira E, Eaton P, Doria G, Miranda A, Gomes I, Quaresma P, Franco R. Gold nanoparticles for the development of clinical diagnosis methods. Analytical Bioanalalytical Chemistry 2008;391(3): 943-950.

20. Silva GA. Introduction to Nanotechnology and Its Applications to Medicine. Surgical Neurology 2004;61(3): 216-220.

21. Baptista PV. Could gold nanoprobes be an important tool in cancer diagnostics?. Expert Reviews of Molecular Diagnostics 2012; 12(6): 541-543.

22. De Jong WH, Borm PJA. Drug delivery and nanoparticles: Applications and hazards. International Journal of Nanomedicine 2008;3(2): 133-149.

23. Ganesh T, Improved Biochemical Strategies for Targeted Delivery of Taxoids. Bioorganic & Medicinal Chemistry 2007;15(11): 3597-3623.

24. Hu CMJ, Zhang LF. Therapeutic Nanoparticles to Combat Cancer Drug Resistance. Current Drug Metabolism 2009;10(8): 836-841.

25. Nakanishi T, Fukushima S, Okamoto K, Suzuki M, Matsumura Y, Yokoyama M, Okano T, Sakurai Y, Kataoka K. Development of the polymer micelle carrier system for doxorubicin. Journal of Controlled Release 2001;74(1-3): 295-302.

26. Ferrari M. Cancer nanotechnology: Opportunities and challenges. Nature Reviews Cancer 2005;5(3): 161-171.

27. Kim BY, Rutka JT, Chan WC. Nanomedicine. The New England Journal of Medicine 2010;363(25): 2434-2443.

28. Riehemann K, Schneider SW, Luger TA, Godin B, Ferrari M, Fuchs H. Nanomedicine – challenge and perspectives. Angewandte Chemie International Edition England 2009;48(5): 872-897.

29. Petros RA, DeSimone JM. Strategies in the design of nanoparticles for therapeutic applications. Nature Reviews Drug Discovery 2010;9(8): 615-627.

30. Bae KH, Chung HJ, Park TG. Nanomaterials for cancer therapy and imaging. Moleculles and Cells 2011;31(4): 295-302.

31.	Taylor A, Wilson KM, Murray P, Fernig DG, Levy R. Long-term tracking of cells using inorganic nanoparticles as contrast agents: are we there yet? Chemical Society Reviews 2012;41(7):2707-2717.

32.	Villalonga-Barber C, Micha-Screttas M, Steele BR, Georgopoulos A, Demetzos C. Dendrimers as biopharmaceuticals: synthesis and properties. Current Topics in Medicinal Chemistry 2008;8(14): 1294-1309.

33.	Clift MJ, Stone V. Quantum dots: an insight and perspective of their biological interaction and how this relates to their relevance for clinical use. Theranostics 2012;2(7): 668-680.

34.	Yamashita T, Yamashita K, Nabeshi H, Yoshikawa T, Yoshioka Y, Tsunoda S, Tsutsumi Y. Carbon nanomaterials: efficacy and safety for nanomedicine. Materials 2012;5(2): 350-363.

35.	Lammers T, Kiessling F, Hennink WE, Storm G. Drug targeting to tumors: principles, pitfalls and (pre-) clinical progress. Journal of Controlled Release 2012;161(2): 175-187.

36.	Walkey CD, Chan WC. Understanding and controlling the interaction of nanomaterials with proteins in a physiological environment. Chemical Society Reviews 2012;41(7): 2780-2799.

37.	Ghosh P, Han G, De M, Kim CK, Rotello VM. Gold nanoparticles in delivery applications. Advanced Drug Delivery Reviews 2008;60(11): 1307-1315.

38.	Cai W, Gao T, Hong H, Sun J. Applications of gold nanoparticles in cancer nanotechnology. Nanotechnology, Science and Applications 2008;1: 17-32.

39.	Pissuwan D, Niidome T, Cortie MB. The forthcoming applications of gold nanoparticles in drug and gene delivery systems. Journal of Controlled Release 2011;149(1): 65-71.

40.	Pathak Y, Thassu D. Drug Delivery Nanoparticles Formulation and Characterization. Rijeka: PharmaceuTech Inc.; 2009. p1-393.

41.	Zhang L, Gu FX, Chan JM, Wang AZ, Langer RS, Farokhzad OC. Nanoparticles in medicine: therapeutic applications and developments. Clinical Pharmacology & Therapeutics 2008;83(5): 761-769.

42.	Davis ME, Chen ZG, Shin DM. Nanoparticle therapeutics: an emerging treatment modality for cancer. Nature Reviews Drug Discovery 2008;7(9): 771-782.

43.	Dinarvand R, de Morais PC, D'Emanuele A. Nanoparticles for Targeted Delivery of Active Agents against Tumor Cells. Journal of Drug Delivery 2012; 2012(2012): 1-2.

44.	Nie S. Understanding and overcoming major barriers in cancer nanomedicine. Nanomedicine 2010;5(4): 523-528.

45.	Allen TM, Cullis PR. Drug delivery systems: entering the mainstream. Science 2004;303(5665): 1818-1822.

46.	Stohrer M, Boucher Y, Stangassinger M, Jain RK. Oncotic pressure in solid tumors is elevated. Cancer Research 2000;60(15): 4251-4255.

47.	Conde J, Rosa J, de la Fuente JM, Baptista PV. Gold-nanobeacons for simultaneous gene specific silencing and intracellular tracking of the silencing events. Biomaterials 2013;34(10): 2516-2523.

48. Gao H Yang Z, Zhang S, Cao S, Shen S, Pang Z, Jiang X. Ligand modified nanoparticles increases cell uptake, alters endocytosis and elevates glioma distribution and internalization. Scientific Reports 2013;3(2534): 1-8.

49. Bartlett DW, Su H, Hildebrandt IJ, Weber WA, Davis ME. Impact of tumor-specific targeting on the biodistribution and efficacy of siRNA nanoparticles measured by multimodality in vivo imaging. PNAS 2007;104(39): 15549-15554.

50. Kirpotin DB, Drummond DC, Shao Y, Shalaby MR, Hong K, Nielsen UB, Marks JD, Benz CC, Park JW. Antibody Targeting of Long-Circulating Lipidic Nanoparticles Does Not Increase Tumor Localization but Does Increase Internalization in Animal Models. Cancer Research 2006;66(13): 6732-6740.

51. Huang X, Jain PK, El-Sayed IH, El-Sayed MA. Gold nanoparticles: interesting optical properties and recent applications in cancer diagnostics and therapy. Nanomedicine 2007;2(5): 681-693.

52. Figueiredo S, Cabral R, Luís D, Fernandes AR, Baptista PV. 2014. Integration of Gold nanoparticles and liposomes for combined anti-cancer drug delivery. In: Seifalian A. (ed.) Nanomedicine; University College London (UK); 2014. Chapter 3; available from http://www.onecentralpress.com/nanomedicine/#.

53. Cobley CM, Chen J, Cho EC, Wang LV, Xia Y. Gold nanostructures: a class of multifunctional materials for biomedical applications. Chemical Society Reviews 2011;40(1): 44-56.

54. Kumar A, Boruah BM, Liang XJ. Gold nanoparticles: Promising nanomaterials for the diagnosis of cancer and HIV/AIDS. Journal of Nanomaterials 2011;2011(2011): 1-17.

55. Baptista PV, Doria G, Quaresma P, Cavadas M, Neves CS, Gomes I, Eaton P, Pereira E, Franco R. Nanoparticles in Molecular Diagnostics. In: Villaverde A. (ed.) Progress in Molecular Biology and Translational Science Elsevier Inc.; 2011. p427-488.

56. Dreaden EC, Austin LA, Mackey MA, El-Sayed MA. Size matters: gold nanoparticles in targeted cancer drug delivery. Therapeutic Delivery 2012;3(4): 457-478.

57. Boisselier E, Astruc D. Gold nanoparticles in nanomedicine: preparations, imaging, diagnostics, therapies and toxicity. Chemical Society Reviews 2009;38(6): 1759-1782.

58. Alexis F, Pridgen E, Molnar LK, Farokhzad OC. Factors affecting the clearance and biodistribution of polymeric nanoparticles. Molecular Pharmaceutics 2008;5(4): 505-515.

59. Knop K, Hoogenboom R, Fischer D, Schubert US. Poly(ethylene glycol) in drug delivery: pros and cons as well as potential alternatives. Angewandte Chemie International Edition England 2010;49(36): 6288-6308.

60. Mieszawska AJ, Mulder WJ, Fayad ZA, Cormode DP. Multifunctional gold nanoparticles for diagnosis and therapy of disease. Molecular Pharmaceutics 2013;10(3): 831-47.

61. Makadia HK, Siegel SJ. Poly lactic-co-glycolic acid (PLGA) as biodegradable controlled drug delivery carrier. Polymers (Basel) 2011;3(3): 1377-1397.

62. Elsabahy M, Wooley KL. Design of polymeric nanoparticles for biomedical delivery applications. Chemical Society Reviews 2012;41(7): 2545-2561.

63. Nicolas J, Mura S, Brambilla D, Mackiewicz N, Couvreur P. Design, functionalisation strategies and biomedical applications of targeted biodegradable/biocompatible polymer-based nanocarriers for drug delivery. Chemical Society Reviews 2013;42(3): 1147-1235.

64. Yazid H, Adnan R, Hamid SA, Farrukh MA. Synthesis and characterization of gold nanoparticles supported on zinc oxide via the deposition-precipitation method. Turkish Journal of Chemistry 2010;34: 639-650.

65. Hainfeld JF, Dilmanian FA, Slatkin DN, Smilowitz HM. Radiotherapy enhancement with gold nanoparticles. Journal of Pharmacy and Pharmacology 2008;60(8): 977-985.

66. Cabral R, Baptista PV. The chemistry and biology of gold nanoparticle-mediated photothermal therapy: promises and challenges. Nano LIFE 2013;3(3): 1330001.

67. Conde J, Ambrosone A, Sanz V, Hernandez Y, Marchesano V, Tian F, Child H, Berry CC, Ibarra MR, Baptista PV, Tortiglione C, de la Fuente JM. Design of Multifunctional Gold Nanoparticles for In Vitro and In Vivo Gene Silencing. ACSNANO 2012;6(9): 8316-8324.

68. Murphy CJ, Gole AM, Hunyadi SE, Stone JW, Sisco PN, Alkilany A, Kinard BE, Hankins P. Chemical sensing and imaging with metallic nanorods. Chemical Communications 2008;5: 544-557.

69. Huang CW, Hao YW, Nyagilo J, Dave DP, Xu LF, Sun XK. Porous Hollow Gold Nanoparticles for Cancer SERS Imaging. Journal of Nano Research 2010;10: 137-148.

70. Sirotkina MA, Elagin VV, Shirmanova MV, Bugrova ML, Snopova LB, Kamensky VA, Nadtochenko VA, Denisov NN, Zagaynova EV. OCT-guided laser hyperthermia with passively tumor-targeted gold nanoparticles. Journal of Biophotonics 2010;3(10-11): 718-727.

71. Cardinal J, Klune JR, Chory E, Jeyabalan G, Kanzius JS, Nalesnik M, Geller DA. Non-invasive Radiofrequency Ablation of Cancer Targeted by Gold Nanoparticles. Surgery 2008;144(2): 125-132.

72. Gannon CJ, Patra CR, Bhattacharya R, Mukherjee P, Curley SA. Intracellular gold nanoparticles enhance non-invasive radiofrequency thermal destruction of human gastrointestinal cancer cells. Journal of Nanobiotechnology 2008;6(2): 1-9.

73. Bhattacharya R, Mukherjee P. Biological properties of "naked" metal nanoparticles. Advanced Drug Delivery Reviews 2008;60(11): 1289-1306.

74. Wang J, Yao K, Wang C, Tang C, Jiang X. Synthesis and drug delivery of novel amphiphilic block copolymers containing hydrophobic dehydroabietic moiety. Journal of Materials Chemistry B 2013;1(17): 2324-2332.

75.   Lim ZZJ, Li JEJ, Ng CT, Yung LYL, Bay BH. Gold nanoparticles in cancer therapy. Acta Pharmacologica Sinica 2011;32(8): 983-990.

76.   Duncan B, Kim C, Rotello VM. Gold nanoparticle platforms as drug and biomacromolecule delivery systems. Journal of Controlled Release 2010;148(1): 122-127.

77.   Schroeder A, Heller DA, Winslow MM, Dahlman JE, Pratt GW, Langer R, Jacks T, Anderson DG. Treating metastatic cancer with nanotechnology. Nature Reviews Cancer 2012;12(1): 39-50.

78.   Hu CMJ, Aryal S, Zhang L. Nanoparticle-assisted combination therapies for effective cancer treatment. Therapeutic Delivery 2010;1(2): 323-334.

79.   Ruoslahti E, Bhatia SN, Sailor MJ. Targeting of drugs and nanoparticles to tumors. Journal of Cell Biology 2010;188(6): 759-768.

80.   Dreaden EC, Mwakwari SC, Sodji QH, Oyelere AK, El-Sayed MA. Tamoxifen-Poly(ethylene glycol)-Thiol Gold Nanoparticle Conjugates: Enhanced Potency and Selective Delivery for Breast Cancer Treatment. Bioconjugate Chemistry 2009;20(12): 2247-2253.

81.   Liu Y, Miyoshi H, Nakamura M. Nanomedicine for drug delivery and imaging: a promising avenue for cancer therapy and diagnosis using targeted functional nanoparticles. International Journal of Cancer 2007;120(12): 2527-2537.

82.   Xie J, Lee S, Chen X. Nanoparticle-based theranostic agents. Advanced Drug Delivery Reviews 2010;62(11): 1064-1079.

83.   Farokhzad OC, Cheng J, Teply BA, Sherifi I, Jon S, Kantoff PW, Richie JP, Langer R. Targeted nanoparticle-aptamer bioconjugates for cancer chemotherapy in vivo. PNAS 2006;103(16): 6315-6320.

84.   Sapra P, Allen TM. Internalizing Antibodies are Necessary for Improved Therapeutic Efficacy of Antibody-targeted Liposomal Drugs. Cancer Research 2002;62(24): 7190-7194.

85.   Park JW, Kirpotin DB, Hong K, Shalaby R, Shao Y, Nielsen UB, Marks JD, Papahadjopoulos D, Benz CC. Tumor targeting using anti-her2 immunoliposomes. Journal of Controlled Release 2001;74(1-3): 95-113.

86.   Chattopadhyay N, Fonge H, Cai Z, Scollard D, Lechtman E, Done SJ, Pignol JP, Reilly RM. Role of antibody-mediated tumor targeting and route of administration in nanoparticle tumor accumulation in vivo. Molecular Pharmaceutics 2012;9(8): 2168-2179.

87.   Kao HW, Lin YY, Chen CC, Chi KH, Tien DC, Hsia CC, Lin MH, Wang HE. Evaluation of EGFR-targeted radioimmuno-gold-nanoparticles as a theranostic agent in a tumor animal model. Bioorganic & Medicinal Chemistry Letters 2013;23(11): 3180-3185.

88.   Ang KK, Berkey BA, Tu X, Zhang HZ, Katz R, Hammond EH, Fu KK, Milas L. Impact of epidermal growth factor receptor expression on survival and pattern of relapse in patients with advanced head and neck carcinoma. Cancer Research 2002;62(24): 7350-7356.

89.   Asadishad B, Vossoughi M, Alemzadeh I. Folate-Receptor-Targeted Delivery of Doxorubicin Using Polyethylene Glycol-Functionalized

Gold Nnanoparticles. Industrial & Engineering Chemistry Research 2010;49(4): 1958-1963.

90. Chithrani BD, Chan WC. Elucidating the mechanism of cellular uptake and removal of protein-coated gold nanoparticles of different sizes and shapes. Nano Letters 2007;7(6): 1542-1550.

91. Kang B, Mackey MA, El-Sayed MA. Nuclear targeting of gold nanoparticles in cancer cells induces DNA damage, causing cytokinesis arrest and apoptosis. Journal of the American Chemical Society 2010;132(5): 1517-1519.

92. de la Fuente JM, Berry CC. Tat Peptide as an Efficient Molecule To Translocate Gold Nanoparticles into the Cell Nucleus. Bioconjugate Chemistry 2005;16(5): 1176-1180.

93. Lu W, Zhang G, Zhang R, Flores LG, 2nd, Huang Q, Gelovani JG, Li C. Tumor site-specific silencing of NF-kappaB p65 by targeted hollow gold nanosphere-mediated photothermal transfection. Cancer Research 2010;70(8): 3177-3188.

94. Lu W, Xiong C, Zhang G, Huang Q, Zhang R, Zhang JZ, Li C. Targeted photothermal ablation of murine melanomas with melanocyte-stimulating hormone analog-conjugated hollow gold nanospheres. Clinical Cancer Research 2009; 15(3): 876-886.

95. Choi CH, Alabi CA, Webster P, Davis ME. Mechanism of active targeting in solid tumors with transferrin-containing gold nanoparticles. Proceedings of the National Academy of Sciences of the United States of America 2010;107(3): 1235-1240.

96. Melancon MP, Lu W, Yang Z, Zhang R, Cheng Z, Elliot AM, Stafford J, Olson T, Zhang JZ, Li C. In vitro and in vivo targeting of hollow gold nanoshells directed at epidermal growth factor receptor for photothermal ablation therapy. Molecular Cancer Therapeutics 2008;7(6): 1730-1739.

97. Park C, Youn H, Kim H, Noh T, Kook YH, Oh ET, Park HJ, Kim C. Cyclodextrin-covered gold nanoparticles for targeted delivery of an anti-cancer drug. Journal of Materials Chemistry 2009;19(16): 2310-2315.

98. Han G, You CC, Kim BJ, Turingan RS, Forbes NS, Martin CT, Rotello VM. Light-Regulated Release of DNA and Its Delivery to Nuclei by Means of Photolabile Gold Nanoparticles. Angewandte Chemie International Edition England 2006;45(19): 3165-3169.

99. Hong R, Han G, Fernández JM, Kim BJ, Forbes NS, Rotello VM. Glutathione-Mediated Delivery and Release Using Monolayer Protected Nanoparticle Carriers. Journal of the American Chemical Society 2006;128(4): 1078-1079.

100. Mayer G, Heckel A. Biologically Active Molecules with a "Light Switch". Angewandte Chemie International Edition England 2006;45(30): 4900-4921.

101. McCoy CP, Rooney C, Edwards CR, Jones DS, Gorman SP. Light-Triggered Molecule-Scale Drug Dosing Devices. Journal of the American Chemical Society 2007;129(31): 9572-9573.

102. Li J, Gupta S, Li C. Research perspectives: gold nanoparticles in cancer theranostics. Quantitative Imaging in Medicine and Surgery. 2013;3(6): 284-291.

103. Agasti SS, Chompoosor A, You C-C, Ghosh P, Kim CK, Rotello VM. Photoregulated Release of Caged Anticancer Drugs from Gold Nanoparticles. Journal of the American Chemical Society 2009;131(16): 5728-5729.

104. You J, Zhang G, Li C. Exceptionally high payload of doxorubicin in hollow gold nanospheres for near-infrared light-triggered drug release. ACS Nano 2010;4(2): 1033-1041.

105. Yavuz MS, Cheng Y, Chen J, Cobley CM, Zhang Q, Rycenga M, Xie J, Kim C, Song KH, Schwartz AG, Wang LV, Xia Y. Gold nanocages covered by smart polymers for controlled release with near-infrared light. Nature Materials 2009;8: 935-939.

106. Gu YJ, Cheng J, Man CW, Wong WT, Cheng SH. Gold-doxorubicin nanoconjugates for overcoming multidrug resistance. Nanomedicine 2012;8(2): 204-211.

107. Brown SD, Nativo P, Smith J-A, Stirling D, Edwards PR, Venugopal B, Flint DJ, Plumb JA, Graham D, Wheate NJ. Gold Nanoparticles for the Improved Anticancer Drug Delivery of the Active Component of Oxaliplatin. Journal of the American Chemical Society 2010;132(13): 4678-4684.

108. Hunt KK, Vorburguer SA. Hurdles and Hopes for Cancer Therapy. Science 2002;297(5580): 415-416.

109. Whitehead KA, Langer R, Anderson DG. Knocking down barriers: advances in siRNA delivery. Nature Reviews Drug Discovery 2009;8(2): 129 138.

110. Giljohann DA, Seferos DS, Prigodich AE, Patel PC, Mirkin CA. Gene regulation with polyvalent siRNA-nanoparticle conjugates. Journal of the American Chemical Society 2009;131(6): 2072-2073.

111. Rosi NL, Giljohann DA, Thaxton CS, Lytton-Jean AK, Han MS, Mirkin CA. Oligonucleotide-modified gold nanoparticles for intracellular gene regulation. Science 2006;312(5776): 1027-1030.

112. Han G, Martin CT, Rotello VM. Stability of Gold Nanoparticle-Bound DNA toward Biological, Physical, and Chemical Agents. Chemical Biology & Drug Design 2006;67(1): 78-82.

113. Han G, Chari NS, Verma A, Hong R, Martin CT, Rotello VM. Controlled Recovery of the Transcription of Nanoparticle-Bound DNA by Intracellular Concentrations of Glutathione. Bioconjugate Chem. 2005;16(6): 1356-1359.

114. Lytton-Jean AK, Langer R, Anderson DG. Five years of siRNA delivery: spotlight on gold nanoparticles. Small 2011;7(14): 1932-1937.

115. Gowda R, Jones NR, Banerjee S, Robertson GP. Use of Nanotechnology to Develop Multi-Drug Inhibitors for Cancer Therapy. Journal of Nanomedicine and Nanotechnology 2013;4(6): 1-16.

116. Conde J, Tian F, Hernández Y, Bao C, Cui D, Janssen K-P, Ibarra MR, Baptista PV, Stoeger T, de la Fuente JM. In vivo tumor targeting via

nanoparticle-mediated therapeutic siRNA coupled to inflammatory response in lung cancer mouse models. Biomaterials 2013;34(31): 7744-7753.

117.   Zhang W, Meng J, Ji Y, Li X, Kong H, Wu X, Xu H. Inhibiting metastasis of breast cancer cells in vitro using gold nanorod-siRNA delivery system. Nanoscale 2011;3(9): 3923-3932.

118.   Ryou SM, Park M, Kim JM, Jeon CO, Yun CH, Han SH, Kim SW, Lee Y, Kim S, Han MS, Bae J, Lee K. Inhibition of xenograft tumor growth in mice by gold nanoparticle-assisted delivery of short hairpin RNAs against Mcl-1L. Journal of Biotechnology 2011;156(2): 89-94.

119.   Brooks PC, Montgomery AM, Rosenfeld M, Reisfeld RA, Hu T, Klier G, Cheresh DA. Integrin alpha v beta 3 antagonists promote tumor regression by inducing apoptosis of angiogenic blood vessels. Cell 1994;79(7): 1157-1164.

120.   Wahid F, Shehzad A, Khan T, Kim YY. MicroRNAs: synthesis, mechanism, function, and recent clinical trials. Biochimica et Biophysica Acta 2010;1803(11): 1231-1243.

121.   Yu Y, Sarkar FH, Majumdar APN. Down-regulation of miR-21 Induces Differentiation of Chemoresistant Colon Cancer Cells and Enhances Susceptibility to Therapeutic egimens. Translational Oncology 2013;6(2):180-186.

122.   Baguley BC. Multiple Drug Resistance Mechanisms in Cancer. Molecular Biotechnology 2010;46(3): 308-316.

123.   Huls M, Russel FGM, Masereeuw R. The Role of ATP Binding Cassette Transporters in Tissue Defense and Organ Regeneration. The Journal of Pharmacology and Experimental Therapeutics 2009;328(1): 3-9.

124.   Wang K, Wu X, Wang J, Huang J. Cancer stem cell theory: therapeutic implications for nanomedicine. International Journal of Nanomedicine 2013;8(1): 899-908.

125.   Ganesh S, Iyer AK, Weiler J, Morrissey DV, Amiji MM. Combination of siRNA-directed Gene Silencing With Cisplatin Reverses Drug Resistance in Human Non-small Cell Lung Cancer. Molecular Therapy-Nucleic Acids 2013;2(7): 1-11.

126.   Susa M, Iyer AK, Ryu K, Choy E, Hornicek FJ, Mankin H, Milane L, Amiji MM, Duan Z. Inhibition of ABCB1 (MDR1) Expression by an siRNA Nanoparticulate Delivery System to Overcome Drug Resistance in Osteosarcoma. PLoS ONE 2010; 5(5): 1-12.

127.   Chaabane W, User SD, El-Gazzah M, Jaksik R, Sajjadi E, Rzeszowska-Wolny J, Los M. Autophagy, Apoptosis, Mitoptosis and Necrosis: Interdependence Between Those Pathways and Effects on Cancer. Archivum Immunologiae et Therapiae Experimentalis 2013;61(1): 43-58.

128.   Amaravadi RK, Thompson CB. The Roles of Therapy-Induced Autophagy and Necrosis in Cancer Treatment. Clinical Cancer Research 2007;13(24): 7271-7279.

129. Saad M, Garbuzenko OB, Minko T. Co-delivery of siRNA and an anticancer drug for treatment of multidrug-resistant cancer. Nanomedicine 2008;3(6): 761-776.

130. Kim D, Yu MK, Lee TS, Park JJ, Jeong YY, Jon S. Amphiphilic polymer-coated hybrid nanoparticles as CT/MRI dual contrast agents. Nanotechnology 2011;22(15): 1-7.

131. Pan Y, Neuss S, Leifert A, Fischler M, Wen F, Simon U, Schmid G, Brandau W, Jahnen-Dechent W. Size-dependent cytotoxicity of gold nanoparticles. Small 2007;3(11): 1941-1949.

132. Yildirimer L, Thanh NTK, Loizidou M, Seifalian AM. Toxicology and clinical potential of nanoparticles. Nano Today 2011;6(6): 585-607.

133. Goodman CM, McCusker CD, Yilmaz T, Rotello VM. Toxicity of gold nanoparticles functionalized with cationic and anionic side chains. Bioconjugate Chemistry 2004;15(4): 897-900.

134. Chen YS, Hung YC, Liau I, Huang GS. Assessment of the In Vivo Toxicity of Gold Nanoparticles. Nanoscale Research Letters 2009;4(8): 858-864.

135. Connor EE, Mwamuka J, Gole A, Murphy CJ, Wyatt MD. Gold nanoparticles are taken up by human cells but do not cause acute cytotoxicity. Small 2005;1(3): 325-327.

136. Tsoli M, Kuhn H, Brandau W, Esche H, Schmid G. Cellular uptake and toxicity of Au55 clusters. Small 2005;1(8-9): 841-844.

137. Ryan JA, Overton KW, Speight ME, Oldenburg CN, Loo L, Robarge W, Franzen S, Feldheim DL. Cellular uptake of gold nanoparticles passivated with BSA-SV40 large T antigen conjugates. Analytical Chemistry 2007;79(23): 9150-9159.

138. Nativo P, Prior IA, Brust M. Uptake and intracellular fate of surface-modified gold nanoparticles. ACS Nano 2008;2(8): 1639-1644.

139. Patra HK, Banerjee S, Chaudhuri U, Lahiri P, Dasgupta AK. Cell selective response to gold nanoparticles, Nanomedice: Nanotechnology, Biology, and Medicine 2007;3(2): 111-119.

140. Chithrani BD, Ghazani AA, Chan WC. Determining the size and shape dependence of gold nanoparticle uptake into mammalian cells. Nano Letters 2006;6(4): 662-668.

141. Jiang W, Kim BY, Rutka JT, Chan WC. Nanoparticle-mediated cellular response is size-dependent. Nature Nanotechnology 2008;3(3): 145-50.

142. Arnida, Malugin A, Ghandehari H. Cellular uptake and toxicity of gold nanoparticles in prostate cancer cells: a comparative study of rods and spheres. Journal of Applied Toxicology 2010;30(3): 212-217.

143. Hauck TS, Ghazani AA, Chan WC. Assessing the effect of surface chemistry on gold nanorod uptake, toxicity, and gene expression in mammalian cells. Small 2008;4(1): 153-159.

144. Sabella S, Brunetti V, Vecchio G, Galeone A, Maiorano G, Cingolani R, Pompa P. Toxicity of citrate-capped AuNPs: an in vitro and in vivo assessment. Journal of Nanoparticle Research 2011;13(12): 6821-6835.

145. Pan Y, Leifert A, Ruau D, Neuss S, Bornemann J, Schmid G, Brandau W, Simon U, Jahnen-Dechent W. Gold nanoparticles of diameter 1.4

nm trigger necrosis by oxidative stress and mitochondrial damage. Small 2009;5(18): 2067-2076.

146.   Li JJ, Hartono D, Ong CN, Bay BH, Yung LYL. Autophagy and oxidative stress associated with gold nanoparticles. Biomaterials 2010;31(23): 5996-6003.

147.   Li JJ, Zou L, Hartono D, Ong CN, Bay BH, Lanry Yung LY. Gold Nanoparticles Induce Oxidative Damage in Lung Fibroblasts In Vitro. Advanced Materials 2008;20(1): 138-142.

148.   Conde J, Larguinho M, Cordeiro A, Raposo LR, Costa PM, Santos S, Diniz MS, Fernandes AR, Baptista PV. Gold-nanobeacons for gene therapy: evaluation of genotoxicity, cell toxicity and proteome profiling analysis. Nanotoxicology 2014;8(5): 521-532.

149.   Hackenberg S, Scherzed A, Kessler M, Hummel S, Technau A, Froelich K, Ginzkey C, Koehler C, Hagen R, Kleinsasser N. Silver nanoparticles: evaluation of DNA damage, toxicity and functional impairment in human mesenchymal stem cells. Toxicology Letters 2011;201(1): 27-33.

150.   Asharani PV, Xinyi N, Hande MP, Valiyaveettil S. DNA damage and p53-mediated growth arrest in human cells treated with platinum nanoparticles. Nanomedicine 2010;5(1): 51-64.

151.   El-Sayed I, Huang X, El-Sayed MA. Selective laser photo-thermal therapy of epithelial carcinoma using anti-EGFR antibody conjugated gold nanoparticles. Cancer Letters 2006;239(1): 129-135.

152.   Schaeublin NM, Braydich-Stolle LK, Schrand AM, Miller JM, Hutchison J, Schlager JJ, Hussain SM. Surface charge of gold nanoparticles mediates mechanism of toxicity. Nanoscale 2011;3(2): 410-420.

153.   Alkilany AM, Nagaria PK, Hexel CR, Shaw TJ, Murphy CJ, Wyatt MD. Cellular uptake and cytotoxicity of gold nanorods: molecular origin of cytotoxicity and surface effects. Small 2009;5(6): 701-708.

154.   Cho WS, Cho M, Jeong J, Choi M, Cho HY, Han BS, Kim SH, Kim HO, Lim YT, Chung BH, Jeong J. Acute toxicity and pharmacokinetics of 13 nm-sized PEG-coated gold nanoparticles. Toxicology and Applied Pharmacology 2009;236(1): 16-24.

155.   Chen YS, Hung YC, Liau I, Huang GS. Assessment of the In Vivo Toxicity of Gold Nanoparticles. Nanoscale Research Letters 2009;4(8): 858-864.

156.   De Jong WH, Hagens WI, Krystek P, Burger MC, Sips AJ, Geertsma RE. Particle size-dependent organ distribution of gold nanoparticles after intravenous administration. Biomaterials 2008;29(12): 1912-1919.

157.   Sonavane G, Tomoda K, Makino K. Biodistribution of colloidal gold nanoparticles after intravenous administration: effect of particle size. Colloids and Surfaces B: Biointerfaces 2008;66(2): 274-280.

158.   Hirn S, Semmler-Behnke M, Schleh C, Wenk A, Lipka J, Schaffler M, Takenaka S, Moller W, Schmid G, Simon U, Kreyling WG. Particle size-dependent and surface charge-dependent biodistribution of gold nanoparticles after intravenous administration. European Journal of Pharmaceutics and Biopharmaceutics 2011;77(3): 407-416.

159. Khlebtsov N, Dykman L. Biodistribution and toxicity of engineered gold nanoparticles: a review of in vitro and in vivo studies. Chemical Society Reviews 2011;40(3): 1647-1671.

160. Longmire M, Choyke PL, Kobayashi H. Clearance properties of nano-sized particles and molecules as imaging agents: considerations and caveats. Nanomedicine (Lond) 2008;3(5): 703-717.

161. Harper SL, Carriere JL, Miller JM, Hutchison JE, Maddux BL, Tanguay RL. Systematic evaluation of nanomaterial toxicity: utility of standardized materials and rapid assays. ACS Nano 2011;5(6): 4688-4697.

162. Pompa P, Vecchio G, Galeone A, Brunetti V, Sabella S, Maiorano G, Falqui A, Bertoni G, Cingolani R. In Vivo toxicity assessment of gold nanoparticles in Drosophila melanogaster. Nano Research 2011;4(4): 405-413.

CHAPTER 6

Gold Nanoparticle and Berberine Entrapped into Hydrogel Matrix as Drug Delivery System

Camila Rufino Souza[1], Henrique R. Oliveira[1], Wagner M. Pinheiro[2], Lubhandwa S. Biswaro[2], Ricardo B. Azevedo[2], Anderson J. Gomes[1]*, Claure N. Lunardi[1]

[1]Laboratory of Nanobiotechnology, Faculty of Ceilandia, University of Brasilia, Brasilia, Brazil
[2]Institute of Biological Sciences, University of Brasilia, Brasilia, Brazil

ABSTRACT

In this study the novel hydrogel loaded with gold nanoparticle (AuNP) enhanced the berberine (BS) release when compared with other formulations of hydrogel. Hydrogels are hydrophilic polymer networks having the capacity to absorb water, ranging from about twenty to thousand times their dry weight. BS is a natural product, a quaternary ammonium salt from the group of isoquinoline alkaloids found in medicinal plants as Berberis Vulgaris. BS has some activity against dysentery, hypertension, inflammation, and liver disease in China and Japan. In this work, BS was used as a model drug to study its association with different types of hydrogel composites of polyvinyl alcohol (BS-PVA 10%); gellan gum (BS-GG 2%), gellan gum-PVA crosslinked with cysteine (cys) (BSGG2%PVA2%cys) and gellan gum-PVA cosslinked with cysteine associated with gold nanoparticles (AuNP-BSGG2%PVA2%cys). Several parameters such as fraction of retained water (Wf), hydration percentage (%H), Swelling (DSw) and time course profile (t = 100%) (TC) were evaluated for all preparations. The results showed that the AuNP-BS-GG2%PVA2%cys was able to retain more water and swelling than the other preparations. The time course of release of the BS to the medium was greater for AuNP-BS-GG2%PVA2%cys making it a candidate to drug delivery studies in biological assays. Also Scanning Electron Microscopy (SEM) images of the surface of these hydrogel were performed. Furthermore, crosslink of the resulting hydrogels were investigated by Fourier Transform Infrared Spectroscopy (FTIR) and differential scanning calorimetry (DSC) and thermogravimetric analysis (TGA). Thus, briefly, the aim of this work was to study three composition of hydrogel loaded with BS and its composition in relation to addition to AuNP and evaluate its profile for further drug delivery application using the Surface Plasmonic Resonance (SPR) as a tool improving the drug release in the new hydrogel.

Keywords: Hydrogel, Berberine, Drug Delivery, AuNPs

1. INTRODUCTION

Berberine (BS) is a natural product, a quaternary ammonium salt from the group of isoquinoline alkaloids, belongsto structural class of protoberberine, which positively-charged ion turns it hydro soluble. It is found in medicinal plants as Berberis Vulgaris [1] -[3] and the salts are derived of ammoniac form, solid, nonvolatile, crystallizable and inodorous when it is oxygenized. Due to its yellow color and characteristics the BS is applied with basic nature pigment [4] . The most interesting aspect of BS is related to its pharmacological properties been reported in the treatment of illness, such as: dysentery, hypertension, diabetes, inflammation, and also cancer [5] - [12] .

Hydrogel is report itself as polymeric network (reticulated structure), which has the feature to swell with water or biological fluids. Moreover, hydrogel storage a large amount of these fluids in its structure without dissolves itself [13] [14] . Besides, it exhibits some particular characteristics, such as hydrophilic behavior, insoluble in water, floppy, elastic and in contact with water, its volume increases considerably, without lose its shape, until achieve physic-chemical equilibrium. Hydrogels are able to mimic nature tissue, being biocompatibility [15] [16] . This feature leads to the amount of storage a large amount of water and special surface property. Hence, in this background, synthetics hydrogels are awaken interest at tissue engineering, cell encapsulate and drug delivery [17] . Another type of nanomaterial that have attracted increasing attention due to their unique properties in multidisciplinary research fields are gold nanoparticles (AuNPs) [2] [3] . The gold core is inert and essentially non-toxic to cells [18] . The particles absorb and resonantly scatter visible and near-infrared light upon excitation of their surface plasmon oscillation (SPR).

Gold nanoparticles associated to hydrogels are based on the behavior of AuNPs properties like the SPR band leading to evaluation of parameters like size among others [19] . Another important characteristic of AuNPs is the ability of dissipating heat when absorbing light, being an interest approach for thermal therapy as a trigger to release of encapsulated drugs from thermosensitive hydrogels [9] [17] . Moreover, there are no reports on the use of BS for preparing crosslinked AuNPs hydrogel nanocomposites, despite its relevance in the context of new formulations for biocompatible hydrogels. The aim of this article was to study the different types of hydrogelbiodegradable composites and the inclusion of gold nanoparticles to this system, evaluating parameters of surface and it's correlation to release time course profile of BS improving the efficiency and versatility to be used in existing therapies.

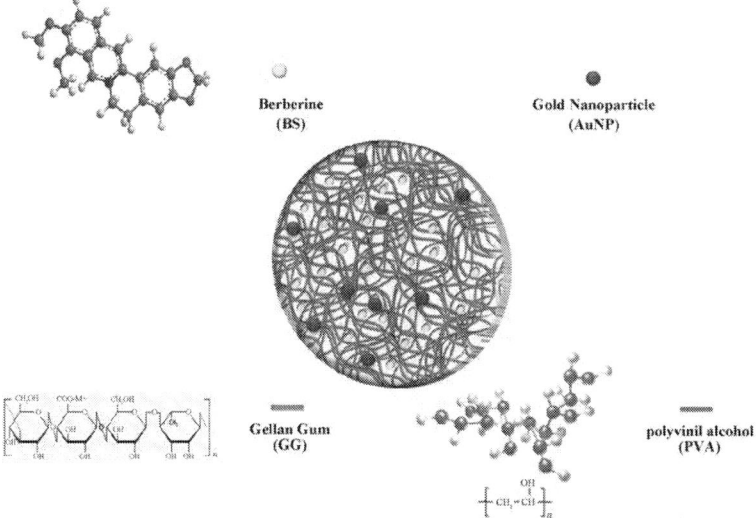

Figure 1. Schematic illustration of possible interaction of BS-GG2%PVA2%cys and AuNPs into the composite.

2. EXPERIMENTAL

2.1. Materials

Berberine (BS from Sigma Chem. Co.); polyvinil alcohol (88% PVA from Sigma Chem. Co.); Deacylated gellan gum GG-LA (Kelcogel from Cpkelco); chloridric acid (from Akros Chem); Au(III) solution (1%) was prepared with HAuCl4, citric acid (from Sigma Chem. Co.); HEPES buffer (from Sigma Chem. Co.); Cisteyne (CYS). Double distilled water was used for the preparation of all solutions.

2.2. Preparation of Hydrogel Composites

PVA 10%. Hydrogel composites were prepared with 10% PVA solution, 7.5% citric acid solution in distilled water,BS was and used as received. All the reagents were of analytical grade. The preparation of the hydrogel has been described elsewhere [20] . In short, 100 mL of a 10% aqueous solution of PVA and citric acid solution and concentrated hydrochloric acid (HCl, 0.05 mL) were mixed, and the resulting dispersion was stirred (using a overhead stirrer at 100 ± 5 rpm) at 70°C for a half-hour to carry out the crosslinked PVA hydrogel production. The thick dispersion so obtained was converted into a membrane by the conventional solution casting method. The resultant membrane was washed thoroughly with distilled water to remove the HCl and was stored in a desiccators at 37°C. The empty hydrogel was designated PVA 10% and the berberine loaded was designated BS- PVA10%.

Gellan Gum. The gellan gum (GG) hydrogel composite were prepared using HEPES (50 mM, pH = 4), following proportion: 2% of GG-LA (89.1 COOH) [21] [22] . In short, 6 mL of HEPES and 20μM of solution CYS and GG-LA were mixed, and the resulting dispersion was stirred (using a overhead stirrer at 100 ± 5 rpm) at 80°C for ten minutes to carry out the crosslinked kelcogell hydrogel production. The thick dispersion so obtained was converted into a membrane by the conventional solution casting method. The empty hydrogel was designated GG2% and the berberine loaded was designated BS-GG2%.

PVA 2% plus Gellan Gum 2%. The hydrogel composite were prepared using HEPES 50 mM, pH = 4), following proportion: 2% of GG-LA (89.1 COOH) and 2% of PVA. In short, 6 mL of HEPES, 20 μM of solution CYS, GG-LA and PVA were mixed, and the resulting dispersion was stirred (using a overhead stirrer at 100 ± 5 rpm) at 80°C for ten minutes to carry out the crosslinked kelcogell hydrogel production. The thick dispersion so obtained was converted into a membrane by the conventional solution casting method. The empty hydrogel was designated GG2%PVA2%cys and the berberine loaded was designated BS-GG2%PVA2%cys.

The BS in micromolar range (10-6M) concentration was added before stirring time to produce hydrogels composites. AuNPs: Gold nanoparticle was prepared by modified sodium citrate reduction method by Turkevich [23] . Placed 0.70 mL of 0.1% HAuCl4 into a 150 mL beaker, then added water to near 60 mL and the solution was heated to boiling. An amount of 2 mL of 1% sodium citrate solution was added dropwhise into beaker solution. The solution became rubbish during stirring for 30 minutes. This solution is called naked AuNPs.

AuNPs-PVA 2% plus Gellan Gum 2%. The nanocomposites were prepared by blending the Au NPs with the PVA 2% plus Gellan Gum 2%. The colloidal AuNPs was added to the BS in micromolar range (10-6 M) concentration was added before stirring time to produce hydrogels composites in the same procedure as described before.

2.3. In Vitro Stability and Swelling of Biodegradable Hydrogel Composite

Approximately 30 mg of each type of composites loaded with 465 μM of BS was deposited into pre-weighed polypropylene sample tubes, weighed, warmed to 37 °C, and combined with 1000 μL of warm HEPES solution at pH 7.4. Samples were incubated at 37°C on a rotary shaker and the HEPPES buffer sampled with total replacement after: 30, 60,120 minutes and daily during 8 days showing stable weight. Recovered buffer was sonicated to disrupt residual hydrogel before Uv-vis analysis (Uv-vis Hitachi U-3900). The swelling ratio, DSw, of each type of composite were determined by accurately weighing each gel sample in the stability study after the HEPES buffer had been removed, correcting for residual buffer, and dividing by the original hydrated sample mass [24] -[26] . To compare the degradation of composites with different swelling characteristics, normalized DSw was defined as:

$$DSw = \frac{m}{m^*}$$

(1)

where m and m^* are the weight of hydrated composite and dried composite respectively.

2.4. Drug Release from Biodegradable Hydrogel Composite

Release profiles of each loaded drug composite were obtained by depositing approximately 30 mg of drug loaded dry hydrogel and adding 1000 µL of HEPES buffer into polypropylene sample tubes, warming to 37°C. Samples were incubated at 37°C on a rotary shaker and the HEPPES buffer sampled with total replacement after 15, 30, 60, 120, 360 minutes and 24 h and daily during 90 days for each preparation. All samples were fresh analyzed by UV-vis measurements were performed on Hitachi U-3900H UV-Vis Spectrophotometer in the wavelength interval of 200 - 800 nm at 37°C using HEPES buffer as blank. The measurements were accomplished when the values of absorbance of BS in the medium in each sample reached the plateau. These values were transformed to percentage of release (100%) and correlated to time (hours) and displayed as time course profile (TC). The fraction of retained water (Wf); hydration percentage (%H); Swelling (DSw) and time course profile (t = 100%) (TC) were evaluated for all hydrogel composite as described in Equations (2)-(4):

$$Wf = \frac{\left(m - m^*\right)}{1}$$

(2)

$$EWC = Wf \times 100$$

(3)

$$H = 100 \times \frac{\left(m - m^*\right)}{m^*}$$

(4)

where m and m^* are the weight of hydrated composite and dried composite respectively.

2.5. Fourier Transform Infrared (FTIR) Spectroscopy Of Biodegradable Hydrogel Composite

The chemical structure of the biodegradable composite materials and the composite itself (unloaded and loaded with BS) were analyzed by FTIR (IR Prestige-21 FRIT-8400S, Shimadzu, Japan) in transmission mode. For that, each composite (1.0 mg) were mixed with KBr (40.0 mg) and then formed into a disc in a manual press. Transmission spectra were recorded using at least 32 scans with 4.0 cm^{-1} resolution, in the spectral range 4000 - 400 cm^{-1}.

2.6. Differential Scanning Calorimetry (DSC) Biodegradable Hydrogel Composite

Thermal characterization of biodegradable hydrogel composite was performed with a Shimadzu DSC-60A. The equipment was calibrated with indium. The sample (approximately 3.0 mg) was heated twice from 25°C to 400°C at 20°C/min in a nitrogen atmosphere (flow rate 30 mL/min). The melting temperature (Tm) was determined from the endothermic peak of the DSC curve recorded in the first heating scan.

2.7. Thermogravimetric Analyses (TGA) of Biodegradable Hydrogel Composite

The composites and its material were analyses. The samples weighting about 3 mg heated from 25°C to 400°C in a platinum pan rate 20°C/min in the nitrogen atmosphere at flow of 30 mL/min. The glass transition temperature (Tg) was determined from the DSC curve recorded in the second heating scan.

2.8. Scanning Electron Microscopy Measurements (SEM)

Scanning electron microscopy (SEM) was used to evaluate the surface of biodegradable composite hydrogel. Samples containing biodegradable composite hydrogel were mounted on aluminum stubs and due to their lack of electrical conductivity they were coated with 50 nm gold coating under an argon atmosphere. The composites were examined and photographed by a Jeol 840 A (Tokyo, Japan) Scanning Electron Microscope operating at 20 kV in the traditional mode (SE1 detector). SEM images were obtained to characterize the surface of the BS-PVA 10%, BS-GG2%cys, BS-GG2%PVA2%cys and AuNPs-BS-GG2%PVA2%cys hydrogel films.

2.9. Statistical Analysis

Experiments were performed in triplicate using freshly prepared samples. Statistical analysis data are expressed as means± standard deviation unless otherwise noted. Comparisons of groups of means were determined by ANOVA and pairs of mean by Student's t-test where appropriate. Significance was assigned at $p < 0.05$.

3. RESULTS AND DISCUSSION

3.1. Hydrogel Composite Preparation

The first attempt to production of biodegradable hydrogel composite with PVA was done in 7.5% w/w however, this hydrogel type dissolved when added to solution of HEPES. As alternative, the concentration was increased to 10%. In this way, we characterized the hydrogel in terms of thickness as displayed for GG2% hydrogel preparation in dried and swollen form (Figure 2) by using a pachymeter. The values obtained for each preparation were GG2% 0.4 mm for dry and 3.1 mm for swollen. Also for PVA10% were 0.2 mm dry and 2.0 swollen; GG2%PVA2%cys were 0.2 dry and 4.0 mm swollen and AuNP-GG2%PVA2%cys were 0.2 dry and 3.5 mm swollen.

3.2. Water Content Parameters And Time Course Release of BS

Water content in the hydrogel composite was affected by the concentration of gellan gum and also by the cysteine content. The values of the fraction of retained water (Wf); hydration percentage (%H); Swelling (DSw) is displayed in Table 1. The hydrogel of BS-PVA10% showed the lower values for all the parameters assayed. Only the time course (TC) of the BS release was higher and this might be due the rigid network during gel formation. Also the others hydrogel composite BS-GG 2%, BS-GG2%PVA2%cys and AuNP-GG2%PVA2%cys preparations showed increased values of water parameters with a faster time course of BS release. This results could be explained by the characteristics of GG, a well known polysaccharide manufactured by microbial fermentation of the Sphingomonas paucimobilis microorganism, that can be dissolved in water, and when heated and mixed with mono or divalent cations its forms gels [13] ; this behavior associated with PVA and cross-linked with cysteine, turned it in the most suitable composite from the preparations evaluated. These features indicate that this preparation is suitable for a drug delivery system for BS in biological systems with high content of water. The citrate-capped AuNPs decreases the intensity of the crosslinking reaction between the PVA/GGcys and BS showing a less cross-linked hydrogels, with reduced swollen profile when compared with BS-GG2%PVA 2%cys.

3.3. Sem Measurements of Hydrogel Composites

In Figure 3 are displayed the SEM of the hydrogels surface of BS-PVA 10%, BS-GG2%, and BS-GG2%PVA2%cys hydrogel films respectively, of dried hydrogels films. The flat and featureless images indicate that the films have a condensed structure and clearly show the appearance of a smooth or fibrous structure. The soft and stiff surface of PVA 10% can explain the lower time course release of the BS from matrix. However, the porous and fibrous like surface of BS-GG2%, and BS-GG2%PVA2%cys could explain the faster BS

Figure 2. The pachymeter for determining the thickness of GG2% hydrogel in dried and swollen form.

Table 1. Biodegradable hydrogel composite: swelling feature and time course release using BS [465 µM].

Hydrogel composite/	Hydrogel properties			
[BS] = 465 µM	Wf	%H	DSw	TC (hs)
BS-PVA 10%	0.4962 ± 0.005	96.9 ± 2.06	1.82 ± 0.22	72 ± 3.90
BS-GG 2%	0.6747 ± 0.438	211.74 ± 40.0	3.1174 ± 0.40	22 ± 0.04
BS-GG 2%PVA2%cys	0.9184 ± 0.013	1150.87 ± 186.4	12.508 ± 1.84	22 ± 0.15
AuNP-BS-GG2% PVA2%cys	0.9623 ± 0.021	1450.87 ± 100.2	8.41 ± 1.105	14 ± 0.50

Wf (fraction of retained water); %H (hydration percentage); DSw (Swelling); TC (Time course).

release and the increased values of physical chemistry parameters analyzed. In addition, 10%PVA hydrogel are less able to hydrate than the types with GG2% explaining its lower time course profile. In these words, GG2% hydrogels composite and associated with PVA are capable of storage a large amount of BS and maintain longer time at thera- peutic line. Both images from Figures 3(B)-(C) displayed more fibrous in topographical surface than the PVA10%. The AuNP-BS-GG2%PVA2%cys displayed in Figure 3(D) shows less fibrous than BS-GG2%PVA2%cys due to less crosslink, but due to thermal characteristics of AuNPs embebed into hydrogel, under the experimental conditions the release was faster (36%) in relation to BS-GG2%PVA2%cys alone.

3.4. FTIR Analysis of Hydrogel Composites

FTIR spectra was obtained with absorption in the region of 400 - 4000 cm^{-1} at room temperature and used to confirm the hydrogels structures. In Figure 4(A) is displayed the PVA 10% hydrogel FTIR spectra with a characteristic peak at 3343 cm^{-1} due typical OH stretching [16]. Also, there is a 2914 cm^{-1} the asymmetric stretching vibrations of methylene group [24]. At region of 1730 cm^{-1} it is possible to observe C = O stretching [10]. The peak in 1431 cm^{-1} and 1135 cm^{-1} indicate the presence of CH-OH [27] also in 1381 cm^{-1} peak is due to CH asymmetrical bending [26], and in the region of 1095 cm^{-1} correspond to CO linkage [24]. In Figure 4(B) is displayed the GG2% hydrogel spectra showing the following peaks in the region 3432 cm^{-1} that correspond the OH group as shown before in the PVA hydrogel; 1620 cm^{-1} due to bending vibration of hydroxyl group, 1462 due to CH asymmetrical bending [28] . In the region of absorption of 1219 cm^{-1} is possible see the sulfate ion. At the region 1219 cm^{-1} has a peak that correspond ester linkages [27] and the peak in the 1043 cm^{-1} indicate the presence of C-O-C group. Both hydrogels of PVA10% and GG2% showed preservation of major groups. These characteristics indicate that have no news groups formed in the blend of this hydrogels. It is possible observe, in Figure 4(C) that the GG2% peaks were preserved as 1620 cm^{-1} shifted to absorption in 1607 cm^{-1}, ester linkages were also preserved, as show in the 1192 cm^{-1}. This phenomenon is shown in the same graphic with PVA10% hydrogel peaks. The C = O stretching in 1730 cm^{-1}, characteristic peak of this hydrogel, shifted to 1721 cm^{-1}. The infrared spectra did not show absorption bands of the typical bridges resulted from the reaction of the hydrogel and BS [10] [29]. It seem just characteristic bands from BS.

3.5. Thermal Analysis of Hydrogel Composites (DSC and TGA)

The DSC heating curves of all hydrogel samples displayed some endothermic peaks due to the melting of the various forms of water that freeze during the cooling stage (Figure 5). The endothermic peak around 40°C - 420°C displayed a peak, which have been ascribed previously to the freezing bound water (the sharp one) [30] -[34] . The thermogram is characterized by peaks crystallization (Tc), melting (Tm) and glass transition (Tg) of all hydrogel composite. The behavior observed for 10%PVA, GG2%, PVAGG2% were very similar showing its Tc in the range of 40°C - 42°C in the first peak and 190°C - 260°C in the following transitions. The biodegradable hydrogel composite BS-GG2%PVA2%cys displayed a similar behavior, however most of the process occurs in exothermic profile. When association occurs, the formation of more structured complex so formed exhibits different characteristics in comparison to components-alone. AuNPs-BS-GG2%PVA2%cys showed similar behavior.

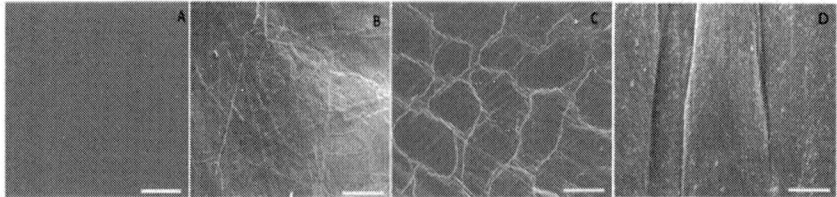

Figure 3. SEM images of hydrogel films. Surface of (a) BS-PVA10%, (b) BS-GG2%, (c) BS-GG2%PVA2%cys and (d) AuNP-BS-GG2%PVA2%cys hydrogel films (270× magnification).

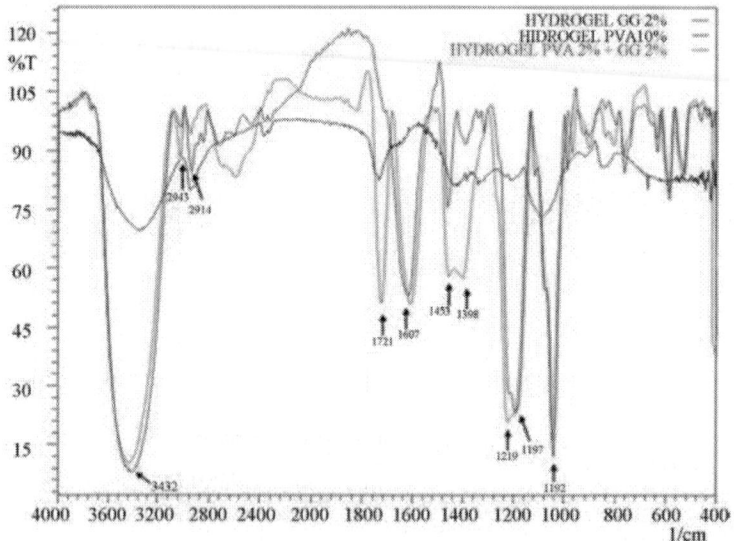

(a)

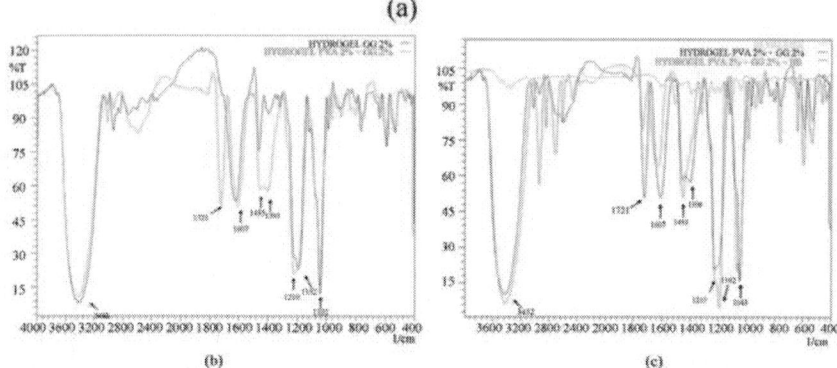

(b)   (c)

Figure 4. FTIR spectra of all hydrogel composite: (a) GG2%; PVA10%; GG2%PVA2%; (b) GG2% , GG2%PVA2%; and (c) BS; GG2%PVA2%; BS-GG2%PVA2%cys.

In Figure 6 is displayed the normalized TGA measurements of all components used in the synthesis biodegradable hydrogel composite. The evaporation process occurs more drastically in the GG2%-(23.0%) and the other component were respectively PVA10%-(5.02%); PVAGG2%-(12.1%); BS-GG2%PVA2%cys-(10.2%) and AuNP-BS-GG2%PVA2%cys-(12.1%). These results indicated that GG2% hydrogel lose more water than PVA10%, also when it is incorporated to PVA and the biodegradable hydrogel loaded with BS, this water lost is reduced indicating that the structure is modified entrapping the water inside the system. The percentage weight loss and the start temperature for all formulation were GG2% (71.5% at 214°C); PVA10%-(90% at 100°C); PVAGG2%-(81.4% at 225°C); BS-GG2%PVA2%cys-(81.2% at 225°C) and AuNP-BS-GG2%PVA2%cys-(80% at 215°C) were observed for all system These results indicate that the GG2% has less mass to lose since its structure is less porous than the network achieved with PVAGG2% BS-GG2%PVA2%cys and AuNP-BS- GG2%PVA2%cys.

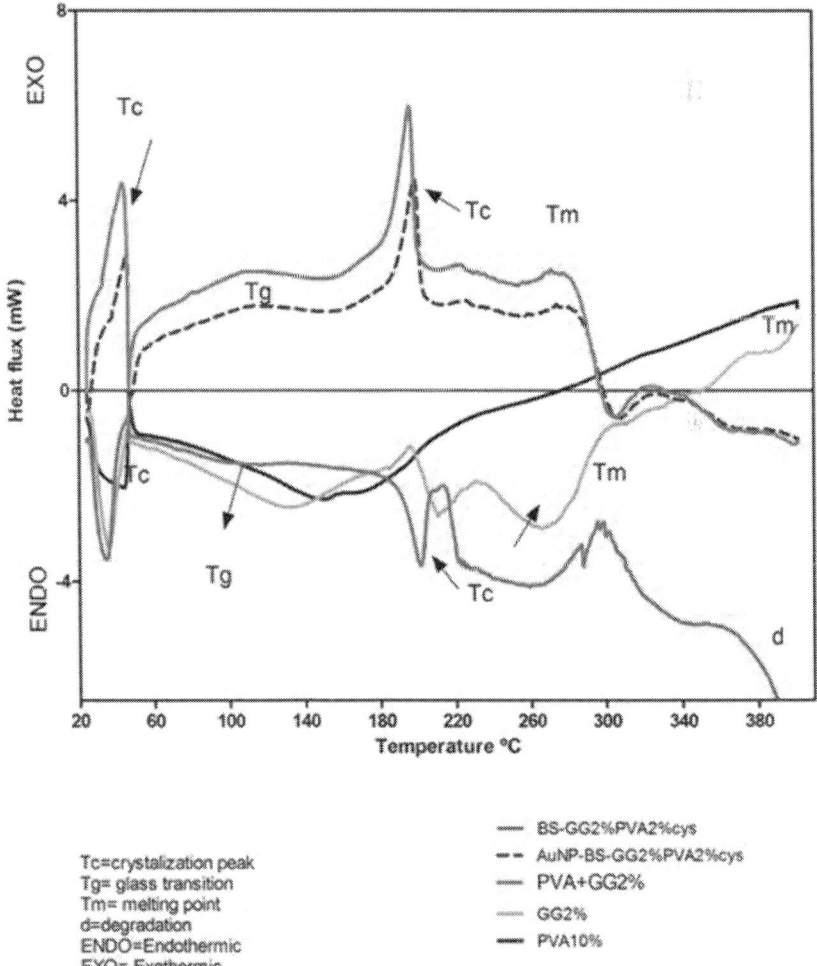

Tc=crystalization peak
Tg= glass transition
Tm= melting point
d=degradation
ENDO=Endothermic
EXO= Exothermic

—— BS-GG2%PVA2%cys
-- AuNP-BS-GG2%PVA2%cys
—— PVA+GG2%
—— GG2%
—— PVA10%

Figure 5. DSC spectra of hydrogel composite: PVA 10%; GG2 %; GG2%PVA2%cys; BS-GG2%PVA2%cys and AuNP-BS-GG2%PVA2%cys.

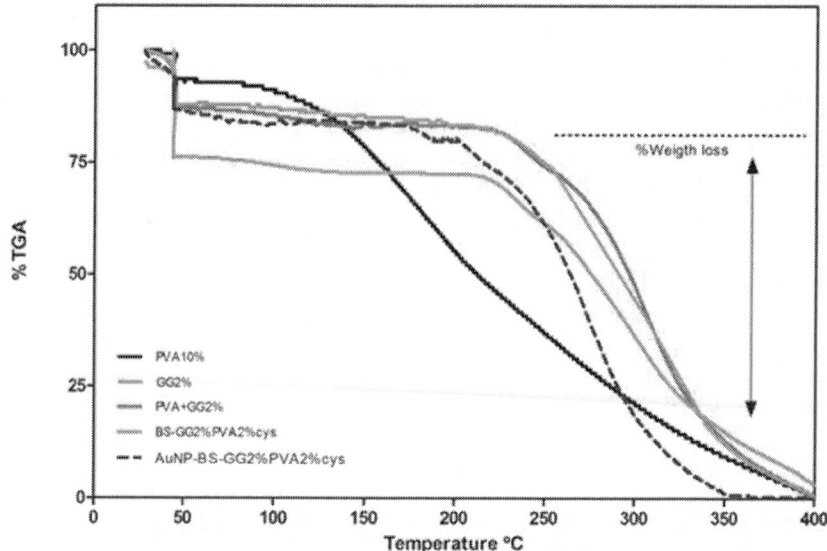

Figure 6. TGA spectra of hydrogel composite: PVA 10%; GG2 %; GG2% PVA2%cys; BS-GG2%PVA2%cys and AuNP-BS-GG2%PVA2%cys.

4. CONCLUSIONS

Polymeric drug delivery system allows the development of innovative concepts that could create more efficient systems to achieve a prolonged and controlled drug delivery. In this study, by the use of the crosslink method, we have prepared several hydrogel composite by varying amount of GG and PVA loaded with BS and its association to gold nanoparticles, and then characterized its water holding capacities, time course release profile; physical-chemical behavior (FTIR, DSC; TGA) and surface analysis (SEM).

The results have demonstrated distinct surface profile of the hydrogel composites and two different time course profiles, which could explain by the feature of hydrogel matrix.

Among all biodegradable hydrogel composites evaluated it was clear that the addition of GG associated to PVA and association to gold nanoparticle, produced the most suitable drug delivery system for BS, due to all characteristics (higher entrapment efficiency) of the matrix structure, making a promising drug delivery system for delivery hydrophilic drugs. Future application of gold nanoparticles (AuNPs) added into the transparent loaded hydrogel can be a promising efficient drug delivery system in skin, by enabling the cell activating capacity once its size is only 1/2000 as large as pores, very easy to penetrate into the dermis. It can carry the drug such as BS deep into the skin to activate skin cells, increases vasodilatation and can enhance the compactness of the fibrous tissue of collagen to turn the skin compact and elastic against aging and other process.

ACKNOWLEDGEMENTS

This material is based upon work supported by the CNPq, FAPDF, FINATEC, DPP-UnB, CAPES.

REFERENCES

1. Huang, Z.J., Zeng, Y., Lan, P., Sun, P.H. and Chen, W.M. (2011) Advances in Structural Modifications and Biological Activities of Berberine: An Active Compound in Traditional Chinese Medicine. Mini-Reviews in Medicinal Chemistry, 11, 1122-1129.
2. Tillhon, M., Guamán Ortiz, L.M., Lombardi, P. and Scovassi, A.I. (2012) Berberine: New Perspectives for Old Remedies. Biochemical Pharmacologyhem Pharmacol, 84, 1260-1267.
3. Xie, W. and Du, L. (2011) Diabetes Is an Inflammatory Disease: Evidence from Traditional Chinese Medicines. Diabetes, Obesity and Metabolism, 13, 289-301.
4. Wang, X.H., Jiang, S.M. and Sun, Q.W. (2011) Effects of Berberine on Human Rheumatoid Arthritis Fibroblast-Like Synoviocytes. Experimental Biology and Medicine (Maywood), 236, 859-866.
5. Chen, X.W., Di, Y.M., Zhang, J., Zhou, Z.W., Li, C.G. and Zhou, S.F. (2012) Interaction of Herbal Compounds with Biological Targets: A Case Study with Berberine. Scientific World Journal, 2012, Article ID: 708292.
6. Chen, Y., Wang, Y., Zhang, J., Sun, C. and Lopez, A. (2011) Berberine Improves Glucose Homeostasis in Streptozotocin-Induced Diabetic Rats in Association with Multiple Factors of Insulin Resistance. ISRN Endocrinology, 2011, Article ID: 519371.
7. Chidambara Murthy, K.N., Jayaprakasha, G.K. and Patil, B.S. (2012) The Natural Alkaloid Berberine Targets Multiple Pathways to Induce Cell Death in Cultured Human Colon Cancer Cells. European Journal of Pharmacology, 688, 14- 21.
8. Tan, W., Li, Y., Chen, M. and Wang, Y. (2011) Berberine Hydrochloride: Anticancer Activity and Nanoparticulate Delivery System. International Journal of Nanomedicine, 6, 1773-1777.
9. Trimarco, V., Cimmino, C.S., Santoro, M., Pagnano, G., Manzi, M.V., Piglia, A., et al. (2012) Nutraceuticals for Blood Pressure Control in Patients with High-Normal or Grade 1 Hypertension. High Blood Pressure & Cardiovascular Prevention, 19, 117-122.
10. Wang, J., Liu, Q. and Yang, Q. (2012) Radiosensitization Effects of Berberine on Human Breast Cancer Cells. International Journal of Molecular Medicine, 30, 1166-1172.

11. Wang, Y., Huang, Y., Lam, K.S., Li, Y., Wong, W.T., Ye, H., et al. (2009) Berberine Prevents Hyperglycemia-Induced Endothelial Injury and Enhances Vasodilatation via Adenosine Monophosphate-Activated Protein Kinase and Endothelial Nitric Oxide Synthase. Cardiovascular Research, 82, 484-492.

12. Zhou, L., Wang, X., Yang, Y., Wu, L., Li, F., Zhang, R., et al. (2011) Berberine Attenuates cAMP-Induced Lipolysis via Reducing the Inhibition of Phosphodiesterase in 3T3-L1 Adipocytes. Biochimica et Biophysica Acta (BBA)—Mole- cular Basis of Disease, 1812, 527-535.

13. Minhas, M.U., Ahmad, M., Ali, L. and Sohail, M. (2013) Synthesis of Chemically Cross-Linked Polyvinyl Alcohol-co- Poly (methacrylic acid) Hydrogels by Copolymerization; A Potential Graft-Polymeric Carrier for Oral Delivery of 5- Fluorouracil. DARU Journal of Pharmaceutical Sciences, 21.

14. Zhang, X.Z., Zhuo, R.X., Cui, J.Z. and Zhang, J.T. (2002) A Novel Thermo-Responsive Drug Delivery System with Positive Controlled Release. International Journal of Pharmaceutics, 235, 43-50.

15. Liu, L., Wang, B., Gao, Y. and Bai, T.C. (2013) Chitosan Fibers Enhanced Gellan Gum Hydrogels with Superior Mechanical Properties and Water-Holding Capacity. Carbohydrate Polymers, 97, 152-158.

16. Moura, M.J., Faneca, H., Lima, M.P., Gil, M.H. and Figueiredo, M.M. (2011) In Situ Forming Chitosan Hydrogels Prepared via Ionic/Covalent Co-Cross-Linking. Biomacromolecules, 12, 3275-3284.

17. Cencetti, C., Bellini, D., Pavesio, A., Senigaglia, D., Passariello, C., Virga, A., et al. (2012) Preparation and Characterization of Antimicrobial Wound Dressings Based on Silver, Gellan, PVA and Borax. Carbohydrate Polymers, 90, 1362-1370.

18. Cobley, C.M., Chen, J., Cho, E.C., Wang, L.V. and Xia, Y. (2011) Gold Nanostructures: A Class of Multifunctional Materials for Biomedical Applications. Chemical Society Reviews, 40, 44-56.

19. Bedford, E.E., Spadavecchia, J., Pradier, C.M. and Gu, F.X. (2012) Surface Plasmon Resonance Biosensors Incorporating Gold Nanoparticles. Macromolecular Bioscience, 12, 724-739.

20. Cristallini, C., Guerra, G.D., Barbani, N. and Bianchi, F. (2007) Biodegradable Bioartificial Materials Made with Chitosan and Poly(vinyl alcohol). Part I: Physicochemical Characterization. Journal of Applied Biomaterials & Biomechanics, 5, 184-191.

21. Cascone, M.G., Barbani, N., Cristallini, C., Giusti, P., Ciardelli, G. and Lazzeri, L. (2001) Bioartificial Polymeric Materials Based on Polysaccharides. Journal of Biomaterials Science, Polymer Edition, 12, 267-281.

22. Oliveira, J.T., Martins, L., Picciochi, R., Malafaya, P.B., Sousa, R.A., Neves, N.M., et al. (2010) Gellan Gum: A New Biomaterial for Cartilage Tissue Engineering Applications. Journal of Biomedical Materials Research Part A, 93, 852- 863.

23. Kimling, J., Maier, M., Okenve, B., Kotaidis, V., Ballot, H. and Plech, A. (2006) Turkevich Method for Gold Nanoparticle Synthesis Revisited. Journal of Physical Chemistry B, 110, 15700-15707.

24. Cai, X., Shao, W., Luan, Y., Pang, J., Li, F. and Li, Z. (2011) Metformin Hydrochloride-Loaded Poly(vinyl alcohol) Composites as Drug Delivery Systems. Journal of Nanoscience and Nanotechnology, 11, 8621-8627.

25. Chen, C.H., Wang, F.Y., Mao, C.F., Liao, W.T. and Hsieh, C.D. (2008) Studies of Chitosan: II. Preparation and Characterization of Chitosan/Poly(vinyl alcohol)/Gelatin Ternary Blend Films. International Journal of Biological Macromolecules, 43, 37-42.

26. Lima, A.C., Sher, P. and Mano, J.F. (2012) Production Methodologies of Polymeric and Hydrogel Particles for Drug Delivery Applications. Expert Opinion on Drug Delivery, 9, 231-248.

27. Hu, Y., Wang, Q. and Tang, M. (2013) Preparation and Properties of Starch-g-PLA/Poly(vinyl alcohol) Composite Film. Carbohydrate Polymers, 96, 384-388.

28. Krauland, A.H., Leitner, V.M. and Bernkop-Schnürch, A. (2003) Improvement in the in Situ Gelling Properties of Deacetylated Gellan Gum by the Immobilization of Thiol Groups. Journal of Pharmaceutical Sciences, 92, 1234-1241.

29. Bashmakova, N., Kutovyy, S., Kornienko, M., Yashchuk, V., Hovorun, D. and Zhurakivsky, R. (2010) Experimental and Calculated by the DFT Method Vibration Spectra of Berberine. Xxii International Conference on Raman Spectroscopy, 1267, 426-427.

30. Ahuja, M., Singh, S. and Kumar, A. (2013) Evaluation of Carboxymethyl Gellan Gum as a Mucoadhesive Polymer. International Journal of Biological Macromolecules, 53, 114-121.

31. Coutinho, D.F., Sant, S.V., Shin, H., Oliveira, J.T., Gomes, M.E., Neves, N.M., et al. (2010) Modified Gellan Gum Hydrogels with Tunable Physical and Mechanical Properties. Biomaterials, 31, 7494-7502.

32. Liu, Y., Geever, L.M., Kennedy, J.E., Higginbotham, C.L., Cahill, P.A. and McGuinness, G.B. (2010) Thermal Behavior and Mechanical Properties of Physically Crosslinked PVA/Gelatin Hydrogels. Journal of the Mechanical Behavior of Biomedical Materials, 3, 203-209.

33. Peng, Z., Kong, L.X., Li, S.D. and Spiridonov, P. (2006) Poly(vinyl alcohol)/Silica Nanocomposites: Morphology and Thermal Degradation Kinetics. Journal of Nanoscience and Nanotechnology, 6, 3934-3938.

34. Wang, H.Y., Lu, S.S. and Lun, Z.R. (2009) Glass Transition Behavior of the Vitrification Solutions Containing Propanediol, Dimethyl Sulfoxide and Polyvinyl Alcohol. Cryobiology, 58, 115-117.

CHAPTER 7

Increasing Possibilities of Nanosuspension

Kumar Bishwajit Sutradhar, Sabera Khatun, and Irin Parven Luna

Department of Pharmacy, Stamford University Bangladesh, 51 Siddeswari Road, Dhaka 1217, Bangladesh

ABSTRACT

Nowadays, a very large proportion of new drug candidates emerging from drug discovery programmes are water insoluble and thus poorly bioavailable. To avoid this problem, nanotechnology for drug delivery has gained much interest as a way to improve the solubility problems. Nano refers to particles size range of 1–1000 nm. The reduction of drug particles into the submicron range leads to a significant increase in the dissolution rate and therefore enhances bioavailability. Nanosuspensions are part of nanotechnology. This interacts with the body at subcellular (i.e., molecular) scales with a high degree of specificity and can be potentially translated into targeted cellular and tissue-specific clinical applications designed to achieve maximal therapeutic efficacy with minimal side effects. Production of drugs as nanosuspensions can be developed for drug delivery systems as an oral formulation and nonoral administration. Here, this review describes the methods of pharmaceutical nanosuspension production including advantages and disadvantages, potential benefits, characterization tests, and pharmaceutical applications in drug delivery.

1. INTRODUCTION

One of the problems facing nanotechnology is the confusion and disagreement among experts about its definition. Nanotechnology is an umbrella term used to define the products, processes, and properties at the nano-microscale that have resulted from the convergence of the physical, chemical, and life sciences. The National Nanotechnology Initiative (NNI) defines, "Nanotechnology as research and development at the atomic, molecular, or macromolecular levels in the sub-100-nm range (w0.1–100 nm) to create structures, devices, and systems that have novel functional properties" [1]. A complete list of the potential applications of nanotechnology is too vast and diverse to discuss in detail, but without doubt, one of the greatest values of nanotechnology will be in the development of new and effective medical treatments [2]. In few words, nanotechnology can be said as "the technology at nanoscale" [3].

2. NANOMEDICINE

Burgeoning interest in the medical applications of nanotechnology has led to the emergence of a new field called nanomedicine, which involves the use of nanotechnology in drug development and offers ever more exciting promises of new diagnoses and cures [4]. It has been defined as "the monitoring, repair, construction, and control of human biological systems at the molecular level, using engineered nanodevices and nanostructures." Therefore, nanomedicine adopts the concepts of nanoscale manipulation and assembly to applications at the clinical level of medical sciences [1]. Most broadly, nanomedicine is the process of diagnosing, treating, preventing disease and traumatic injury, relieving pain, and preserving and improving human health, using molecular tools and molecular knowledge of the human body. In short, nanomedicine is the application of nanotechnology to medicine [5].

Applications of nanotechnology in medicine are potentially enormous. It is recognized that as particles get smaller, the surface area increases with a greater proportion of atoms/molecules found at the surface compared to those inside [4].

2.1. Nanosuspension

A nanosuspension is a submicron colloidal dispersion of drug particles. A pharmaceutical nanosuspension is defined as very finely colloid, biphasic, dispersed, and solid drug particles in an aqueous vehicle, size below 1 μm, without any matrix material, stabilized by surfactants and polymers, and prepared by suitable methods for drug delivery applications, through various routes of administration like oral, topical, parenteral, ocular and pulmonary routes [6]. The particle-size distribution of the solid particles in nanosuspensions is usually less than one micron with an average particle size ranging between 200 and 600 nm [7]. A nanosuspension not only solves the problem of poor solubility and bioavailability but also alters the pharmacokinetics of drug and that improves drug safety and efficacy. In case of drugs that are insoluble in both water and in organic media instead of using lipidic systems, nanosuspensions are used as a formulation approach. Nanosuspension formulation approach is most suitable for the compounds with high log P value, high melting point, and high dose. The use of nanotechnology to formulate poorly water-soluble drugs as nanosuspension offers the opportunity to address nature of the deficiency associated with this class of drugs. Nanosuspension has been reported to enhance absorption and bioavailability; it may help to reduce the dose of the conventional oral dosage forms. Drug particle size reduction leads to an increase in surface area and consequently in the rate of dissolution as described by the Nernst-Brunner and Levich modification of the Noyes-Whitney equation. In addition, an increase in saturation solubility is postulated by particle size reduction due to an increased dissolution pressure explained by the Ostwald-Freundlich equation. An increasing amount of amorphous drug fraction could induce higher saturation solubility. In nanosuspension technology, the drug is maintained in the required crystalline state with reduced particle size, leading to an increased dissolution rate and therefore improved bioavailability.

Drugs encapsulated within nanosuspensions exist in pharmaceutically acceptable crystalline or amorphous state. Nanosuspensions can successfully formulate the brick dust molecules for improved dissolution and good absorption [6].

2.2. Potential Benefits of Nanosuspension Technology for Poorly Soluble Drugs [6, 8–10]

We have the following.(1)Reduced particle size, increased drug dissolution rate, increased rate and extent of absorption, increased bioavailability of drug, area under plasma versus time curve, onset time, peak drug level, reduced variability, and reduced fed/fasted effects. Due to the particle size reduction, the penetration capability of topical nanosuspension preparations increases significantly (Figure 1).(2)Nanosuspensions can be used for compounds that are water insoluble but which are soluble in oil. On the other hand, nanosuspensions can be used in contrast with lipidic systems, and successfully formulate compounds that are insoluble in both water and oils.(3)Nanoparticles can adhere to the gastrointestinal mucosa, prolonging the contact time of the drug and thereby enhancing its absorption.(4)A pronounced advantage of nanosuspension is that there are many administration routes for nanosuspensions, such as oral, parenteral, pulmonary, dermal and ocular.(5)Nanosuspension of nanoparticles (NPs) offers various advantages over conventional ocular dosage forms, including reduction in the amount of dose, maintenance of drug release over a prolonged period of time, reduction in systemic toxicity of drug, enhanced drug absorption due to longer residence time of nanoparticles on the corneal surface, higher drug concentrations in the infected tissue, and suitability for poorly water-soluble drugs, and smaller particles are better tolerated by patients than larger particles; therefore, nanoparticles may represent auspicious drug carriers for ophthalmic applications.(6)Nanosuspension has low incidence of side effects by the excipients.(7)Nanosuspensions overcome delivery issues for the compounds by obviating the need to dissolve them and by maintaining the drug in a preferred crystalline state of size sufficiently small for pharmaceutical acceptability.(8)Increased resistance to hydrolysis and oxidation and increased physical stability to settling.(9)Reduced administration volumes, essential for intramuscular, subcutaneous, and ophthalmic use.(10)Finally, nanosuspensions can provide the passive targeting.

2.3. Ingredients Used in the Formulation of Nanosuspension

We have the following: stabilizer; organic solvents; cosurfactants; other additives like buffers, salts, polyols, osmogents, and cryoprotectants [11].

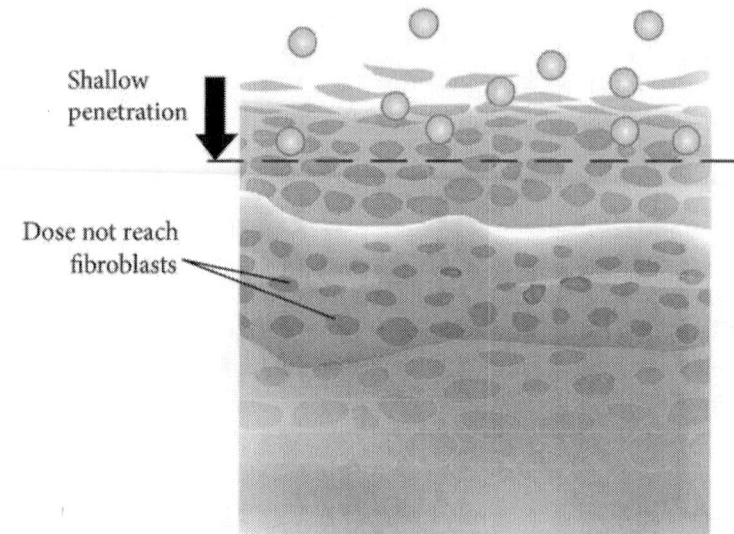

Shallow penetration

Dose not reach fibroblasts

Large, conventional topical preparation

Figure 1. Penetration capability of nanosuspension over conventional topical preparation.

3. PREPARATION METHODS OF NANOSUSPENSIONS

For manufacturing nanosuspensions, there are two converse methods "bottom-up" and the "top-down" technologies. Conventional methods of precipitation are called "bottom-up technology". The "top-down technologies" are the disintegration methods and are preferred over the precipitation methods. These include media milling (Nanocrystals), high-pressure homogenization in water (Dissocubes), high-pressure homogenization in nonaqueous media (Nanopure) and combination of precipitation and high-pressure homogenization (Nanoedge).

Techniques like emulsion as templates and microemulsion as templates are also used for preparing nanosuspensions [12].

3.1. Different Methods of Nanosuspension Preparation [12]

See Figure 2.

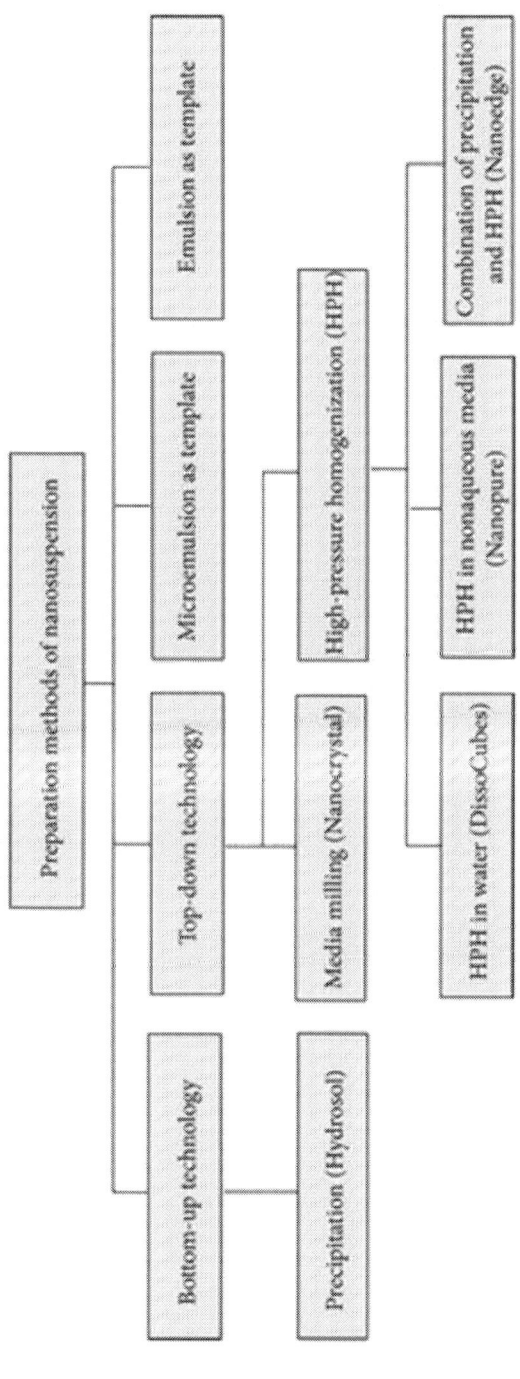

Figure 2. Schematic diagram of preparation methods of nanosuspension

3.2. Media Milling (Nanocrystals)

This method was discovered by Liversidge et al. in 1992 and first patented by "Nanosystems" group, and now this patent transferred to "Elan drug delivery."

Here, the particle size is reduced by the high shear rate. And the total process is performed under controlled temperature. Otherwise, at high shear rate, some temperature will build up which will degrade some of the ingredients in the dosage form.

This equipment is known as high shear media milling or pearl mills (Figure 3).

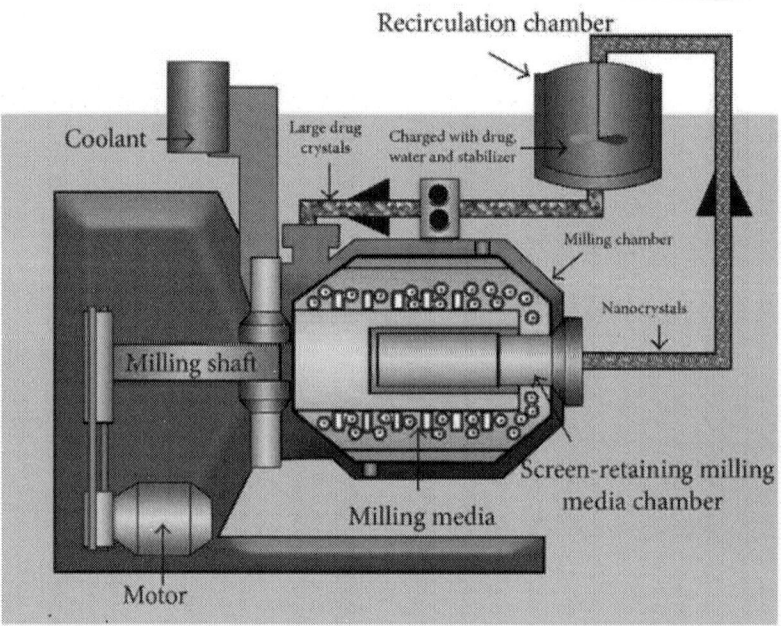

Figure 3. Schematic representation of the media milling process [11].

This mill consists of three major columns:(a)milling chamber;(b)milling shaft;(c)recirculation chamber.

3.2.1. Principle
Here, the main principle involved in the size reduction is "impaction". By this shear, the microparticles are braked down into nanoparticles. And it is connected to the recirculating chamber so that continuous production will be carried out. It is suitable for both batch operation and continuous operation. By this, we can reduce the particle size up to <200 nm in 30–60 min only [13].

3.3. High-Pressure Homogenization (Dissocubes)

The process was developed by R. H. Muller, and first patent was taken by DDS Gmbh. Later, patent was transferred to Skype pharmaceuticals.

Commonly used homogenizer is the APV Micron Lab 40 (APV Deutschland Gmbh, Lubeck, Germany). And another type is piston-gap homogenizers. And it is manufactured by Avestin (Avestin Inc., Ottawa, Canada). And another one is Stansted (Stansted Fluid Power Ltd. Stansted, UK).

The main principle is high pressure that is 100–1500 bars. By this pressure we can easily convert the micron size particle, into nanosize particle. And it initially needs the micron range particle that is <25 micrometer, so that we have to get the sample from the jet mill because by using jet mill we can reduce the particle size up to <25 micrometer (Figure 4).

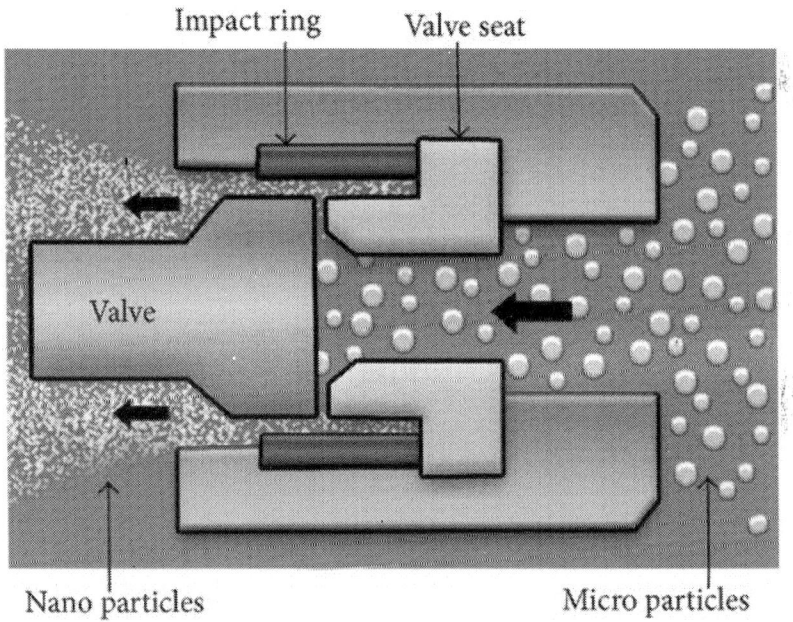

Figure 4. Schematic representation of the high-pressure homogenization process [11].

And we can use this equipment for both batch and continuous operations. Capacity is also 40 mL to thousand litres. Here, first, we have to convert the particles into presuspension form (after jet milling).

3.3.1. Principle
High shear and high pressure are due to particle collisions; the particle size will be reduced. Here, we have to add viscosity enhancers to increase the viscosity of nanosuspension. In this methods we have to mainly concentrate on two

parameter called pressure and homogenization cycles (depending on particle hardness analyzed by particle size and polydispersibility index) [13].

3.4. Emulsion as Template [13]

These emulsions are also useful for the preparation of nanosuspensions. The drugs which were insoluble in volatile organic solvents or partially soluble in water are prepared by this method.

This method is done by two types: as shown in Figure 5.

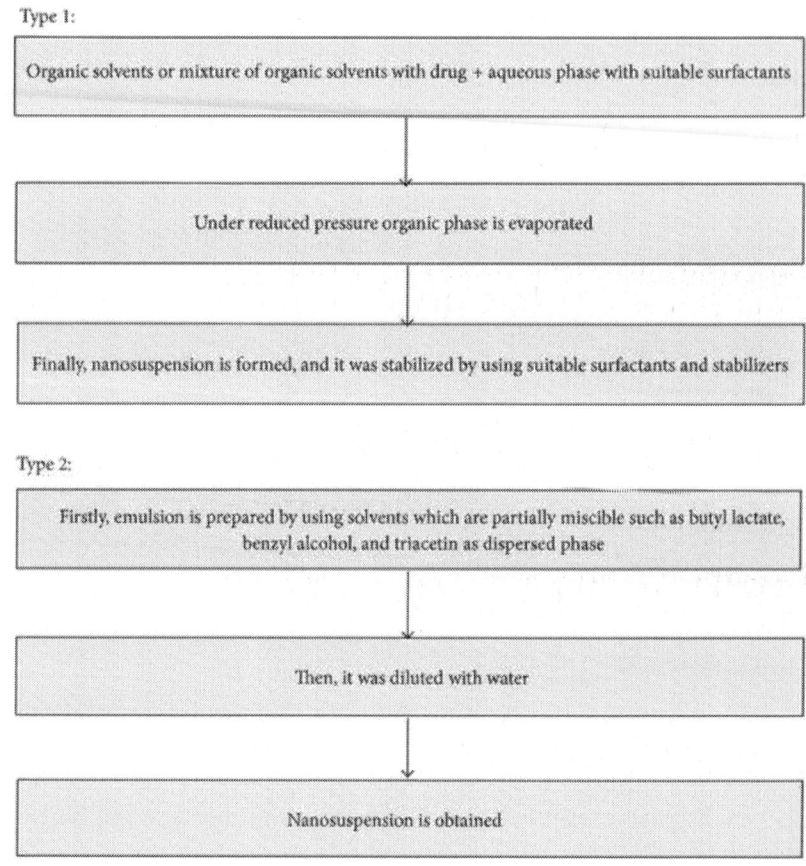

Type 1:

Organic solvents or mixture of organic solvents with drug + aqueous phase with suitable surfactants

Under reduced pressure organic phase is evaporated

Finally, nanosuspension is formed, and it was stabilized by using suitable surfactants and stabilizers

Type 2:

Firstly, emulsion is prepared by using solvents which are partially miscible such as butyl lactate, benzyl alcohol, and triacetin as dispersed phase

Then, it was diluted with water

Nanosuspension is obtained

Figure 5. Schematic representation of the emulsion as template process.

3.5. Microemulsion as Template

Microemulsions are thermodynamically stable and isotropically clear dispersion of the two immiscible liquids such as oil and water, and they were stabilized by an interfacial film of surfactant and cosurfactant. In this, firstly, the microemulsion was prepared the dug solution was mixed to that prepared emulsion and drug loading efficiency was tested [13].

3.6. Precipitation Method

Precipitation has been applied for years to prepare submicron particles within the last decade, especially for the poorly soluble drugs. Typically, the drug is firstly dissolved in a solvent. Then, this solution is mixed with a miscible antisolvent in the presence of surfactants. Rapid addition of a drug solution to the antisolvent (usually water) leads to sudden supersaturation of drug in the mixed solution and generation of ultrafine crystalline or amorphous drug solids (Figure 6). This process involves two phases: nuclei formation and crystal growth. When preparing a stable suspension with the minimum particle size, a high nucleation rate but low growth rate is necessary. Both rates are dependent on temperature: the optimum temperature for nucleation might lie below that for crystal growth, which permits temperature optimization [6].

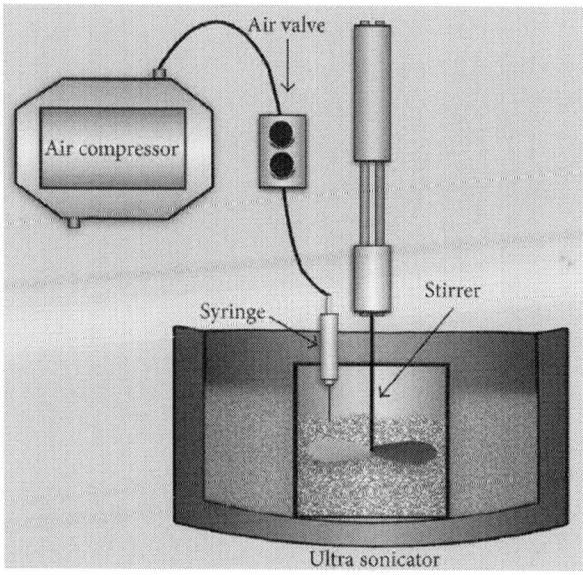

Figure 6: Schematic representation of the precipitation method [14].

3.7. Dry Cogrinding

Recently, nanosuspensions can be obtained by dry milling techniques. Successful work in preparing stable nanosuspensions using dry grinding of poorly soluble drugs with soluble polymers and copolymers after dispersing in a liquid media has been reported [15]. Colloidal particle formation of many poorly water soluble drugs like like griseofulvin, glibenclamide and nifedipine obtained by grinding with polyvinylpyrrolidone (PVP) and sodium dodecylsulfate (SDS)

[16]. Many soluble polymers and copolymers such as PVP, polyethylene glycol (PEG), hydroxypropyl methylcellulose (HPMC), and cyclodextrin derivatives have been used. Physicochemical properties and dissolution of poorly water-soluble drugs were improved by cogrinding because of an improvement in the surface polarity and transformation from a crystalline to an amorphous drug. Dry cogrinding can be carried out easily and economically and can be conducted without organic solvents. The cogrinding technique can reduce particles to the submicron level and a stable amorphous solid can be obtained [12].

3.8. Some Other Methods for Nanosuspension Preparation [19, 20]

We have the following.
(i) Laser fragmentation.
(ii) Nanojet technology.
(iii) Emulsion solvent diffusion method.
(iv) Melt emulsification method.
(v) Supercritical fluid method.

Table 1 describes different preparation methods of nanosuspension technologies with their advantages and disadvantages.

4. CHARACTERIZATION TESTS [8]

Chemical
(i). Active ingredient
(ii). Degradation products
(iii). Moisture (for lyophilized and solid dosage forms)
(iv). Preservatives
(v). pH.

Physical
(i). Particle-size distribution
(ii). Particle-size distribution in response to accelerated ageing and shipping (freeze/thaw, mechanical agitation, and centrifugation)
(iii). Drainability (from sides of container)
(iv). Syringeability, and injectability
(v). Resuspendability
(vi). Dissolution in water or biorelevant medium
(vii).   Compatibility after admixture
(viii).   Zeta potential (electrostatic self-repulsion of particles).

Biological
(i) Sterility
(ii) Pyrogenicity
(iii) In vivo pharmacokinetics.

Table 1. Advantages and disadvantages of different preparation methods of nanosuspension technologies [7, 13, 17, 18]

Preparation methods	Advantages	Disadvantages
Precipitation	(i) Simple process (ii) Economical production (iii) Ease of scale-up	(i) Drug has to be soluble at least in one solvent and that this solvent needs to be miscible with a nonsolvent (ii) Growing of crystals needs to be limited by surfactant addition
High-pressure homogenization	(i) General applicability to most drugs (ii) For dilute and high concentrated nanosuspensions preparation (iii) Simple technique (iv) Sterile products preparation (v) Drugs which belong to BCS CLASS II and IV (vi) Ease of scale-up and little batch-to-batch variation	(i) High number of homogenization cycles (ii) Prerequisite micronized drug particles (iii) Possible contamination of product could occur from metal ions coming off from the wall of the homogenizer (iv) Presuspension is required (v) High number of homogenization cycles
Emulsion/microemulsion template	(i) High drug solubilization (ii) Long shelf life (iii) Large-scale preparation (iv) Low cost (v) Simple manufacturing method (vi) Some organic solvents like ethyl acetate and ethyl formate can be used	(i) Used organic solvents are much unsuitable as human health cost (ii) Use of high amount of surfactant and stabilizers
Media milling	(i) Little batch-to-batch variation (ii) Ease of handling large quantities of drugs	(i) Generation of residue of milling media (ii) Time-consuming process (iii) Prolonged milling may induce the formation of amorphous leading to instability (iv) Scale-up is not easy due to mill size and weight
Dry cogrinding	(i) Easy process (ii) No organic solvent (iii) Requiring short grinding time	Generation of residue of milling media

The essential characterization parameters for nanosuspensions are as follows.

4.1. Mean Particle Size and Particle-Size Distribution

The mean particle size and the width of particle-size distribution are important characterization parameters as they govern the saturation solubility, dissolution velocity, physical stability and even biological performance of nanosuspensions. It has been indicated by Müller and Peters (1998) that saturation solubility and dissolution velocity show considerable variation with the changing particle size of the drug [21]. Photon correlation spectroscopy (PCS) (B. W. Müller and R. H. Müller, 1984) can be used for rapid and accurate determination of the mean particle diameter of nanosuspensions [22].

Moreover, PCS can even be used for determining the width of the particle-size distribution (polydispersity index (PI)). The PI is an important parameter that governs the physical stability of nanosuspensions and should be as low as possible for the long-term stability of nanosuspensions. A PI value of 0.1–0.25 indicates a fairly narrow size distribution whereas a PI value greater than 0.5 indicates a very broad distribution. No logarithmic normal distribution can definitely be attributed to such a high PI value. Although PCS is a versatile technique, because of its low measuring range (3 nm to 3 μm) it becomes difficult to determine the possibility of contamination of the nanosuspension by microparticulate drugs (having particle size greater than 3 μm). Hence, in addition to PCS analysis, laser diffractometry (LD) analysis of nanosuspensions should be carried out in order to detect as well as quantify the drug microparticles that might have been generated during the production process. Laser diffractometry yields a volume size distribution and can be used to measure particles ranging from 0.05 to 80 μm, and in certain instruments, particle sizes up to 2000 μm can be measured. The typical LD characterization includes determination of diameter 50% LD (50) and diameter 99% LD (99) values, which indicate that either 50 or 99% of the particles are below the indicated size. The LD analysis becomes critical for nanosuspensions that are meant for parenteral and pulmonary delivery. Even if the nanosuspension contains a small number of particles greater than 5-6 μm, there could be a possibility of capillary blockade or emboli formation, as the size of the smallest blood capillary is 5-6 μm. It should be noted that the particle size data of a nanosuspension obtained by LD and PCS analysis are not identical as LD data are volume based and the PCS mean diameter is the light intensity weighted size. The PCS mean diameter and the 50 or 99% diameter from the LD analyses are likely to differ, with LD data generally exhibiting higher values. The nanosuspensions can be suitably diluted with deionized water before carrying out PCS or LD analysis. For nanosuspensions that are intended for intravenous administration, particle-size analysis by the Coulter counter technique is essential in addition to PCS and LD analysis. Since the Coulter counter gives the absolute number of particles per volume unit for the different size classes, it is a more efficient and appropriate technique than LD analysis for quantifying the contamination of nanosuspensions by microparticulate drugs [11].

4.2. Crystalline State and Particle Morphology

The assessment of the crystalline state and particle morphology together helps in understanding the polymorphic or morphological changes that a drug might undergo when subjected to nanosizing. Additionally, when nanosuspensions are prepared drug particles in an amorphous state are likely to be generated. Hence, it is essential to investigate the extent of amorphous drug nanoparticles generated during the production of nanosuspensions. The changes in the physical state of thedrug particles as well as the extent of the amorphous fraction can be determined by X-ray diffraction analysis [23, 24] and can be supplemented by differential scanning calorimetry [25]. In order to get an actual idea of particle morphology, scanning electron microscopy is preferred [23].

4.3. Particle Charge (Zeta Potential)

The determination of the zeta potential of a nanosuspension is essential as it gives an idea about the physical stability of the nanosuspension. The zeta potential of a nanosuspension is governed by both the stabilizer and the drug itself. In order to obtain a nanosuspension exhibiting good stability, for an electrostatically stabilized nanosuspension, a minimum zeta potential of 30 mV is required, whereas in the case of a combined electrostatic and steric stabilization, a minimum zeta potential of 20 mV is desirable [26].

4.4. Saturation Solubility and Dissolution Velocity

The determination of the saturation solubility and dissolution velocity is very important as these two parameters together help to anticipate any change in the in vivo performance (blood profiles, plasma peaks, and bioavailability) of the drug. As nanosuspensions are known to improve the saturation solubility of the drug, the determination of the saturation solubility rather than an increase in saturation solubility remains an important investigational parameter. The saturation solubility and dissolution velocity of the drug nanosuspensions in different physiological buffers as well as at different temperatures should be assessed according to methods reported in the pharmacopoeia. The investigation of the dissolution velocity of nanosuspensions reflects the advantages that can be achieved over conventional formulations, especially when designing the sustained release dosage forms based on nanoparticulate drugs [11].

4.5. In Vivo Biological Performance

The establishment of an in vitro/in vivo correlation and the monitoring of the in vivo performance of the drug are an essential part of the study, irrespective of the route and the delivery system employed. It is of the utmost importance in the case of intravenously injected nanosuspensions since the in vivo behavior of the drug depends on the organ distribution, which in turn depends on its surface properties, such as surface hydrophobicity and interactions with plasma proteins.

In fact, the qualitative and quantitative composition of the protein absorption pattern observed after the intravenous injection of nanoparticles is recognized as the essential factor for organ distribution [27–31]. Hence, suitable techniques have to be used in order to evaluate the surface properties and protein interactions to get an idea of in-vivo behavior. Techniques such as hydrophobic interaction chromatography can be used to determine surface hydrophobicity [32], whereas 2D PAGE [27] can be employed for the quantitative and qualitative measurement of protein adsorption after intravenous injection of drug nanosuspensions in animals [11].

5. APPLICATIONS OF NANOSUSPENSION

Applications of nanosuspensions had landmarking history. Among these few applications are given below.

5.1. Oral Drug Delivery

Most of the time the oral route is preferred because it has numerous well-known advantages. Some orally administered antibiotics such as atovaquone and Buparvaquone reflect this problem very well. Nanosizing of such drugs can lead to a dramatic increase in their oral absorption and subsequently bioavailability. The oral administration of naproxen nanoparticles lead to an area under the curve (AUC) (0–24 h) of 97.5 mg-h/L compared with just 44.7 mg-h/L for naprosyn suspensions and 32.7 mg-h/L for anaprox tablets. Oral administration of the gonadotropin inhibitor Danazol as a nanosuspension leads to an absolute bioavailability of 82.3 and the conventional dispersion (Danocrine) only to 5.2%. A nanosuspension of amphotericin B showed a significant improvement in its oral absorption in comparison with the conventional commercial formulation [6].

5.2. Bioavailability Enhancement

The poor oral bioavailability of the drug may be due to poor solubility, poor permeability, or poor stability in the gastrointestinal tract (GIT). Nanosuspensions resolve the problem of poor bioavailability by solving the twin problems of poor solubility and poor permeability across the membrane. Bioavailability of poorly soluble oleanolic acid, a hepatoprotective agent, was improved using a nanosuspension formulation. The therapeutic effect was significantly enhanced, which indicates higher bioavailability. This was due to the faster dissolution (90% in 20 min) of the lyophilized nanosuspension powder when compared with the dissolution from a coarse powder (15% in 20 min) [18].

5.3. Parenteral Drug Delivery

Nanosuspensions can be administered via different parenteral administration routes ranging from intra-articular via Intraperitoneal to intravenous injection. For administration by the parenteral route, the drug either has to be solubilized or has particle/globule size below 5 μm to avoid capillary blockage. In addition, nanosuspensions have been found to increase the efficacy of parenterally administered drugs. Paclitaxel nanosuspensions revealed their superiority over taxol in reducing the median tumour burden [33]. Similarly, aphidicolin, a poorly water soluble new anti parasitic lead molecule, when administered as a nanosuspension resulted in an improvement in EC50 in comparison to DMSO-dissolved drug [34]. Clofazimine nanosuspension, a poorly water-soluble antileprotic drug, revealed an improvement in stability and efficacy over the liposomal clofazimine in M. avium-infected female mice [35]. Rainbow and coworkers reported an intravenous itraconazole nanosuspension enhanced efficacy of antifungal activity relative to a solution formulation in rats [7]. Intrathecal delivery of nanosuspension busulfan to a mouse model of neoplastic meningitis led to a significant increase in survival [36].

5.4. Pulmonary Drug Delivery

Nanosuspensions may prove to be an ideal approach for delivering drugs that exhibit poor solubility in pulmonary secretions. Aqueous nanosuspensions can be nebulized using mechanical or ultrasonic nebulizers for lung delivery. Because of their small size, it is likely that in each aerosol droplet at least one drug particle is contained, leading to a more uniform distribution of the drug in lungs. The nanoparticulate nature of the drug allows the rapid diffusion and dissolution of the drug at the site of action. At the same time, the increased adhesiveness of the drug to mucosal surfaces offers a prolonged residence time for the drug at the absorption site. This ability of nanosuspensions offers quick onset of action initially, and then controlled release of the active moiety is highly beneficial and is required by most pulmonary diseases. Budesonide drug nanoparticles were successfully nebulized using an ultrasonic nebulizer [6].

5.5. Ocular Drug Delivery

Nanosuspensions can prove to be a boon for drugs that exhibit poor solubility in lachrymal fluids [6]. The protective barriers of the eye make drug delivery difficult without tissue damage. Poor drug absorption and penetration of drugs to intraocular tissues limit the delivery of drugs. Use of nanoparticles and nanosuspensions for drug delivery to the intraocular tissues is being developed. One example is cross-linked polymer nanosuspensions of dexamethasone, which show enhanced anti-inflammatory activity in a model of rabbit eye irritation [20].

5.6. Targeted Drug Delivery

Nanosuspensions can be used for targeted delivery as their surface properties and in vivo behavior can easily be altered by changing either the stabilizer or the milieu. The engineering of stealth nanosuspensions (analogous to stealth liposomes) by using various surface coatings for active or passive targeting of the desired site is the future of targeted drug delivery systems [11]. Kayser formulated a nanosuspension of aphidicolin to improve drug targeting against Leishmania-Infected macrophages. He stated that the drug in the conventional form had an effective concentration (EC 50) of 0.16 mcg/mL, whereas the nanosuspension formulation had an enhanced activity with an (EC 50) of 0.003 mcg/mL [61].

Scholer et al. showed an improved drug targeting to the brain in the treatment of toxoplasmic encephalitis in a new murine model infected with Toxoplasma gondii using a nanosuspension formulation of atovaquone [62].

5.7. Mucoadhesion of the Nanoparticles

Nanoparticles orally administered in the form of a suspension diffuse into the liquid media and rapidly encounter the mucosal surface. The direct contact of the particles with the intestinal cells through a bioadhesive phase is the first step before particle absorption. The adhesiveness of the nanosuspensions not only helps to improve bioavailability but also improves targeting of the parasites persisting in the GIT, for example, Cryptosporidium parvum. Mucoadhesive Buparvaquone nanosuspensions, because of their prolonged residence at the infection site, revealed a 10-fold reduction in the infectivity score of Cryptosporidium parvum as compared to the Buparvaquone nanosuspensions without mucoadhesive polymers [6].

5.8. Nanosuspension: Breaking the Barrier of the Skin

Drug nanoparticles can be incorporated into creams and water-free ointments. The nanocrystalline form leads to an increased saturation solubility of the drug in the topical dosage form and thus enhancing the diffusion of the drug into the skin (Figure 1) [7].

5.9. Central Nervous System

Nanosuspensions afford a means of administering increased concentrations of poorly water-soluble drugs to the brain with decreased systemic effects. Significant efficacy has been shown with microparticulate busulfan in mice administered intrathecally. The work has advanced to Phase I in patients afflicted with neoplastic meningitis, administered via an Ommaya reservoir for intraventricular delivery and via lumbar puncture. The drug was well tolerated and resulted in delayed progression of disease. Epidural injection of a 10% butamben suspension for cancer pain was well tolerated in dogs and humans. Future work will probably also involve less invasive routes, utilizing either passive targeting (via PEGylation, as has been done for liposomes) or active targeting to the brain following intravenous administration of nanosuspensions. In these latter publications, it was found that use of the agent Polysorbate 80 in the formulation led to deposition of apolipoprotein E on the nanoparticles, which facilitated brain uptake by receptors on the brain endothelial cells [8].

5.10. Nanosuspension Formulations for Treating Bioweapon-Mediated Diseases

Several concepts of targeting of nanosuspension dosage forms for treatment of bioweapon-mediated diseases have been developed at the Baxter Healthcare Corporation. Alterations of pharmacokinetic profiles of existing antibiotics can lead to enhanced efficacy with reduced side effects. This has been shown for a nanosuspension formulation of the antifungal agent itraconazole. Secondly, viral sanctuaries breed resistance and often include the brain and lymphatics. These may be targeted by loading nanoparticulate drug into macrophages which target these organs, increasing antiviral drug concentration in these typically inaccessible regions. Finally, a strategy for dendritic cell vaccines has been developed for use against bioweapons [63].

Table 2 summarizes successfully manufactured nanosuspension of different drugs for different delivery routes with their manufacturing method and indication.

Table 2. Summary of drug nanosuspensions.

Drug	Drug delivery route	Manufacturing method	Indication
Silybin	Oral, IV	HPH	Human prostate cancer [37, 38]
All-trans retinoic acid (ATRA)	IV, oral, and skin	Modified precipitation method	Antiproliferative drug against tumor [14, 39]
Mitotane	Oral	Emulsion as template	Symptomatic treatment of advanced adrenocortical carcinoma (ACC) [38, 40, 41]
Clofazimine	Intravenous	High-pressure homogenization	Murine *Mycobacterium avium* infection [42]
Cyclosporin A	Inhalation	Antisolvent precipitation	In immunosuppression [43, 44]
Amphotericin B	Ocular	Solvent displacement process	Management of ophthalmic fungal infections [45]
Olmesartan medoxomil	Oral	Media milling	Antihypertensive agent [46]
Simvastatin	Oral	Nanoprecipitation	Lipid-lowering agent [47]
Azithromycin	Oral	HPH	Antimicrobial [38, 41]
Nifedipine	Oral	Dry cogrinding	Treatment of vascular diseases [48–50]
Salbutamol sulfate	Pulmonary inhalation	HPH	Antiasthmatic [51]
Diclofenac	Transdermal	Emulsification	NSAID [52]
Oridonin	IV	HPH	Antitumor [53, 54]
Albendazole	Oral	HPH	Lipophilic anthelmintic drug [55]
Loviride	IV	Milling	Antivirotic [56, 57]
Naproxen	I.P	Precipitation	Analgesic activity [58]
Paclitaxel	IV	HPH	Anticancer [8, 36]
Nevirapine	Parenteral	HPH	Antiretroviral [57]
Ibuprofen	Ocular	Emulsion solvent diffusion method	Ocular anti-inflammatory activity [6, 59]
Megestrol acetate	Oral	Laser fragmentation and media milling	Anorexia and cachexia [19, 60]

6. RECENT TREND OF NANOSUSPENSION

In recent years, nanosuspension technology has been successfully applied to tackle the formulation issues of poorly soluble drugs [64]. Most recently, nanopowders have been used as a delivery system for oral administration to enhance the dissolution rates of poorly soluble drugs. Tween 80/poloxamer 188 stabilised nanosuspension of a hydrophobic antiretroviral drug; loviride was prepared on a laboratory scale by media milling, and sucrose cofreeze-dried nanopowders were obtained [65]. Pulmonary products are essentially feasible. Nanosuspensions can be aerosolized using commercial nebulizers, but no products have been created. The reason may be commercial and not technical. It makes little sense to replace a well-selling product with a nanosuspension simply because pulmonary deposition might be superior. The cost of market introduction is too high. Even with a new molecule, an established routine delivery technology is preferable [66].

In addition, injection of poorly water-soluble nanosuspension drugs is an emerging and rapidly growing field that has drawn increasing attention due to its benefits in reducing toxicity and increasing drug efficacy through elimination of cosolvent in the formulation [38]. The current approaches for parenteral delivery include salt formation, solubilization using cosolvents, micellar solutions, complexation with cyclodextrin and recently liposomes. However, there are limitations on the use of these approaches because of the limitations on their solubilization capacity and parenteral acceptability. In this regard, liposomes are much more tolerable and versatile in terms of parenteral delivery. However, they often suffer from problems such as physical instability, high manufacturing cost, and difficulties in scale-up. Nanosuspensions would be able to solve the problems mentioned above [33].

Some recent studies based on stability have proved that nanosuspension could significantly improve the chemical and photo stability of Quercetin compared with the solution stored in the same conditions. Nanosuspension technology would be an effective route to improve the stabilization of the chemical labile drugs [64]. Tam et al. [67, 68] have recently attempted to achieve stable nanosuspensions via a novel design of flocs structure called "open flocs." Thin film freezing was used to produce BSA nanorods with aspect ratio of approximately 24. These BSA nanorods were found to be highly stable when dispersed into hydrofluoroalkane (HFA) propellant, with no apparent sedimentation observed for 1 year. Due to the high aspect ratio of BSA nanorods and relatively strong attractive van der Waals (vdW) forces at the contact sites between the particles, primary nanorods were locked together rapidly as an open structure upon addition of HFA, inhibiting collapse of the flocs [68].

A nanosuspension of indinavir has been loaded into bone marrow derived macrophages and injected into HIV-1-challenged humanized mice. The targeted delivery system significantly reduced numbers of virus-infected cells in plasma, lymph nodes, spleen, liver, and lung and led to CD4 (+) T-cell protection [69]. Spironolactone (SP) is a mineralocorticoid widely prescribed in pediatric population. It is a poor water-soluble drug characterized by incomplete oral bioavailability, bitter taste, and tendency to destabilize in aqueous media. Regarding the good solubility of Spironolactone in lipid materials, lipid nanoparticles seemed to be an excellent way to overcome these issues [70].

To overcome the problems associated with oral absorption and bioavailability issues, various strategies have been utilized [71], and nanosuspension is emerged as a promising strategy for the efficient delivery of hydrophobic drugs nowadays [11].

7. CONCLUSION

The formulation of poorly soluble drugs has always been a challenging problem faced by pharmaceutical scientists. In this case, nanosuspension formulations can be considered as a promising candidate. Various techniques described in this review alone or in combination can be successfully used to solve the poor bioavailability problem of hydrophobic drugs and drugs which are poorly soluble in aqueous and organic solutions. Nanosuspensions can be administered through oral, parenteral, ophthalmic, pulmonary, and topical routes. It can play a very important role for human betterment as the technology is simple, requires less excipients, and increases dissolution velocity and saturation solubility. By emphasizing this technology, our society will be benefited financially also. Thus, nanosuspension technology is able enough to bring enormous immediate benefits and will revolutionize the research and practice of medicine in the field of pharmacy.

REFERENCES

1. K. J. Morrow, R. Bawa, and C. Wei, "Recent advances in basic and clinical nanomedicine," Medical Clinics of North America, vol. 91, no. 5, pp. 805–843, 2007.
2. S. K. Sahoo, S. Parveen, and J. J. Panda, "The present and future of nanotechnology in human health care," Nanomedicine, vol. 3, no. 1, pp. 20–31, 2007.
3. J. Ramsden, "What is nanotechnology?" in Nanotechnology: An Introduction, pp. 1–14, Elsevier, New York, NY, USA, 2011.
4. V. S. W. Chan, "Nanomedicine: an unresolved regulatory issue," Regulatory Toxicology and Pharmacology, vol. 46, no. 3, pp. 218–224, 2006.
5. R. A. Freitas, "What is nanomedicine?" Nanomedicine, vol. 1, no. 1, pp. 2–9, 2005.
6. C. Prabhakar and K. B. Krishna, "A review on nanosuspensions in drug delivery," International Journal of Pharma and Bio Sciences, vol. 2, no. 1, pp. 549–558, 2011.
7. J. Chingunpituk, "Nanosuspension technology for drug delivery," Walailak Journal of Science & Technolog, vol. 4, no. 2, pp. 139–153, 2007.
8. B. E. Rabinow, "Nanosuspensions in drug delivery," Nature Reviews Drug Discovery, vol. 3, no. 9, pp. 785–796, 2004.
9. P. Liu, X. Rong, J. Laru et al., "Nanosuspensions of poorly soluble drugs: preparation and development by wet milling," International Journal of Pharmaceutics, vol. 411, no. 1-2, pp. 215–222, 2011.

10. H. M. Ibrahim, H. R. Ismail, A. E. A. Lila, et al., "Formulation and optimization of ocular poly-D, L-lactic acid nano drug delivery system of amphotericin-B using box behnken design," International Journal of Pharmacy and Pharmaceutical Sciences, vol. 4, no. 2, pp. 342–349, 2012.

11. V. B. Patravale, A. A. Date, and R. M. Kulkarni, "Nanosuspensions: a promising drug delivery strategy,"Journal of Pharmacy and Pharmacology, vol. 56, no. 7, pp. 827–840, 2004.

12. H. Banavath, K. S. Raju, M. T. Ansari, M. S. Ali, and G. Pattnaik, "Nanosuspension: an attempt to enhance bioavailability of poorly soluble drugs," International Journal of Pharmaceutical Sciences and Research, vol. 1, no. 9, pp. 1–11, 2010.

13. G. A. Reddy and Y. Anilchowdary, "Nanosuspension technology: a review," IJPI's Journal of Pharmaceutics and Cosmetology, vol. 2, no. 8, pp. 47–52, 2012.

14. X. Zhang, Q. Xia, and N. Gu, "Preparation of all-trans retinoic acid nanosuspensions using a modified precipitation method," Drug Development and Industrial Pharmacy, vol. 32, no. 7, pp. 857–863, 2006.

15. A. Wongmekiat, Y. Tozuka, T. Oguchi, and K. Yamamoto, "Formation of fine drug particles by cogrinding with cyclodextrins. I. The use of β-cyclodextrin anhydrate and hydrate," Pharmaceutical Research, vol. 19, no. 12, pp. 1867–1872, 2002.

16. K. Itoh, A. Pongpeerapat, Y. Tozuka, T. Oguchi, and K. Yamamoto, "Nanoparticle formation of poorly water-soluble drugs from ternary ground mixtures with PVP and SDS," Chemical and Pharmaceutical Bulletin, vol. 51, no. 2, pp. 171–174, 2003.

17. G. P. Kumar and K. G. Krishna, "Nanosuspensions: the solution to deliver hydrophobic drugs,"International Journal of Drug Delivery, vol. 3, no. 4, pp. 546–557, 2011.

18. A. Vaghela, M. Jain, H. Limbachiya, and D. P. Bharadia, "Nanosuspension technology," International Journal of Universal Pharmacy and Life Sciences, vol. 2, no. 2, pp. 306–317, 2012.

19. J. P. Sylvestre, M. C. Tang, A. Furtos, G. Leclair, M. Meunier, and J. C. Leroux, "Nanonization of megestrol acetate by laser fragmentation in aqueous milieu," Journal of Controlled Release, vol. 149, no. 3, pp. 273–280, 2011.

20. J. McMillan, E. Batrakova, and H. E. Gendelman, "Cell delivery of therapeutic nanoparticle," in Progress in Molecular Biology and Translational Science, vol. 104, pp. 571–572, Elsevier, New York, NY, USA, 2011.

21. R. H. Müller and K. Peters, "Nanosuspensions for the formulation of poorly soluble drugs I: preparation by a sizereduction technique," International Journal of Pharmaceutics, vol. 160, pp. 229–237, 1998.

22. B. W. Müller and R. H. Müller, "Particle size analysis of latex suspensions and microemulsions by photon correlation spectroscopy," Journal of Pharmaceutical Science, vol. 73, no. 7, pp. 915–918, 1984.

23. R. H. Müller and B. H. L. Böhm, "Emulsions and nanosuspensions for the formulation of poorly soluble drugs," in Nanosuspensions, R. H. Müller, S. Benita, and B. H. L. Böhm, Eds., pp. 149–174, Medpharm Scientific Publishers, Stuttgart, Germany, 1998.

24. R. H. Müller and M. J. Grau, "Increase of dissolution velocity and solubility of poorly water soluble drugs as nanosuspension," in Proceedings of the World Meeting APGI/APV, vol. 2, pp. 623–624, Paris, France, 1998.

25. T. R. Shanthakumar, S. Prakash, R. M. Basavraj, et al., "Comparative pharmacokinetic data of DRF-4367 using nanosuspension and HP_CD formulation. Proceedings of the International Symposium on Advances in Technology and Business Potential of New Drug Delivery Systems," B. V. Patel Educational Trust and B. V. Patel PERD Centre, vol. 5, abstract 55, p. 75, 2004.

26. R. H. Müller and C. Jacobs, "Production and characterization of a budesonide nanosuspension for pulmonary administration," Pharmaceutical Ressearch, vol. 19, pp. 189–194, 2002.

27. T. Blunk, D. F. Hochstrasser, J. C. Sanchez, B. W. Muller, and R. H. Muller, "Colloidal carriers for intravenous drug targeting: plasma protein adsorption patterns on surface-modified latex particles evaluated by two-dimensional polyacrylamide gel electrophoresis," Electrophoresis, vol. 14, no. 12, pp. 1382–1387, 1993.

28. T. Blunk, M. Lück, A. Calvör, et al., "Kinetics of plasma protein adsorption on model particles for controlled drug delivery and drug targeting," European Journal of Pharmaceutics and Biopharmaceutics, vol. 42, pp. 262–268, 1996.

29. M. Lück, W. Schroder, S. Harnisch, et al., "Identification of plasma proteins facilitated by enrichment on particulate surfaces: analysis by two-dimensional electrophoresis and N-terminal microsequencing,"Electrophoresis, vol. 18, pp. 2961–2967, 1997.

30. M. Lück, B. R. Paulke, W. Schröder, T. Blunk, and R. H. Müller, "Analysis of plasma protein adsorption on polymeric nanoparticles with different surface characteristics," Journal of Biomedical Materials Research, vol. 39, no. 3, pp. 478–485, 1997.

31. R. H. Müller, "Differential opsonization: a new approach for the targeting of colloidal drug carriers,"Archiv der Pharmazie, vol. 322, p. 700, 1989.

32. K. H. Wallis and R. H. Müller, "Determination of the surface hydrophobicity of colloidal dispersions by mini-hydrophobic interaction chromatography," Pharmazeutische Industrie, vol. 55, pp. 1124–1128, 1993.

33. E. Merisko-Liversidge, G. G. Liversidge, and E. R. Cooper, "Nanosizing: a formulation approach for poorly-water-soluble compounds," European Journal of Pharmaceutical Sciences, vol. 18, no. 2, pp. 113–120, 2003.

34. O. Kayser, "Nanosuspensions for the formulation of aphidicolin to improve drug targeting effects against Leishmania infected macrophages," International Journal of Pharmaceutics, vol. 196, no. 2, pp. 253–256, 2000.

35. K. Peters, S. Leitzke, J. E. Diederichs et al., "Preparation of a clofazimine nanosuspension for intravenous use and evaluation of its therapeutic efficacy in murine Mycobacterium avium infection,"Journal of Antimicrobial Chemotherapy, vol. 45, no. 1, pp. 77–83, 2000.

36. B. E. Rabinow, "Nanosuspensions for parenteral delivery," in Nanoparticulate Drug Delivery Systems, pp. 33–49, Informa Healthcare, London, UK, 2007.

37. D. Zheng, Y. Wang, D. Zhang et al., "In vitro antitumor activity of silybin nanosuspension in PC-3 cells," Cancer Letters, vol. 307, no. 2, pp. 158–164, 2011.

38. L. Wu, J. Zhang, and W. Watanabe, "Physical and chemical stability of drug nanoparticles," Advanced Drug Delivery Reviews, vol. 63, no. 6, pp. 456–469, 2011.

39. All trans Retinoic Acid, http://www.thehamner.org/docs/pbpk_11/Day3.Exercise1.ATRA.pdf.

40. London Cancer New Drugs Group-APC/DTC Briefing, "Mitotane for the adjuvant treatment of adrenocortical carcinoma," September 2011, http://www.nelm.nhs.uk/en/Download/?file%3DMDs3NjY0MTM7L 3VwbG9hZC9QaXJmZW5pZG9uZV9EZWMgMjAxMS5wZGY_.pdf.

41. X. Pu, J. Sun, M. Li, and Z. He, "Formulation of nanosuspensions as a new approach for the delivery of poorly soluble drugs," Current Nanoscience, vol. 5, no. 4, pp. 417–427, 2009.

42. K. Peters, S. Leitzke, J. E. Diederichs et al., "Preparation of a clofazimine nanosuspension for intravenous use and evaluation of its therapeutic efficacy in murine Mycobacterium avium infection,"Journal of Antimicrobial Chemotherapy, vol. 45, no. 1, pp. 77–83, 2000.

43. J. M. Tam, J. T. McConville, R. O. Williams, and K. P. Johnston, "Amorphous cyclosporin nanodispersions for enhanced pulmonary deposition and dissolution," Journal of Pharmaceutical Sciences, vol. 97, no. 11, pp. 4915–4933, 2008.

44. R. Calne, "Cyclosporine as a milestone in immunosuppression," Transplantation Proceedings, vol. 36, no. 2, pp. 13S–15S, 2004.

45. S. Das and P. K. Suresh, "Nanosuspension: a new vehicle for the improvement of the delivery of drugs to the ocular surface. Application to amphotericin B," Nanomedicine, vol. 7, no. 2, pp. 242–247, 2011.

46. H. P. Thakkar, B. V. Patel, and S. P. Thakkar, "Development and characterization of nanosuspensions of olmesartan medoxomil for bioavailability enhancement," Journal of Pharmacy and Bioallied Sciences, vol. 3, no. 3, pp. 426–434, 2011.

47. V. M. Pandya, J. K. Patel, and D. J. Patel, "Effect of different stabilizer on the formulation of simvastatin nanosuspension prepared by nanoprecipitation technique," Research Journal of Pharmaceutical, Biological and Chemical Sciences, vol. 1, no. 4, pp. 910–917, 2010.

48. L. Zhao, Y. Wei, Y. Yu, and W. Zheng, "Polymer blends used to prepare nifedipine loaded hollow microspheres for a floating-type oral drug delivery system: In vitro evaluation," Archives of Pharmacal Research, vol. 33, pp. 443–450, 2010.

49. G. P. Kumar and K. G. Krishna, "Nanosuspensions: the solution to deliver hydrophobic drugs," International Journal of Drug Delivery, vol. 3, pp. 546–557, 2011.

50. J. Hecq, M. Deleers, D. Fanara, H. Vranckx, and K. Amighi, "Preparation and characterization of nanocrystals for solubility and dissolution rate enhancement of nifedipine," International Journal of Pharmaceutics, vol. 299, no. 1-2, pp. 167–177, 2005.

51. Bhavna, F. J. Ahmad, R. K. Khar, S. Sultana, and A. Bhatnagar, "Techniques to develop and characterize nanosized formulation for salbutamol sulfate," Journal of Materials Science: Materials in Medicine, vol. 20, supplement 1, pp. S71–S76, 2009.

52. H. Piao, N. Kamiya, A. Hirata, T. Fujii, and M. Goto, "A novel solid-in-oil nanosuspension for transdermal delivery of diclofenac sodium," Pharmaceutical Research, vol. 25, no. 4, pp. 896–901, 2008.

53. G. Lei, Z. Dianrui, C. Minghui, Z. Tingting, and W. Shumei, "Preparation and characterization of an oridonin nanosuspension for solubility and dissolution velocity enhancement," Drug Development and Industrial Pharmacy, vol. 33, no. 12, pp. 1332–1339, 2007.

54. X. Qi, D. Zhang, X. Xu, et al., "Oridonin nanosuspension was more effective than free oridonin on G2/M cell cycle arrest and apoptosis in the human pancreatic cancer PANC-1 cell line," International Journal of Nanomedicine, vol. 7, pp. 1793–1804, 2012.

55. R. Ravichandran, "Preparation and characterization of albendazole nanosuspensions for oral delivery," International Journal of Green Nanotechnology, vol. 2, no. 1, pp. B1–B24, 2010.

56. R. Dhanapal and J. V. Ratna, "Nanosuspensions technology in drug delivery-a review," International Journal of Pharmacy Review & Research, vol. 2, no. 1, pp. 46–52, 2012.

57. R. Shegokar, K. K. Singh, and R. H. Müller, "Nevirapine nanosuspension: comparative investigation of production methods," Nanotechnology Development, vol. 1, no. e4, pp. 16–22, 2011.

58. "Enhanced Analgesic activity of Polymeric or Lipidic Nanosuspension of Naproxen,"http://www.aapsj.org/abstracts/AM_2002/AAPS2002-002064.pdf.

59. P. Lakshmi and G. A. Kumar, "Nanosuspension technology: a review," International Journal of Pharmacy and Pharmaceutical Sciences, vol. 2, supplement 4, pp. 35–40, 2010.

60. E. Merisko-Liversidge and G. G. Liversidge, "Nanosizing for oral and parenteral drug delivery: a perspective on formulating poorly-water soluble compounds using wet media milling technology,"Advanced Drug Delivery Reviews, vol. 63, no. 6, pp. 427–440, 2011.

61. O. Kayser, "Nanosuspensions for the formulation of aphidicolin to improve drug targeting effects against Leishmania infected macrophages," International Journal of Pharmaceutics, vol. 196, no. 2, pp. 253–256, 2000.

62. N. Scholer, K. Krause, O. Kayser, et al., "Atovaquone nanosuspensions show excellent therapeutic effect in a new murine model of reactivated toxoplasmosis," Antimicrobial Agents and Chemotherapy, vol. 45, pp. 1771–1779, 2001.

63. K. K. Jain, "Miscellaneous applications," in The Handbook of Nanomedicine, p. 325, Humana Press, New York, NY, USA, 2008.

64. L. Gao, G. Liu, X. Wang, F. Liu, Y. Xu, and J. Ma, "Preparation of a chemically stable quercetin formulation using nanosuspension technology," International Journal of Pharmaceutics, vol. 404, no. 1-2, pp. 231–237, 2011.

65. E. Ojewole, I. Mackraj, P. Naidoo, and T. Govender, "Exploring the use of novel drug delivery systems for antiretroviral drugs," European Journal of Pharmaceutics and Biopharmaceutics, vol. 70, no. 3, pp. 697–710, 2008.

66. R. H. Müller and C. M. Keck., "Twenty years of drug nanocrystals: where are we, and where do we go?"European Journal of Pharmaceutics and Biopharmaceutics, vol. 80, pp. 1–3, 2012.

67. J. M. Tam, J. D. Engstrom, D. Ferrer, R. O. Williams, and K. P. Johnston, "Templated open flocs of anisotropic particles for pulmonary delivery with pressurized metered dose inhalers," Journal of Pharmaceutical Sciences, vol. 99, no. 7, pp. 3150–3165, 2010.

68. J. D. Engstrom, J. M. Tam, M. A. Miller, R. O. Williams, and K. P. Johnston, "Templated open flocs of nanorods for enhanced pulmonary

delivery with pressurized metered dose inhalers," Pharmaceutical Research, vol. 26, no. 1, pp. 101–117, 2009.

69. P. P. Constantinides, M. V. Chaubal, and R. Shorr, "Advances in lipid nanodispersions for parenteral drug delivery and targeting," Advanced Drug Delivery Reviews, vol. 60, no. 6, pp. 757–767, 2008.

70. Z. Bourezg, S. Bourgeois, S. Pressenda, T. Shehada, and H. Fessi, "Redispersible lipid nanoparticles of Spironolactone obtained by three drying methods," Colloids and Surfaces A, vol. 413, no. 5, pp. 191–199, 2012.

71. N. Saffoon, R. Uddin, N. H. Huda, and K. B. Sutradhar, "Enhancement of oral bioavailability and solid dispersion: a review," Journal of Applied Pharmaceutical Science, vol. 1, no. 7, pp. 13–20, 2011.

CHAPTER 8

Nanoparticles for Dermal and Transdermal Drug Delivery

Okoro Uchechi, John D. N. Ogbonna and Anthony A. Attama

Drug Delivery Research Unit, Department of Pharmaceutics, University of Nigeria, Nsukka, Enugu State, Nigeria

1. INTRODUCTION

The term "nanoscale" refers to particle size range from ~ 1 to 100 nm [1], but for the purpose of drug delivery, nanoparticles in the range of 50 – 500 nm are acceptable depending on the route of administration. The method by which a drug is delivered can have a significant effect on its efficacy. Some drugs have an optimum concentration range within which maximum benefit is derived and concentrations above or below this range can be toxic or produce no therapeutic benefit. The slow progress in the efficacy of the treatment of several diseases has suggested a growing need for a multidisciplinary approach to the delivery of therapeutics to target tissues [2]. Transdermal drug delivery systems (TDDS) or patches are controlled-release devices that contain the drug either for localized treatment of tissues underlying the skin or for systemic therapy after topical application to the skin surface [3]. TDDS are available for a number of drugs, although the formulation matrices of these delivery systems differ. They differ from conventional topical formulations in the following ways:

- they have an impermeable occlusive backing film that prevents intensive water loss from the skin beneath the patch;
- the formulation matrix of the patch maintains the drug concentration gradient within the device after application so that drug delivery to the interface between the patch and the skin is sustained; and
- TDDS are kept in place on the skin surface by an adhesive layer ensuring drug contact with the skin and continued drug delivery [4].

Topical or transdermal drug delivery is challenging because the skin acts as a natural and protective barrier. TDDS were introduced into the US market in the late 1970s [5], but transdermal delivery of drugs had been used for a very long time. There have been previous reports about the use of mustard plasters to alleviate chest congestion and belladonna plasters as analgesics. The mustard plasters were homemade as well as available commercially where mustard seeds were ground and mixed with water to form a paste, which was in turn used to form a dispersion type of delivery system. Several methods have been examined to increase the permeation of therapeutic molecules into and through the skin and one such approach is use of nanoparticulate delivery system.

The skin has been an important route for drug delivery when topical, regional, or systemic effects are desired. Nevertheless, skin constitutes an excellent barrier and presents difficulties for the transdermal delivery of therapeutic agents, since few drugs possess the characteristics required to permeate across the stratum corneum in sufficient quantities to reach a therapeutic concentration in the blood [6]. In order to enhance drug transdermal absorption, different methodologies have been investigated, developed, and patented. Improvement in physical permeation-enhancement technologies has led to renewed interest in transdermal drug delivery. Some of these novel advanced transdermal permeation-enhancement technologies include iontophoresis, electroporation, ultrasound, microneedles to open up the skin, and more recently the use of transdermal nanocarriers.

2. THE HUMAN SKIN

The potential of using the intact skin as the port of drug administration to the human body has been recognized for several decades. However, the skin is a very difficult barrier to the ingress of materials allowing only small quantities of a drug to penetrate over a period of time. In order to design a drug delivery system, one must first understand the skin anatomy and its implication of drug-of choice and method of delivery.

The human skin is the largest organ in our body with surface area of 1.8-2.0 m². It is composed of three main layers; the epidermis, dermis and hypodermis (subcutaneous layer) (Fig. 1). The skin is a well energized organ that protects the organism against environmental factors and regulates heat and water loss from the body.

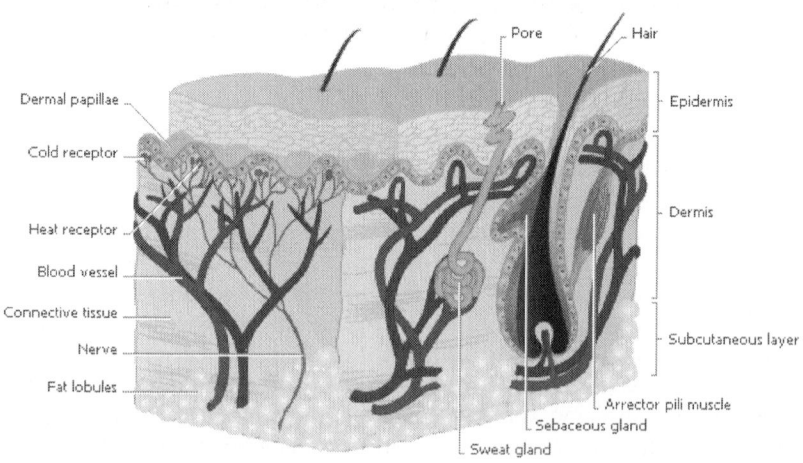

Figure 1. Structure of the skin

3. ROUTES OF DRUG PENETRATION THROUGH THE SKIN

The permeation of drugs through the skin involves the diffusion through the intact epidermis through the skin appendages (hair follicles and sweat glands). These skin appendages form shunt pathways through the intact epidermis, occupying only 0.1% of the total human skin. It is known that drug permeation through the skin is usually limited by the stratum corneum (Fig. 2). Three main penetration routes are recognized (Fig. 3).

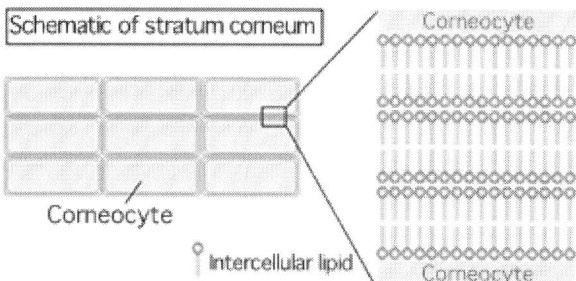

Figure 2. The stratum corneum

3.1. THE INTERCELLULAR LIPID ROUTE

Interlamellar regions in the stratum corneum, including linker regions, contain less ordered lipids and more flexible hydrophobic chains. This is the reason for the nonplanar spaces between crystalline lipid lamellae and their adjacent cells' outer membrane. Fluid lipids in skin barrier are crucially important for transepidermal diffusion of the lipidic and amphiphilic molecules, occupying those spaces for the insertion and migration through intercellular lipid layers of such molecules [7]. The hydrophilic molecules diffuse predominantly "laterally" along surfaces of the less abundant water-filled interlamellar spaces or through such volumes; polar molecules can also use the free space between a lamella and a corneocyte outer membrane to the same end.

3.2. The Transcellular Route

Intracellular macromolecular matrix within the stratum corneum abounds in keratin, which does not contribute directly to the skin diffusive barrier but supports mechanical stability and thus intactness of the stratum corneum. Transcellular diffusion is practically unimportant for transdermal drug transport [8]. The narrow aqueous transepidermal pathways have been observed using confocal laser scanning microscopy. Here, regions of poor cellular and intercellular lipid packing coincide with wrinkles on skin surface and are simultaneously the sites of lowest skin resistance to the transport of hydrophilic entities. This lowest-resistance pathway leads between clusters of corneocytes at the locations where such cellular groups show no lateral overlap. The contribution to transdermal drug transport can increase with pathway widening or multiplication, e.g., that which is caused by exposing the stratum corneum to a strong electrical (electroporation/iontophoresis), mechanical (sonoporation/sonophoresis), or thermal stimulus, or suitable skin penetrants.

3.3. Follicular Penetration

Recently, follicular penetration has become a major focus of interest due to the fact that drug targeting to the hair follicle is of great interest in the treatment of skin diseases. However, follicular orifices occupy only 0.1% of the total skin surface area. For this reason, it was assumed to be a nonimportant route for drug penetration. But a variety of studies have shown that hair follicles could be an interesting option for drug penetration through the skin [6]. Such follicular pathways have also been proposed for topical administration of polystyrene nanoparticles. They were investigated in porcine skin (*ex vivo*) and human skin (*in vivo*). Surface images revealed that polystyrene nanoparticles accumulated preferentially in the follicular openings. This distribution was increased in a time-dependent manner, and the follicular localization was favored by the smaller particle size. The study also confirmed similarity in the penetration between both membranes (porcine and human skin). In other investigations, the influence of microparticle size in skin penetration has been shown by differential stripping. Nanoparticles can act as efficient drug carriers through the follicle or can be utilized as follicle blockers to stop the penetration of topically applied substances.

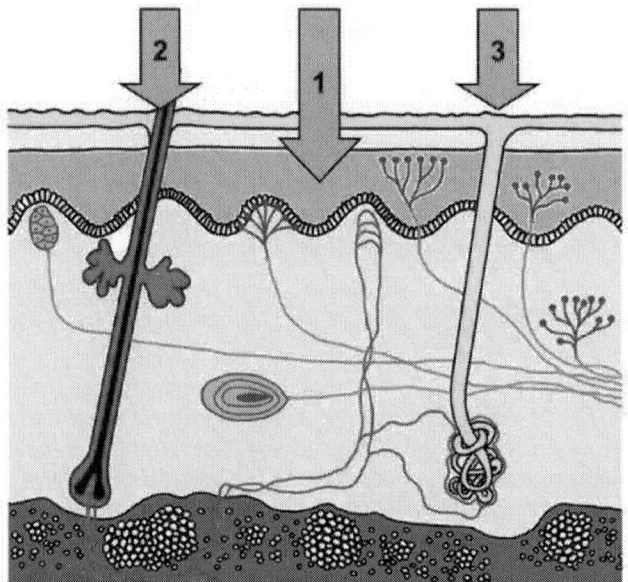

Figure 3. Structure of the skin showing routes of penetration: (1) across the intact horny layer, (2) through the hair follicles with the associated sebaceaous glands, or (3) via the sweat glands

4. MAIN FACTORS FOR NANO-BASED DELIVERY SYSTEM

4.1. Particle Size, Size Distribution and Zeta Potential

Particle size and shape affect drug release, physical stability and cellular uptake of the nanoparticulate materials. The yield and size distribution of each system are affected by certain in-process operations and conditions such as stirring rate, temperature, type and amount of dispersing agent as well as the viscosity of the organic and aqueous phases [9,10]. Zeta potential of a dispersion is necessary for dispersion stability [11].

4.2. Surface Properties

The attachment of nanoparticles to cell membrane is affected by the surface charge of the particles. Variation of the particle surface charge could potentially control binding to the tissue and direct nanoparticles to cellular compartments both *in vitro* and *in vivo*. Cellular surfaces are dominated by negatively charged sulphated proteoglycans molecules that play pivotal roles in cellular proliferation, migration and motility [12]. Cell surface proteoglyans consist of a core protein anchored to the membrane and linked to one or more glycosaminoglycan side chains (heparan, dermatan, keratan or chondrotine sulfates) to produce a structure that extends away from the cell surface.

Nanoparticles show a high affinity for cellular membrane mainly due to electrostatic interactions [12]. It is known that cell membranes have large negatively charged domains, which should repel negatively charged nanoparticles. The high cellular uptake of negatively charged nanoparticles is related first to the non-specific process of nanoparticles adsorption on the cell membrane and second to formation of nanoparticle clusters [13]. The adsorption of the negatively charged particles at the positively charged sites via electrostatic interaction can lead to localized neutralization and a subsequent bending of the membrane favouring in turn endocytosis for cellular uptake [14]. Thus the formulation of nanoparticles with different surface properties can influence their cellular uptake and intracellular distribution and it is possible to localize the nanoparticles to specific intracellular targets (lysosomes, mitochondria, cytoplasm, etc) by modifying their surface charge [15].

There are some investigations that showed the effect of surface charge, for example polymer charge density of dendrimers was found to significantly impact membrane permeability. The most densely charged polymer facilitates the transport of dye molecule across the membrane [16]. Other investigation showed that lipid coating of ionically charged nanoparticles was able to increase endothelial cell layer crossing 3 or 4 fold compared with uncoated particles, whereas nanoparticles coating of neutral particles did not significantly alter their permeation characteristics across the endothelial cell monolayer [13]. Transdermal drug administration systems have been limited to certain drugs of a range of molecular weight and lipophilicity, and of certain charge preference. For instance, cationic compounds have a positive effect on skin permeation, since the skin carries a negative surface charge due to phosphatidylcholine [17] and carbohydrates found in mammalian cells contain negatively charged groups. Therefore, nanoparticles with predominant positive charge would promote transdermal permeation.

5. DERMATOPHARMACOKINETICS

Dermatopharmacokinetics describe the pharmacokinetics of topically applied drugs in the stratum corneum with pharmacodynamic effects. The smart techniques (tape stripping and microdialysis) use in dermatopharmacokinetic methodology assesses the cutaneous drug concentration at the site of application. Various studies have shown dermatopharmacokinetics to be a reliable and reproducible method for determining bioequivalence, and have indicated that it is applicable for all topical dermatological drug products. Dermatopharmacokinetics refer to the determination of stratum corneum concentration-time curves for topical actives. This is analogous to plasma/urine concentration-time curves for systemically or orally administered drugs, and the concept is clearly adaptable to microdialysis, where drug is determined in the skin compartment in which the microdialysis fibre is positioned (Fig. 4).

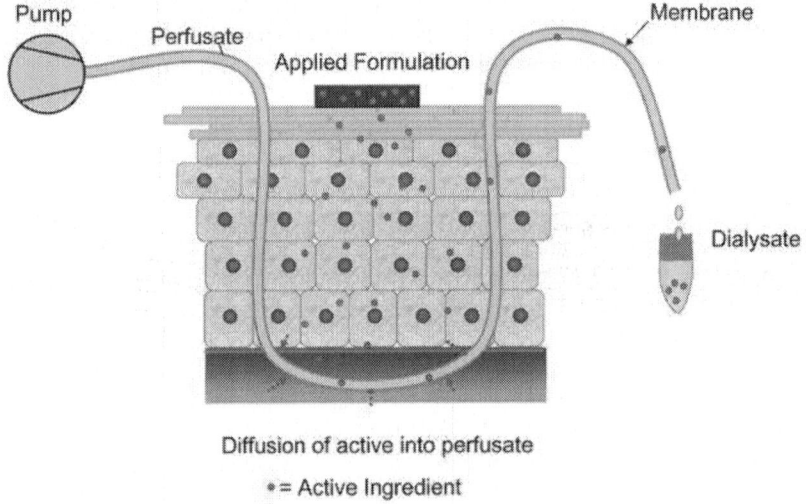

Diffusion of active into perfusate

• = Active Ingredient

Figure 4. Sampling in the skin by microdialysis

Although, this procedure is invasive, it is a method of great potential offering information of high value and relevance. There could be sampling in a compartment within the skin. It is a technically demanding procedure, however, requiring experimental dexterity of high order. The potential for use on diseased skin is a unique and considerable advantage over other techniques, but real challenges remain with respect to reproducibility, sensitivity, applicable drugs, etc.

Stratum corneum tape-stripping is a minimally invasive method for determining drug levels in human stratum corneum *in vivo*. It involves repeated application of adhesive tapes on a site that has been treated with a topical formulation and determination of drug levels in stratum corneum collected on tape strips.

The dermatopharmacokinetics approach suggested by the Food and Drug Administration (FDA) proposes to evaluate the level of a topically applied drug in the stratum corneum during its uptake and clearance so as to calculate classic pharmacokinetic parameters [18]. The assumption is that stratum corneum concentration-time curves are directly related to concentration-time curves in the epidermis and dermis.

When applied to diseased skin, topical drug products induce one or more therapeutic responses, where onset, duration, and magnitude depend on the relative efficiency of three sequential processes, namely:

- the release of the drug from the dosage form
- penetration of the drug through the skin barrier, and
- generation of the desired pharmacological effect.

Because topical products deliver the drug directly to or near the intended site of action, measurement of the drug uptake into and drug elimination from the stratum corneum can provide a dermatopharmacokinetics means of assessing the bioequivalence of two topical drug products [19,20]. Presumably, two formulations that produce comparable stratum corneum concentration-time curves may be bioequivalence, just as two oral formulations are judged bioequivalent if they produce comparable plasma concentration-time curves. Even though the target site for topical dermatologic drug products in some instances may not be the stratum corneum, the topical drug must still pass through the stratum corneum, except in instances of damage, to reach deeper sites of action [21]. In certain instances, the stratum corneum itself is the site of action. For example, in fungal infections of the skin, fungi reside in the stratum corneum and therefore dermatopharmacokinetic measurement of an antifungal drug in the stratum corneum represents direct measurement of drug concentration at the site of action [22]. In instances where the stratum corneum is disrupted or damaged, *in vitro* drug release may provide additional information toward the bioequivalent assessment. In this context, the drug release rate may reflect drug delivery directly to the dermal skin site without passage through the stratum corneum. For antiacne drug products, target sites are the hair follicles and sebaceous glands. In this setting, the drug diffuses through the stratum corneum, epidermis, and dermis to reach the site of action. The drug may also follow follicular pathways to reach the sites of action. The extent of follicular penetration depends on the particle size of the active ingredient if it is in the form of a suspension [21, 23-25]. Under these circumstances, the dermatopharmacokinetic approach is still expected to be applicable because studies indicate a positive correlation between the stratum corneum and follicular concentrations. Although the exact mechanism of action for some dermatological drugs is unclear, the dermatopharmacokinetic approach may still be useful as a measure of bioequivalence because it has been demonstrated that the stratum corneum functions as a reservoir, and stratum corneum concentration is a predictor of the amount of drug absorbed [26].

For reasons thus cited, dermatopharmacokinetic principles should be generally applicable to all topical dermatological drug products including antifungal, antiviral, antiacne, antibiotic, corticosteroid, and vaginally applied drug products. The dermatopharmacokinetic approach can thus be the primary means to document bioavailability/bioequivalence. Generally, bioequivalence determinations using dermatopharmacokinetic studies are performed in healthy subjects because skin where disease is present demonstrates high variability and changes over time. Use of healthy subjects is consistent with similar use in bioequivalence studies for oral drug products.

A dermatopharmacokinetic approach is not generally applicable when:

- a single application of the dermatological preparation damages the stratum corneum
- for otic preparations except when the product is intended for otic inflammation of the skin; and
- for ophthalmic preparations because the cornea is structurally different from the stratum corneum.

6. IDEAL DRUGS FOR DERMAL AND TRANSDERMAL DELIVERY

Owing to the selective nature of the skin barrier, only a small pool of drugs can be delivered systemically at therapeutically relevant rates [27]. Few drugs constitute the whole segment of the transdermal drug market. Besides great potency, the physicochemical drug characteristics often evoked as favourable for percutaneous delivery include moderate lipophilicity and low molecular weight [28]. However, a large number of pharmaceutical agents do not fulfill these criteria. This is especially true for macromolecules, such as insulin, human growth hormone or cyclosporine, which are very challenging from the drug delivery point of view. The physicochemical properties of ideal drug for transdermal delivery include:

- Molecular weight less than approximately 1000 Daltons.
- Affinity for both lipophilic and hydrophilic phases. Extreme partitioning characteristics not ideal.
- Low melting point.
- Should be potent, with short half life and be non-irritating.

Overcoming low skin permeability to xenobiotics can be achieved by a variety of approaches, and is an active field of research. Their effectiveness and applicability will vary from drug to drug depending on the physicochemical nature of the compound. New drug discovery is still a complicated process and generally requires substantial time and monetary investment. Technologies for formulation change provide the benefit of improving pharmaceutical product efficacy and safety as well as patient convenience; these technologies provide a relatively simple approach to creating new pharmaceuticals compared with new drug discovery because the active compounds used in the formulation have already been approved [29-31]. Nnamani *et al* [32] developed and evaluated the antimicrobial activities of an alternative non-invasive, convenient and cost-effective transdermal drug delivery system (TDDS) containing gentamicin in biodegradable polyester-based matrices. Other drugs which have been formulated for dermal and transdermal delivery are nitroglycerin, nicotine, scopolamine, clonidine, fentanyl, 17-β-estradiol, testosterone, Boswellic acid (*Boswellia serrata*) and curcumin (*Curcuma longa*).

7. ADVANTAGES OF DERMAL AND TRANSDERMAL DRUG DELIVERY

Transdermal delivery provides convenient and pain-free self-administration for patients. It eliminates frequent dosing administration and plasma level peaks and valleys associated with oral dosing and injections to maintain constant drug concentrations, and a drug with a short half-life can be delivered easily. All this leads to enhanced patient compliance, especially when long-term treatment is required, as in chronic pain treatment and smoking cessation therapy [3,33,34].

- Avoidance of hepatic first-pass metabolism and the gastrointestinal (GI) tract for poorly bioavailable drugs is another advantage of transdermal delivery. Elimination of the first-pass effect allows the amount of drug administered to be lower, and hence, safer in hepato-compromised patients, resulting in the reduction of adverse effects.

- Transdermal systems are generally inexpensive when compared with other therapies on a monthly cost basis, as patches are designed to deliver drugs from 1 to 7 days.

- The other advantage of transdermal delivery is that multiple dosing, on-demand or variable-rate delivery of drugs is possible with the latest programmable systems, adding more benefits to the conventional patch dosage forms.

- The general acceptability of transdermal products by patients is very high, which is also evident from the increasing market for transdermal products.

- Transdermal route permits the use of a relatively potent drug with minimal risk of system toxicity [35,36].

- In case of toxicity, the transdermal patch can easily be removed by the patient [37].

8. DISADVANTAGES OF DERMAL AND TRANSDERMAL DELIVERY SYSTEMS

Even though dermal and transdermal delivery systems have a lot of advantages over conventional topical formulation, it still suffer from a lot of limitations. The disadvantages of dermal and transdermal delivery systems according to Ranade and Cannon [38] are that:

- Not all drugs are suitable for transdermal delivery.
- Drugs that require high blood levels cannot be administered.
- The adhesive used may not adhere well to all types of skin.
- Drugs or drug formulation may cause sensitization or irritation which must be evaluated fairly early in the development process.
- The patches may/can be uncomfortable to wear.
- The manufacture requires specialized equipments which results in the formulation being more expensive to manufacture than conventional dosage forms thus the formulation will not be economical for most patients.
- There is always a lag time for drug to penetrate through the skin to the systemic circulation, therefore TDDS is not suitable for drugs requiring rapid onset of action.
- There is a requirement for low dose/high permeable drug. In general a drug with molecular weight less than 400, $logP_{o/w}$=2-3 and dose less than 10 mg will be the best candidate for transdermal delivery.

9. CHARACTERIZATION OF DERMAL AND TRANSDERMAL DELIVERY SYSTEMS AND THEIR PERFORMANCE

Dermal and transdermal delivery systems are characterized using different methods.

9.1. Drug Solubility Determination

The determination of solubility of the drug in the transdermal/dermal matrix early in the formulation process can avoid crystallization problem, which is one of the instabilities in transdermal drug delivery systems (TDDS). This instability in the matrix which could be due to supersaturation makes the formulation metastable and upon storage results in changes in the liberation/release rate of the drug from the formulation.

9.2. Micromeritic Measurements

9.2.1. Particle-Size, Shape and Zeta Potential Analysis

Light scattering is an important way of characterizing colloidal and macromolecular dispersions and could be useful in assessing properties of particulate TDDS e.g. ethosomes. The particle size and size distribution are primarily measured using wet laser diffraction sizing otherwise called dynamic light scattering (DLS) [39]. Size of formulation can also be determined using dynamic light scattering (e.g. using a Zetasizer). This is necessary to ascertain the possible effect of the size on drug release and penetration across barriers in transdermal and dermal delivery as well as to monitor stability over time. The zeta potential of a formulation is very important. It is determined using Zetsizer or by other means, and gives information on the charge of the particles and the tendency of the particles in a formulation to aggregate or to remain discrete.

9.2.2. Specific Surface Area

An important parameter of bulk powders is the specific surface area expressed per unit weight. The specific surface area measurement includes the cracks, crevices, nooks, and crannies present in the particles. To include these features in the surface-area measurement, methods have been developed to probe these convoluted surfaces through adsorption by either a gas or a liquid [40-42]. The most widely used surface area measurement technique is the adsorption of a monolayer of gas, typically krypton or nitrogen as the adsorbate gas in helium as an inert diluent, using the method developed by Brunauer, Emmett, and Teller known as the BET method. Surface area affects spreading and occlusivity of TDDS.

9.3. Visualization by Transmission Electron Microscopy

A combination of transmission electron microscopy (TEM) and freeze fracturing otherwise referred to as freeze fracture electron microscopy (FFEM) could be used to visualize skin structures and certain perturbations in the skin. A micrograph image is generated by transmitting a beam of electrons through a specimen appropriately treated to enhance the visualization of skin structural details. High resolution of TEM makes it possible to visualize both structures and transition processes in the epidermis. Using different techniques, epidermal granules [43], Langerhans cells [44] and the lipids in stratum corneum and epidermis [45], amongst others, have been observed. Samples preparation in FFTEM involves freezing the sample and subsequent longitudinal fracturing approximately parallel to the original skin surface under high vacuum [46]. Further treatment could be done on the sample after which the fracture is viewed under high voltage. This visualization method can provide information on the interaction between the nanoparticle formulation and the skin. Since the fracture will always run along the plane of least resistance, FFEM micrographs of treated stratum corneum often show the lipid coated surfaces of corneocytes or the lipid lamellae.

9.4. Stability

Physical and chemical instabilities of carrier systems often limit their widespread use in medical applications [47]. Instabilities in ethosomes and other nanocarrier formulations are caused by hydrolysis or oxidation of the phospholipid molecules and are indicated by leakage of the encapsulated drug and alterations in vesicle size due to fusion and aggregation [48,49]. Changes in size and size distribution, entrapment efficiency and aggregation of vesicles are very important parameters in monitoring stability. These parameters can be assessed by EM or DLS repeatedly over time at varying storage conditions. It has recently been found that although multilamellar and large unilamellar benzocaine-loaded ethosome vesicles remained substantially stable with time, in terms of drug entrapment yield and particle dimensions, small unilamellar vesicles showed high tendency to form aggregates due to increased surface area exposed to the medium [10]. Such vesicle aggregation indicates instability. In addition, changes in storage conditions led to marked decrease in particle dimensions and drug-entrapping yield with less regular morphology for frozen-and-thawed multilamellar ethosome dispersions, while the untreated multilamellar and unilamellar vesicular dispersions remained homogenous and stable with regard to those parameters assessed over the period [50]. Temperature of formulation and storage conditions affect physical stability of nanoparticle preparations [10,51].

Optical characteristics, viscosity and physical changes such as cracking or creaming are also important in assessing stability of ethosomes. Ethosomes are colloidal disperse systems therefore, cracking and creaming may be observed during storage as in water-in-oil emulsions. The use of an innovative optical analyzer, Turbiscan Lab® Expert, in studying the influence of optical characteristics on long-term stability of vesicular colloidal delivery systems has been advocated [52]. The principle of this measurement is based on the variation of the droplet volume fraction (migration) or mean size (coalescence), thus resulting in the variation of backscattering and transmission signals as a function of time. No variation of particle size occurs when the backscattering profile is within the interval ± 2 %. Variations greater than 10 % either as a positive or negative value in the graphical scale of backscattering are representative of an unstable formulation.

9.5. High-Pressure Liquid Chromatography (HPLC)

It is used to monitor the stability of pure drug substance and drugs in formulation with quantitation of degradation product. A liquid mobile phase is pumped under pressure through a stainless steel column containing particles of stationary phase with a diameter of 3-10 μm. The analyte is loaded onto the head of the column via a loop valve and separation of a mixture occurs according to the relative lengths of time spent by its components in the stationary phase. All components in a mixture spend more or less the same time in the mobile phase in order to exit the column. The column effluent can be monitored with a variety of detectors.

The combination of high-pressure liquid chromatography (HPLC) with monitoring by UV/Visible detection provides an accurate, precise and robust method for quantitative analysis of pharmaceutical products and is the industry standard method for this purpose. The two principal mechanisms which produce retardation of a compound passing through a column are straight-phase packing where adsorption of polar groups of a molecule onto the polar groups of a stationary phase occur and reverse-phase packing which is due to partitioning of the lipophilic portion of a molecule into the stationary phase.

9.6. Liquid Chromatography–Mass Spectrometry (LC/MS)

Mass spectrometry in conjunction with liquid chromatography provides a method for characterizing impurities in drugs and formulation excipients [53]. It provides highly sensitive and specific methods for determining drugs and their metabolites in biological fluids and tissues.

9.7. Fourier Transform Infra Red (Ftir) Spectroscopy

FTIR spectroscopic properties are used to determine the chemical stability of the drug in a TDDS. FTIR spectra of formulations, the starting materials and pure drug sample are normally obtained at a range of 4000-400 cm^{-1} and the spectra obtained on infrared spectrophotometer using potassium bromide of spectroscopic grade.

Detailed insights into the organization of the stratum corneum can be gained through the study of the vibrations of amide, amine and carboxylic groups and the frequencies of the methylene stretching, scissoring and rocking vibrations. FTIR is used to study the lateral lipid organization of the intercellular lipid matrix in stratum corneum, which is essential for the barrier function of stratum corneum, as more densely organized membranes are less permeable to substances. The stretching vibrations are used to determine whether lipids are in an ordered (hexagonal or orthorhombic lateral packing) or disordered packing (liquid phase), while the scissoring and rocking vibration provide detailed information on the presence of orthorhombic phases. By performing measurements at different temperatures, also the thermotropic behaviour of the lipids can be determined.

9.8. Attenuated Total Reflectance FTIR (ATR-FTIR)

Attenuated total reflectance FTIR (ATR-FTIR) is a modification of FTIR. In this technique, IR radiation is not transmitted through the sample but reflected by the sample. With this technique, it is possible to perform measurements on stratum corneum *in vivo*, because the skin can be placed on the ATR crystal. The IR radiation beam penetrates only to a limited extent into stratum corneum. In order to detect substances in the stratum corneum, it is necessary to remove stratum corneum layers, by tape-stripping, which makes it also possible to generate a penetration profile of an applied substance in stratum corneum [54-56]. ATR-

FTIR has been used to determine effects of topically applied substances on the lipid organization in the stratum corneum [54,57]. ATR-FTIR can be combined with tape-stripping to determine the penetration profile of hydrophilic and lipophilic substances in stratum corneum in addition to the water profile of the stratum corneum.

9.9. Differential Scanning Calorimetry (DSC)

This technology is used to evaluate the degree of perturbation of the skin lipids as a result of penetration of a formulation or drug through skin. The free intercellular lipid bilayers of the stratum corneum have a unique composition compared to other epithelial lipid bilayers and consist of ceramides (50%), cholesterol (25%), and fatty acids (10-20%, highly enriched in linoleic acid). These common skin lipids are detected at different transition temperatures when the skin is subjected to DSC studies.

9.10. Small Angle X-Ray Diffraction (SAXD)

This technique is used to analyse the long range order of the crystalline structure of lipids. Stratum corneum is a very thin layer of about 10 μm and composed of corneocytes and an intercellular lipid matrix. The ordered structure of the intercellular lipid matrix plays an important role in skin barrier function. Structural analyses of intercellular lipids in mammalian stratum corneum by X-ray diffraction have shown more detailed lipid structure models. The X-ray pattern of a lamellar phase is characterized by a series of sequential maxima, which are positioned at equal interpeak distances at increasing scattering angle [58]. The sequential peaks are referred to as the 1st order (positioned at distance Q1), the 2nd order (Q2), the 3rd order (Q3), etc, in which Q is directly related to the scattering angle. The repeat distance (d) of a lamellar phase can be directly calculated from the peak positions $d=2\pi\pi/Q1=4\ \pi\ \pi\ /Q2=6\ \pi\ \pi\ /Q3$, etc. In skin research SAXD is used to study the lamellar organization of the lipids in the intercellular matrix of stratum corneum of humans and other mammals. Furthermore, SAXD measurements using lipid mixtures of ceramides, cholesterol and free fatty acids have revealed the role of the various lipid classes in the lamellar phases. Additionally, it has been used to study effects of topically applied substances [59] or physical stratum corneum perturbation methods [54]. SAXD is also used to study the effects of hydrophilic and lipophilic agents like nanoparticles on the lamellar organization of isolated stratum corneum.

9.11. Dermal Irritation Assay

If a new drug is intended to be applied to the skin or eyes, one of the first tests to be conducted would be to determine if the drug, or the formulation containing the drug, will cause irritation of the skin or eyes. Even if a drug is intended only for dermal application, eye irritation testing may also be required because of the possibility of inadvertent exposure to the eyes. The test is conducted as follows:

Six male albino rabbits are to be clipped free of hair on the back. One area of skin is left intact, whereas another is abraded in a tic-tac-toe pattern with the point of a hypodermic needle so as to incise the superficial epidermis layer without causing bleeding. The test material, 0.5 ml of liquid or 0.5 g of solid or semisolid is applied to each site under a 1 × 1 inch gauze pad. The entire trunk of the animal is wrapped with an impervious material and held in place with tape for 24 h. The patches are then removed and excessive material wiped off. The skin reactions are scored at 24 and 72 h after the initial application according to a scheme such as that listed in Table 1.

Table 1. Dermal irritation scoring system

Skin reaction	Value
Erythema and eschar formation	
No erythema	0
Very slight erythema (barely perceptible)	1
Well-defined erythema	2
Moderate to severe erythema	3
Severe erythema (beet redness) to slight eschar formation (injuries in depth)	4
Edema formation	
No edema	0
Very slight edema (barely perceptible)	1
Slight edema (edges of area well defined by definite raising)	2
Moderate edema (raised approximately 1 mm)	3
Severe edema (raised more than 1 mm and extending beyond the area of exposure)	4

The mean values of the six rabbits for erythema and eschar formation at 24 and 72 h for both intact and abraded skin (four values) are added. The mean values of the six rabbits for edema at 24 and 72 h (four values) are also added. The total of eight values is divided by 4 to give the primary irritation index. Values of 5 or greater are considered indicative of a positive irritant [60].

9.12. Occlusivity

It is usually the aim of cosmetic chemists to maintain the skin's softness and freshness and it is considered important to retain moisture in the stratum corneum. The degree to which a formulation retains or promotes the loss of moisture from the stratum corneum is termed the occlusivity. The occlusivity of formulations for topical application is determined *in vivo* by measuring the suppression of transepidermal water loss (TEWL) of the skin. The occlusive effect of the formulation also depends on the characteristics of the skin such as the lipid level and prevailing environmental condition. The occlusivity of films formed by nanoparticles varies with time, type of formulation, coating amount, physical form, size of particles etc. It is necessary to determine the occlusivity of a nanoparticle formulation for topical application as it directly affects liberation and penetration of the encapsulated drug. Under occlusive conditions, the skin is more hydrated and transport of drug could be higher.

9.13. Spreadability

Pharmaceutical semisolid preparations include ointments, pastes, creams, emulsions, gels, and rigid foams. Their common property is the ability to cling to the application surface for a reasonable period of time before they are washed off or worn off [61]. They usually serve as vehicles for topically applied drugs, as emollients, or as protective or occlusive dressings, or they may be applied to the skin and membranes such as the rectal, buccal, nasal, and vaginal mucosa, urethral membrane, external ear lining, or the cornea [62]. These preparations are widely used as a means of altering the hydration state of the substrate (i.e., the skin or the mucous membrane) and for delivering the drugs (topical or systemic) by means of the topical–mucosal route. Nanoparticles for transdermal application could be formulated as gels, creams, emulsions, foams etc, or dispersed in ointment bases. This makes the spreadability characteristics of the formulation very pertinent in achieving the desired objective.

The efficacy of topical therapy depends on the patient spreading the formulation in an even layer to deliver a standard dose. The optimum consistency of such a formulation helps ensure that a suitable dose is applied or delivered to the target site. This is particularly important with formulations of potent drugs. A reduced dose would not deliver the desired effect, and an excessive dose may lead to undesirable side effects. The delivery of the correct dose of the drug depends highly on the spreadability of the formulation. Spreadability, in principle, is related to the contact angle of the drop of a liquid or a semisolid preparation on a standardized substrate and is a measure of lubricity, which is directly related to the coefficient of friction [63]. Spreadability is subjectively assessed at shear rates varying from 10^2 to 10^5 s^{-1}. The rate of shear during spreading, $\gamma\gamma$ s^{-1}, is calculated using the following equation for plane laminar flow between two parallel plates:

$$\gamma = \frac{v}{d}$$

(1)

in which v is the relative velocity of the plates (cm s^{-1}) and d is the distance between them (cm); that is, a measure of thickness of the film between the skin surfaces [64].

To assess the spreadability of a topical or a mucosal semisolid preparation, the important factors to consider include hardness or firmness of the formulation, the rate and time of shear produced upon smearing, and the temperature of the target site [64]. The rate of spreading also depends on the viscosity of the formulation, the rate of evaporation of the solvent, and the rate of increase in viscosity with concentration that results from evaporation [65].

small mean diameters (<100 nm). Their large inner volume allows the loading of small biomolecules while their outer surface can be chemically modified to render themselves various novel features that can be used to load proteins and genes for effective drug delivery [146], even through the skin.

Fullerenes (Fig. 7) are 1-nm scale carbon spheres of 60 carbon atoms. Although fullerenes are hydrophobic, they can be organically functionalized by attaching hydrophilic moiety and become water-soluble and capable of carrying genes, proteins and other biomolecules for delivery purposes [146]. Their small size, spherical shape and hollow interior all provide therapeutic opportunities and have been proposed for use in cosmetic products like sunscreens, moisturizers, long lasting makeup, etc.

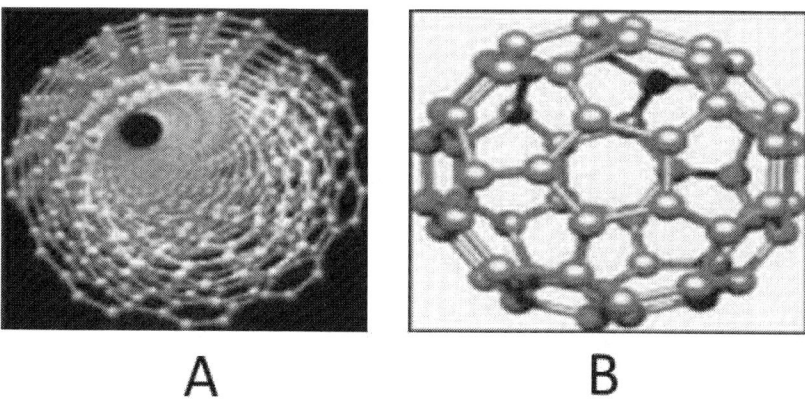

A   **B**

Figure 7. Carbon nanotube (A) and fullerene (B)

11. INFLUENCE OF NANOPARTICULATE FORMULATIONS ON BIOCHEMICAL PROCESSES OF THE SKIN

Apart from environmental protection against radiation, functions of the skin include heat regulation, immune response, biochemical synthesis, sensory detection, regulation of absorption/loss of water and electrolytes. The stratum corneum formed from nonviable corneocytes plays the major role.

A crucial question for the investigation of nanoparticulate drug delivery carriers is the site of drug release from the particles, i.e., does the release occur in suspension or on the skin surface leaving the carrier particles outside or do the particles penetrate the skin to release the drug within the tissue? Nanoparticles formulations for dermal delivery or transdermal delivery influenced some of the traditional functions of the skin. Once applied to the skin, enzymes activated by body heat led to the formation of an active ingredient (allyl isothiocynate). Transport of the active drug component took place by passive diffusion across the skin-the very basis of transdermal drug delivery [147,148]. The alcohol in ethosomes initiates the process of transdermal permeation and drug release by its permeation enhancing effect [149]. The major hindrance to TDDS is the stratum corneum layer that forms a strong barrier and limiting factor to skin penetration and permeation of many drugs.

The processes involved in drug delivery from ethosomes through the skin are illustrated in Fig. 8 [46]. The alcohol makes the vesicles to be packed loosely and the vesicle membranes to become softer and malleable [13]. It also causes reversible perturbations in the deeper layers of SC and penetrates intercellular lipid layers of skin cell membranes making them more fluid and less dense [150]. The vesicles then squeeze through the intercellular spaces into the deeper layers of skin. It has been shown that drug particles are concentrated more on the inside wall than in the core of vesicles [151]. In this position, release of the vesicular content is thermodynamically favored. Owing to the increased affinity, due to its lipid content, the vesicle fuses with the lipid contents of the skin layers and releases its content which then diffuses into deeper layers of the skin or membrane and into systemic circulation. Other mechanisms, such as the free drug diffusion, may be involved in penetration.

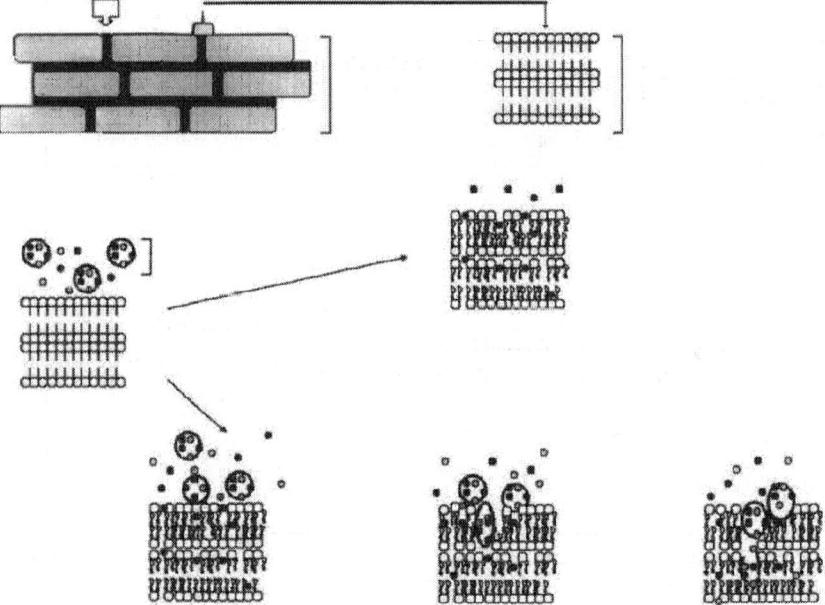

Figure 8. Mechanism of drug delivery from ethosomal vesicular carriers through the skin [46,115]. Note the initial fluidization of the skin architecture.

12. REGULATIONS ON DERMAL AND TRANSDERMAL DELIVERY SYSTEMS

Safety and toxicological issues are the most important issues for a drug delivery system. Safety is an obvious concern for the fast growth of nanoparticles mediated drug delivery [152]. Governmental regulatory agencies such as the United States Food and Drug Agency (USFDA) have established guidelines describing the kind of safety tests that should be conducted in animals in order to have a new drug approved for use in clinical trials and in order to get approval of a new drug application (NDA) for marketing. The rationale and circumstances for conducting reproductive, mutagenicity, carcinogenicity, irritation, and sensitization studies have already been mentioned. The requirements for acute, subacute, and chronic toxicity studies for pharmaceutical products intended for use in humans as described according to are the requirements of the United States, Japan, and Europe because these areas represent the largest pharmaceutical markets in the world today. These requirements have been developed at the International Conference on Harmonization to provide uniformity among the three regions [153]. Phases I, II, and III refer to the different phases of human clinical trials. Phase I refers to the initial trials, limited to one or a few doses to determine absorption, pharmacokinetics, and an initial estimate of safety. Phase II refers to larger scale studies to establish safety and to get an initial estimate of clinical efficacy. Phase III refers to the final, large-scale, multicenter trials aimed at establishing efficacy.

The Food and drug agency (FDA) paradigm for regulation of new products is based on the concepts of risk management, which includes identification, analysis and control of risk [154]. The regulation and approval by the FDA is on a ''product by product'' basis, with the overall regulation process falling into three stages: premarket approval, premarket acceptance and post-market surveillance.

Premarket approval: Prior to market introduction of any new pharmaceuticals, high-risk medical devices, food additives, colors, and biologicals, FDA approval is required. The producer/sponsor of the product is responsible for identifying and assessing the risks presented by the product. This party will also be responsible for indicating means to minimize the risks in a product application.

Premarket acceptance: This category refers to products that are often copies of similar products that were approved previously or are products prepared according to approved specifications. For these products, the FDA receives and reviews some form of notice that the products will be marketed and the products undergo a more rapid review process than premarket approval.

Postmarket surveillance: In this category, FDA manages the risks of GRAS products like foods, cosmetics, radiation emitting electronic products and materials such as food additives and food packaging. For products in this category, market entry, and distribution are at the discretion of the manufacturer/producer. These products are generally regulated by the application of good manufacturing practices. FDA takes regulatory action if adverse events that threaten public or individual health occur.

The FDA coordinates policies within itself and with other government agencies. As and when new toxicological risks that derive from the new materials and/or new conformations of existing materials are identified, the FDA will require new tests.

The FDA regulations are for products, not technologies. In addition, the FDA regulates only the claims made by the product sponsor. If the manufacturer makes no nanotechnology claims regarding the manufacture or performance of the product, the FDA may be unaware at the time that the product under review employed nanotechnology. Finally, the FDA has only limited authority over some potentially high-risk products, such as cosmetics. Many products are regulated only if they cause adverse health-related events in use. To date there have been few resources available to assess the risks of these products.

13. DERMAL AND TRANSDERMAL FORMULATIONS ON THE MARKET

A lot of dermal and transdermal drug delivery systems have been licensed for manufacture after passing through the regulatory approval and trials as specified by different countries example FDA (United States of America). Some of the drugs currently available on the market are presented in Table 3.

Table 3. Currently available medications for transdermal delivery [155,156].

Drug	Trade name	Type of transdermal patch	Manufacturer	Indication
Fentanyl	Duragesic	Reservoir	Alza/Janssen Pharmaceutica	Moderate/ Severe pain
Nitroglycerine	Deponit	Drug in adhesive	Schwarz Pharma	Angina Pectoris
	Minitran	Drug in adhesive	3M Pharmaceuticals	
	Nitrodisc	Micro reservoir	Searle, USA	
	Nitrodur	Matrix	Key Pharmaceuticals	
	TransdermNitro	Reservoir	Alza/Novartis	
	Nitroderm TTS	Face	Novartis	
	Diafusor	Matrix	Schering-Plough	
	Transdermal-NTG	Rim	Warner Chilcott Lab	
	Nitrocine	Rim	Kremer Urban	
	Nitro patch	Rim	Adria Lab	

	NTS patch	Rim	Bolar, Major, Qualitest, Bio-Line, Goldline, Geneva, Rugby WarnerChilcott Lab	
Isosorbide dinitrate	Frandol Tape	Matrix	Toaeiyo, Yamanouchi Pharm.	
Nicotine	Prostep	Reservoir	ElanCorp/Lederie Labs	Smoking Cessation
	Nicotrol	Drug in adhesive	Cygnus Inc./McNeil Consumer Products Ltd.	
	Nicotinell	Matrix	Novartis	
	Nikofrenon	Matrix	Novartis	
	Habitraol	Drug in adhesive	Novartis	
Testosterone	Androderm	Reservoir	Thera Tech/ GlaxoSmithKline	Hypogonadism in males
	Testoderm TTS	Reservoir	Alza	
Clonidine	Catapres-TTS	Membrane matrix hybrid type	Alza/Boehinger Ingelheim	Hypertension
Lidocaine	Lidoderm	Drug in adhesive	Cerner Multum, Inc.	Anesthetic
Scopolamine	Transderm Scop	Membrane matrix hybrid type	Alza/Novartis	Motion sickness
Hyoscine	Trasiderm-Scop	Matrix	Novartis	
	Kimite-patch	Matrix	Myun Moon Pharm. Co.	
Minoxidil 4%	Nanominox		Sinere, Germany	Hair growth promoter

Acyclovir	Supravir cream		Trima, Israel	herpes infection.
Many ingredients	Cellutight EF		Hampden Health, USA	Topical cellulite
Estradiol	Climara	Drug in adhesive	3M Pharmaceuticals/ Berlex Labs	Postmenstrual Syndrome
			Noven Pharma/Novartis	
	Vivelle	Drug in adhesive	Alza/Novartis.	
	Estraderm	Reservoir	Women First Healthcare, Inc	
	Esclim	Drug in adhesive	Johnson & Johnson	
Ethinyl Estradiol	Ortho Evra	Drug in adhesive		

14. DERMAL AND TRANSDERMAL DELIVERY OF PHYTOPHARMACEUTICALS

Novel drug delivery system is a novel approach to drug delivery that addresses the limitations of the traditional drug delivery systems. Phytopharmaceuticals are pharmaceuticals using traditional compounds derived from botanicals instead of chemicals. Because these natural ingredients are more easily and more readily metabolized by the body they produce fewer if any side effects and provide increased absorption in the bloodstream resulting in more thorough and effective treatments unlike pharmaceuticals produced from chemical compounds which are prone to adverse side effects [157]. The formulation of dermal and transdermal delivery of phytopharmaceuticals is gaining interest owing to the benefits accruable from it. One of the first few attempts to utilize TDDS containing phytopharmaceuticals was investigation aimed to formulate transdermal films incorporating herbal drug components such as boswellic acid (*Boswellia serrata*) and curcumin (*Curcuma longa*), which utilizes skin as a site for continuous drug administration into the systemic circulation [157]. TDDS avoids first pass metabolism of the drug without the pain associated with injection; moreover the system provides a sustained drug delivery with infrequent dosing via zero-order kinetics and the therapy can easily be terminated at any time. For the local action of the drug at the site of administration of TDDS, turmeric are used which is considered a new version of ayuverdic turmeric*poultice* or *lepa* [158].

Application of vesicular encapsulation holds great promise in the development and use of phytomedicines considering the difficulties of their formulation into stable dosage forms. Certain physicochemical properties of many herbal extracts make their formulation difficult due to stability and processing challenges. By using appropriate techniques, vesicular products of herbal extracts with enhanced stability and efficacy have been produced. A new drug delivery device known as phytosome, composed of phosphatidylcholine, has been developed to overcome the poor absorption of flavonoids, a challenge due mainly to their large molecular sizes and poor miscibility with the lipid contents of cell membrane linings [159]. Phytosomes are well absorbed when taken orally.

Evaluations of phytosomes indicate that a bond is formed between a flavonoid and a phosphatidylcholine molecule to form a hybrid that is highly lipid-miscible. The development and applications of a variety of novel vesicular herbal formulations such as liposomes, phytosomes, transfersomes and ethosomes have been reported [160,161]. Ethosomes, by virtue of their special characteristics, may circumvent the hindrances to successful delivery of phytomedicines. Both soluble and insoluble phytomedines can be encapsulated in ethosomes. Ethosomes also offer protection from premature degradation and increased biodistribution, which would make for improved bioavailability and more beneficial therapeutic outcome for TDDS.

CONCLUSION

From the myriad published studies involving nanoparticles, it is clear that nanoparticles have the potential to effectively deliver drugs across the skin barrier. Conventional liposomes, flexible liposomes, ethosomes, niosomes and ultradeformable liposomes, etc offer potential value as dermal and transdermal drug delivery systems in addition to other lipid nanoparticles.

REFERENCE

1.   Hatto P (2011) ISO concensus definitions relevant to nanomaterials and nanotechnologies. 4th Annual Nano Safety for Success Dialogue. ISO TC 229and BSI NTI/1 Nanotechnologies Standardization Committees. 29th and 30th March 2011. Brussels. Available from: http://ec.europa.eu/health/ nanotechnology [Last accessed 15 August 2012].

2.   Devi VK, Saisivam S, Maria GR, Deepti PU (2003) Design and evaluation of matrix diffusion controlled transdermal patches of verapamil hydrochloride. Drug Dev. Ind. Pharm. 29:495-503.

3.   Valenta C, Auner BG (2004) The use of polymers for dermal and transdermal delivery. Eur. J. Pharm. Biopharm. 58: 279–289.

4.   Block HL (2010) Biopharmaceutics and drug delivery systems In: Comprehensive Pharmacy Review. Eds. Leon Shargel, Alan H Mutnick, Paul F Souney, Larry N Swanson, Lippincott Williams and Wilkins, USA. pp. 83-96.

5.   Prausnitz MR, Langer R (2008) Transdermal drug delivery. Nat. Biotech. 26(11):1261–1268.

6.   Escobar-Chávez JJ, Díaz-Torres R, Rodríguez-Cruz IM, Domínguez-Delgado CL, Morales RS, Ángeles-Anguiano E, Melgoza-Contreras LM (2012) Nanocarriers for transdermal drug delivery. Research and Reports in Transdermal Drug Deliv. 1: 3–17.

7.   Geinoz S, Guy RH, Testa B, Carrupt PA (2004) Quantitative structure–permeation relationships (QSPeRs) to predict skin permeation: a critical evaluation. Pharm Res. 21:83–92.

8.   Cevc G, Vier IU (2010) Nanotechnology and the transdermal route. A state of the art review and critical appraisal. J. Control. Rel.141:277–299.

9.   Pinto Reis C, Nuefeld RJ, Ribeiro AJ, et al., (2006) Nanoencapsulation 1. Methods for preparation of drug-loaded polymeric nanoparticles. Nanomedicine. 2: 8-21.

10.   Maestrelli F, Capasso G, Gonzalez-Rodriguez ML, et al., (2009) Effect of preparation technique on the properties and in vivo efficacy of benzocaine-loaded ethosomes. J Liposome Res 19(4): 253-60.

11.   Attama AA, Schicke BC, Paepenmüller T, Müller-Goymann CC (2007) Solid lipid nanodispersions containing mixed lipid core and a polar heterolipid: Characterization. Eur. J. Pharm. Biopharm. 67: 48-57.

12.   Bernfild M, Gotte M, Park PW, Reizes O, Fitzgerald ML, Lincecum J, Zako M (1999) Functions of cell surface heparan sulphate proteoglycans. Annu. Rev. Biochem. 68: 729-777.

13.   Honary S and Zahir F (2013) Effect of Zeta Potential on the Properties of Nano-Drug Delivery Systems - A Review (Part 1).Trop. J. Pharm Res. 12 (2): 255-264

14.   Win KY, Feng SS (2005) Effects of particle size and surface coating on cellular uptake of polymeric nanoparticles for oral delivery of anticancer drugs. Biomaterials 26: 2713–2722.

15.   Patila S, Sandberg A, Heckert E, Self W, Sea S (2007) Protein adsorption and cellular uptake of cerium oxide nanoparticles as a function of zeta potential. Biomaterials 28: 4600–4607.

16.   Wolinsky JB, Grinstaff MW (2008) Therapeutic and diagnostic applications of dendrimers for cancer treatment. Adv. Drug Deliv. Rev. 60: 1037–1055.

17.   Chang JH, Cho MA, Son HH, Lee CK, Yoon MS, Cho HH, Seo DS, Kim KJ (2006a). Characterization and formation of phospholipid nanoemulsion coatings on Mg-modified sericite surface. J. Ind. Eng. Chem, 12: 635-638.

18.   FDA (1998) Guidance for Industry: topical dermatological drug product NDAs and ANDAs- in vivo bioavailability, bioequivalence, in vitro release and associated studies (Draft).

19. Shah, V. P. and H. I. Maibach (eds.), 1993, Topical Drug Bioavailability, Bioequivalence and Penetration, Plenum Press.

20. Shah VP, Flynn GL, Yacobi A, et al., (1998) "Bioequivalence of topical dermatological dosage forms - Methdods of evaluation of bioequivalence," Pharm. Res. 15: 167-171.

21. Schaefer, H. and T. E. Redelmeir (eds.), 1996, Skin Barrier, Principles of Percutaneous Absorption, Eds, Karger Publishers.

22. Pershing LK, Lambert L, Wright ED, et al., (1994) "Topical 0.05% betamethasone dipropionate: Pharmacokinetic and pharmacodynamic dose-response studies in humans," Arch Dermatol, 130:740-747.

23. Allec A, Chatelus A, Wagner N (1997) "Skin distribution and pharmaceutical aspects of adapalene gel," J. Am. Acad. Dermatol. 36: S119-125.

24. Hueber F, Schaefer H, Wepierre J (1994) Role of transepidermal and transfollicular routes in percutaneous absorption of steroids: in vitro studies on human skin, Skin Pharmacology, 7: 237-244.

25. Illel B, Schaefer H, Wepierre J, Doucet O (1991) Follicles play an important role in percutaneous absorption, J. Pharm. Sci. 80: 424-427.

26. Rougier A, Rallis M, Kiren P, Lotte C (1990) "In vivo percutaneous absorption: A key role for stratum corneum/vehicle partitioning," Arch Dermatol Res. 282: 498-505.

27. Kalpana SP, Mikolaj M, Courtney LS, Nicole KB, Priyanka G, Audra LS (2010) Challenges and opportunities in dermal/transdermal delivery. Ther. Deliv. 1(1): 109–131.

28. Bos JD, Meinardi MMHM (2000) The 500 Dalton rule for the skin penetration of chemical compounds and drugs. Ex. Dermatol. 9(3): 165–169.

29. Tsutomu I, Tohru M (2010) Techniques for efficient entrapment of pharmaceuticals in biodegradable solid micro/nanoparticle. Expert Opinion Drug Deliv. 7(6): 1-11.

30. Moghimi SM, Hunter AC, Murray JC (2005) Nanomedicine: current status and figure prospects. FASEB J. 19: 311-330.

31. Marcato PD and Duran N (2008) New aspects of nanopharmaceutical delivery systems. J Nanosci Nanotechnology. 6: 2216-2229.

32. Nnamani PO, Kenechukwu FC, Dibua EU, Ogbonna CC, Momoh MA, Ogbonna JDN, Okechukwu DC, Olisemeke AU and Attama AA (2013). Bioactivity of gentamicin contained in novel transdermal drug delivery systems (TDDS) formulated with biodegradable polyesters; Afr. J. Pharm. Pharmacol. 7(28): 1987-1993.

33. Dnyanesh NT, Vavia PR (2003) Acrylate-based transdermal therapeutic system of nitrendipine. Drug Dev. Ind. Pharm. 29: 71–78.

34. Chandak AR, Verma PRP (2008) Development and evaluation of HPMC based matrices for transdermal patches of tramadol. Clin. Res. Reg. Affairs. 25: 13–30.

35. Mundargi RC, Patil SA, Agnihotri SA, Aminabhavi TM (2007) Evaluation and controlled release characteristics of modified xanthan films for transdermal delivery of atenolol. Drug Dev. Ind. Pharm. 33:79–90.

36. Mutalik S, Udupa N (2004) Glibenclamide transdermal patches: Physicochemical, pharmacodynamic and pharmacokinetic evaluations. J. Pharm. Sci. 93: 1577–1594.

37. Chang HI, Perrie Y, Coombes AGA (2006) Delivery of the antibiotic gentamicin sulphate from precipitation cast matrices of polycaprolactone. J. Control. Rel. 110:414-421.

38. Ranade VV, Cannon JB (2011) Drug Delivery Systems, Third Edition, Taylor and Francis, Boca Raton.

39. Touitou E, Dayan N, Bergelson L, et al., (2000). Ethosomes-- novel vesicular carriers for enhanced delivery: characterization and skin penetration properties. J. Control. Rel. 65: 403-18.

40. Martin A, Bustamante P (1993) Physical Pharmacy: Physical Chemical Principles in the Pharmaceutical Sciences, 4th edn, Lippincott Williams & Wilkins, Philadelphia, PA, pp. 436–439.

41. Newman A.W (1995) Micromeritics in Physical Characterization of Pharmaceutical Solids, (ed. H.G. Brittain), Marcel Dekker, New York, NY, Ch. 9, 254–264p.

42. Chikazawa M, Takei T (1997) Specific Surface Area in Powder Technology Handbook (eds K. Gotoh, H. Masuda & K. Higashitani), 2nd edn, Marcel Dekker, Inc., New York, NY, Ch. III. 8, pp. 337–349.

43. Ishida-Yamamoto A, Simon M, Kishibe M, Miyauchi Y, Takahashi H, Yoshida S, O'Brien TJ, Serre G, Iizuka H (2004) 'Epidermal lamellar granules transport different cargoes as distinct aggregates'. Journal of Investigative Dermatology 122(5): 1137-1144.

44. Demarchez M, Asselineau D, Regnier M, Czernielewski J (1992) 'Migration of Langerhans Cells into the Epidermis of Human Skin Grafted onto Nude-Mice'. Journal of Investigative Dermatology, 99(5): S54-S55.

45. Holman BP, Spies F, Bodde HE (1990) 'An Optimized Freeze-Fracture Replication Procedure for Human Skin'. Journal of Investigative Dermatology 94(3):332-335.

46. Norlen L, Al-Amoudi A, Dubochet J (2003) 'A cryotransmission electron microscopy study of skin barrier formation'. Journal of Investigative Dermatology. 120(4): 555-560.

47. Stark B, Pabst G, Prassl R (2010) Long-term stability of sterically stabilized liposomes by freezing and freeze-drying: effects of cryoprotectants on structure. Eur. J. Pharm. Sci. 41: 546-555.

48. Lopez-Pinto JM, Gonzalez-Rodriguez ML, Rabasco AM (2005) Effect of cholesterol and ethanol on dermal delivery from DPPC liposomes. Int. J. Pharm. 298: 1-12.

49. Nakhla T, Marek M, Kovalcik T (2000) Issues associated with large-scale production of liposomal formulations. Drug Deliv. Technol. 2:1-6.

50. Maestrelli F, Gonza´lez-Rodrı´guez ML, Rabasco AM, et al., (2006). Effect of preparation technique on the properties of liposomes encapsulating ketoprofen-cyclodextrin complexes aimed for transdermal delivery. Int. J. Pharm. 312:53-60.

51. Attama AA, Müller-Goymann CC (2008) Effect of beeswax modification on the lipid matrix and solid lipid nanoparticle crystallinity. Colloids and Surfaces A: Physicochemical and Engineering Aspects. 315: 189-195.

52. Celia C, Trapasso E, Cosco D, et al (2009) Turbiscan Lab® Expert analysis of the stability of ethosomes and ultradeformable liposomes containing a bilayer fluidizing agent. Colloids Surf B Biointerfaces 72: 155-60.

53. Watson DG (1999) Pharmaceutical Analysis. A textbook for Pharmaceutical students and Pharmaceutical Chemists. 1st Ed. Churchil Livingstone, UK.

54. Coderch L, de Pera M, Perez-Cullell N, Estelrich J, de la Maza A, Parra JL (1999) 'The effect of liposomes on skin barrier structure'. Skin Pharmacol. Applied Skin Physiol., 12(5): 235-246.

55. Curdy C, Naik A, Kalia YN, Alberti I, Guy RH (2004) 'Non-invasive assessment of the effect of formulation excipients on stratum corneum barrier function in vivo'. Int. J. Pharm. 271(1-2): 251-256.

56. Honeywell-Nguyen PL, Gooris GS, Bouwstra JA (2004) 'Quantitative assessment of the transport of elastic and rigid vesicle components and a model drug from these vesicle formulations into human skin in vivo'. J. Investig. Dermatol. 123(5): 902-910.

57. Jadoul A, Doucet J, Durand D, Preat V (1996) 'Modifications induced on stratum corneum structure after in vitro iontophoresis: ATR-FTIR and X-ray scattering studies'. J. Control. Rel. 42(2):165-173.

58. Attama AA, Müller-Goymann CC (2006) A critical study of novel physically structured lipid matrices composed of a homolipid from Cupra hircus and theobroma oil. Int. J. Pharm. 322, 67-78.

59. Brinkmann I, Müller-Goymann CC (2005) An attempt to clarify the influence of glycerol, propylene glycol, isopropyl myristate and a combination of propylene glycol and isopropyl myristate on human stratum corneum. Pharmazie. 60: 215-220.

60. Oilman MR (1982). Skin and Eye Testing in Animals. Principles and Methods of Toxicology; Hayes, A.W., Ed.; Raven Press: New York. 209–222p.

61. Garg A, Aggarwal D, Garg S, Singla AK (2002). Spreading of Semisolid Formulations An Update. Pharmaceutical Technology, Sept.

62. Idson B, Lazarus J.J (1987) Semisolids. In The Theory and Practice of Industrial Pharmacy, L. Lachman, H.A. Lieberman, and J.L. Kanigs, Eds. Lea and Febiger, Philadelphia, PA, 2d ed. pp. 215–244.

63. Duggin G (1996) Softening skin with emollient ingredients. Manufacturing Chemist 67 (6): 27–31.

64. Barry BW, Grace AJ (1972) Sensory testing of spreadability: Investigation of rheological conditions operative during application of topical preparations," J. Pharm. Sci. 61(3): 335–341.

65. Rance RW (1973) Studies of the factors controlling the action of hair sprays. Part I: The spreading of hair spray resin solutions on hair. J. Soc. Cosm. Chem. 24 (7): 501–522.

66. Islam MT, Rodríguez-Hornedo N, Ciotti S, Ackermann C (2004) Rheological characterization of topical Carbomer gels neutralized to different pH. Pharm. Res. 21(7): 1192-1199.

67. Lawrence MJ, Rees GD (2000) Microemulsion-based media as novel drug delivery systems. Adv Drug Deliv Rev. 45: 89-121.

68. Kreilgaard M (2002) Influence of microemulsions on cutaneous drug delivery. Adv Drug Deliv Rev. 54: S77-S98.

69. He W, Tan Y, Tian Z, et al., (2011). Food protein-stabilized nanoemulsions as potential delivery systems for poorly water-soluble drugs: preparation, in vitro characterization and pharmacokinetics in rats. Int J Nanomedicine. 6: 521-533.

70. Heuschkel S, Goebel A, Neubert RHH (2008) Microemulsions — modern colloidal carrier for dermal and transdermal drug delivery. J Pharm Sci. 97: 603-631.

71. Zhao X, Liu J, Zhang X, et al., (2006) Enhancement of transdermal delivery of theophylline using microemulsion vehicle. Int. J. Pharm. 327: 58-64.

72. Neubert RHH (2011) Potentials of new nanocarriers for dermal and transdermal drug delivery. Eur J Pharm Biopharm. 77: 1-2.

73. Grampurohit N, Ravikumar P, Mallya R (2011) Microemulsions for topical use–a review. Ind J Pharm Edu Res. 45: 100-107.

74. Valenta C, Schultz K (2004) Influence of carrageenan on the rheology and skin permeation of microemulsion formulations. J. Control. Rel. 95: 257-265.

75. Sonneville-Aubrun O., Simonnet J.T., L'Alloret F (2004) Nanoemulsions: a new vehicle for skincare products. Adv Colloid Interface Sci. 108–109:145–149.

76. Seiden MV, Muggia F, Astrow A, et al (2004) A phase II study of liposomal lurtotecan (OSI-211) in patients with topotecan resistant ovarian cancer. Gynecol Oncol. 93: 229–232.

77. Kuo F, Subramanian B, Kotyla T, Wilson TA, Yoganathan S, Nicolosi RJ (2008) Nanoemulsions of an anti-oxidant synergy formulation containing gamma tocopherol have enhanced bioavailability and anti-inflammatory properties. Int. J. Pharm. 363:206–213.

78. Wu H, Ramachandran C, Bielinska AU, et al (2001) Topical transfection using plasmid DNA in a water-in-oil nanoemulsion. Int. J. Pharm. 221:23–34.

79. Subramanian B, Kuo F, Ada E, et al (2008) Enhancement of anti-inflammatory property of aspirin in mice by a nano-emulsion preparation. Int. Immunopharmacol. 2008;8:1533–1539.

80. Mou D, Chen H, Du D, et al (2008) Hydrogel-thickened nanoemulsion system for topical delivery of lipophilic drugs. Int. J. Pharm. 353: 270–276.

81. Wu H, Ramachandran C, Weiner ND, Roessler BJ (2001) Topical transport of hydrophilic compounds using water-in-oil nanoemulsions. Int. J. Pharm. 220: 63–75.

82. Alves MP, Scarrone AL, Santos M, Pohlmann AR, Guterres SS (2007) Human skin penetration and distribution of nimesulide from hydrophilic gels containing nanocarriers. Int. J. Pharm. 341: 215–220.

83. Kumar VS, Asha K (2011) Herbosome a Novel carrier for herbal drug delivery. Int. J Current Pharm. Res. 3(3): 36-41.

84. Jung S, Otberg N, Thiede G, Richter H, Sterry W, Panzner S, Lademann J (2006) Innovative liposomes as a transfollicular drug delivery system: penetration into porcine hair follicles. J. Invest. Dermatol. 126: 1728–1732.

85. Honeywell-Nguyen PL, Wouter Groenink HW, de Graaff AM, Bouwstra JA (2003) The in vivo transport of elastic vesicles into human skin: effects of occlusion, volume and duration of application. J Control. Rel. 90: 243–255.

86. Honeywell-Nguyen PL, de Graaff AM, Groenink HW, Bouwstra JA (2002) The in vivo and in vitro interactions of elastic and rigid vesicles with human skin. Biochim. Biophys. Acta. 1573: 130–140.

87. Verma DD, Verma S, Blume G, Fahr A (2003) Liposomes increase skin penetration of entrapped and non-entrapped hydrophilic substances into human skin: a skin penetration and confocal laser scanning microscopy study. Eur. J. Pharm. Biopharm. 55: 271–277.

88. Cui Z, Han S, Padinjarae D, Huang L (2005) Immunostimulation mechanism of LPD nanoparticles as a vaccine carrier. Mol. Pharm. 2: 22–28.

89. Herffernan M, Murthy N (2005) Polyketal nanoparticles: a new pH-sensitive biodegradable drug delivery vehicle. Bioconjug. Chem. 16: 1340–1342.

90. Dubey V, Mishra D, Dutta T, Nahar M, Saraf DK, Jain NK (2007) Dermal and transdermal delivery of an anti-psoriatic agent via ethanolic liposomes. J. Control. Rel. 123: 148–154.

91. Manosroi A, Kongkaneramit L, Manosroi J (2004) Stability and transdermal absorption of topical amphotericin B liposome formulations. Int. J. Pharm. 270: 279–286.

92. Maestrelli F, González-Rodríguez ML, Rabasco AM, Mura P (2005) Preparation and characterisation of liposomes encapsulating ketoprofen–cyclodextrin complexes for transdermal drug delivery. Int. J. Pharm. 298: 55–67.

93. Essa EA, Bonner MC, Barry BW (2004) Electrically assisted skin delivery of liposomal estradiol; phospholipid as damage retardant. J. Control. Rel. 95: 535–546.

94. Sharma BB, Jain SK, Vyas SP (1994) Topical liposome system bearing local anaesthetic lignocaine: preparation and evaluation. J. Microencapsul. 11: 279–286.

95. Uchegbu IF, Florence AT (1995) Non-ionic surfactant vesicles (niosomes): physical and pharmaceutical chemistry. Adv. Colloid Interface Sci. 58: 1-55.

96. Vora B, Khopade AJ, Jain NK (1998) Proniosome based transdermal delivery of levonorgestrel for effective contraception. J. Control. Rel. 54: 149-165.

97.   Alsarra IA, Bosela AA, Ahmed SM, Mahrous GM (2005) Proniosomes as a drug carrier for transdermal delivery of ketorolac. Eur. J. Pharm. Biopharm. 59: 485–490.

98.   Muzzalupo R, Tavano L, Cassano R, Trombino S, Ferrarelli T, Picci N (2011) A new approach for the evaluation of niosomes as effective transdermal drug delivery systems. Eur. J. Pharm. Biopharm. 79: 28–35.

99.   Manconi M, Caddeo C, Sinico C, et al (2011) Ex vivo skin delivery of diclofenac by transcutol containing liposomes and suggested mechanism of vesicle–skin interaction. Eur. J. Pharm. Biopharm. 78: 27–35.

100.   Mura S, Manconi M, Sinico C, Valenti D, Fadda AM (2009) Penetration enhancer containing vesicles (PEVs) as carriers for cutaneous delivery of minoxidil. Int. J. Pharm. 380: 72–79.

101.   Guinedi AS, Mortada ND, Mansour S, Hathout RM (2005). Preparation and evaluation of reverse-phase evaporation and multilamellar niosomes as ophthalmic carriers of acetazolamide. Int. J. Pharm. 306: 71–82.

102.   Balakrishnana P, Shanmugama S, Lee WS, et al (2009) Formulation and in vitro assessment of minoxidil niosomes for enhanced skin delivery. Int. J. Pharm. 377: 1–8.

103.   Junyaprasert VB, Singhsa P, Suksiriworapong J, Chantasart D (2012) Physicochemical properties and skin permeation of Span 60/Tween 60 niosomes of ellagic acid. Int. J. Pharm. 423: 303–311.

104.   Jain S, Jain P, Umamaheshwari RB, Jain NK (2003) Transfersomes—A Novel Vesicular Carrier for Enhanced Transdermal Delivery: Development, Characterization, and Performance Evaluation. Drug Dev. Ind. Pharm. 29 (9): 1013-1026.

105.   Planas ME, Gonzalez P, Rodriguez S, Sanchez G, Cevc G (1992) Non-invasive percutaneous induction of topical analgesia by a new type of drug carrier and prolongation of the local pain intensity by liposomes. Anesth. Analge. 95: 615–621.

106.   Cevc G, Schatzlein A, Blume G (1995) Transdermal drug carrier basic properties, optimization and transfer efficiency in the case of epicutaneously applied peptides. J. Control. Rel. 36: 3–16.

107.   Paul A, Cevc G, Bachhawat BK (1998). Transdermal immunization with an integral membrane component gap junction protein, by means of ultradeformable drug carriers, transfersomes. Vaccine 16: 188–195.

108.   Mbah CC, Builders PF, Attama AA (2014). Nanovesicular carriers as alternative drug delivery systems: ethosomes in focus Expert Opin. Drug Deliv. 11(1):1-15.

109.   Touitou E (1996). Composition of applying active substances to or through the skin. US5716638.

110.   Jain S, Tiwary AK, Sapra B, Jain NK (2007). Formulaion and evaluation of thosomes for transdermal delivery of lamivudine. AAPS PharmSciTech. 8(4): Article 111.

111. Coderch L., Fonollosa J., De Pera M., et al (2000) Influence of cholesterol on liposome fluidity by EPR: relationship with percutaneous absorption. J Control. Rel. 68: 85-95.

112. Serikawa T, Kikuchi A, Sugaya S, et al., (2006) In vitro and in vivo evaluation of novel cationic liposomes utilized for cancer gene therapy. J. Control. Rel. 113(3): 255-260.

113. Barry BW (2004) Breaching the skin' barrier to drugs. Nat. Biotechnol. 22: 165-167.

114. Honeywell-Nguyen PL, Bouwstra JA (2005) Vesicles as a tool for transdermal and dermal delivery. Drug Discov. Today. 2: 67-74.

115. Elsayed MMA, Abdallah OY, Naggar VF, et al., (2007a). Lipid vesicles for skin delivery of drugs: reviewing three decades of research. Int. J. Pharm. 332: 1-16.

116. Ainbinder D, Touitou E (2005) Testosterone ethosomes for enhanced transdermal delivery. Drug Deliv. 12: 297-303.

117. Paolino D, Lucania G, Mardente D, et al., (2005) Ethosomes for skin delivery of ammonium glycyrrhizinate: in vitro percutaneous permeation through human skin and in vivo anti-inflammatory activity on human volunteers. J. Control. Rel. 106: 99-110.

118. Dayan N, Touitou E (2000) Carriers for skin delivery of trihexyphenidyl HCl: ethosomes vs. liposomes. Biomaterials. 21: 1879-85.

119. Elsayed MM, Abdallah OY, Naggar VF, et al., (2007b) Deformable liposomes and ethosomes as carriers for skin delivery of ketotifen. Pharmazie 62: 133-137

120. Li G, Fan Y, Fan C, et al. (2012) Tacrolimus-loaded ethosomes: physicochemical characterization and in vivo evaluation. Eur. J. Pharm. Biopharm. In press 2012.

121. Chourasia MK, Kang L, Chan SY (2011) Nanosized ethosomes bearing ketoprofen for improved transdermal delivery. Results Pharm Sci. 1: 60–67.

122. Fang YP, Huang YB, Wua PC, Tsai YH (2009) Topical delivery of 5-aminolevulinic acid-encapsulated ethosomes in a hyperproliferative skin animal model using the CLSM technique to evaluate the penetration behavior. Eur J Pharm Biopharm. 73: 391–398.

123. Paolino D, Lucania G, Mardente D, Alhaique F, Fresta M (2005) Ethosomes for skin delivery of ammonium glycyrrhizinate: in vitro percutaneous permeation through human skin and in vivo anti-inflammatory activity on human volunteers. J Control. Rel. 106: 99–110.

124. Touitou E, Dayan N, Bergelson L, Godina B, Eliaz M (2000) Ethosomes – novel vesicular carriers for enhanced delivery: characterization and skin penetration properties. J Control. Rel. 65: 403–418.

125. Maheshwari RGS, Tekade RK, Sharma PA, et al (2012) Ethosomes and ultradeformable liposomes for transdermal delivery of clotrimazole: a comparative assessment. Saudi Pharm J. 20: 161–170.

126. Dayan N, Touitou E (2000) Carriers for skin delivery of trihexyphenidyl HCl: ethosomes vs liposomes. Biomaterials. 21: 1879–1885.

127. Parekh HS (2007) The advance of dendrimers– a versatile targeting platform for gene/drug delivery. Curr. Pharm Des. 13, 2837–2850.

128. Esfand R, Tomalia DA (2001) Poly(amidoamine) (PAMAM) dendrimer: from biomimicry to drug delivery and biomedical applications. Drug Discov. Today 6:427–436.

129. D'Emanuele A, Attwood D (2005) Dendrimer-drug interactions. Adv. Drug Deliv. Rev. 57: 2147–2162.

130. Wang Z, Itoh Z, Hosaka Y, et al (2003) Novel transdermal drug delivery system with polyhydroxyalkanoate and starburst polyamidoamine dendrimer. J Biosci Bioeng. 95: 541–543.

131. Chauhan AS, Sridevi S, Chalasani KB, et al (2003) Dendrimer-mediated transdermal delivery: enhanced bioavailability of indomethacin. J. Control. Rel. 90: 335–343.

132. Yiyun C, Na M, Tongwen X, et al (2007). Transdermal delivery of nonsteroidal anti-inflammatory drugs mediated by polyamidoamine (PAMAM) dendrimer. J. Pharm. Sci. 96: 595–602.

133. Venuganti VVK, Perumal OP (2008) Effect of poly(amidoamine) (PAMAM) dendrimer on skin permeation of 5-fluorouracil. Int. J. Pharm. 361: 230–238.

134. Niederhafner P, Šebestík J, Ježek J (2005) Peptide dendrimers. J. Peptide Sci. 11: 757–788.

135. Mehnert W, Mäder K (2001) Solid lipid nanoparticles. Production, characterization and applications. Adv Drug Del Rev. 47: 165–96.

136. Müller RH, Keck CM (2004) Challenges and solutions for the delivery of biotech drugs – a review of drug nanocrystal technology and lipid nanoparticles. J Biotech. 113: 151–70.

137. Castelli F, Puglia C, Sarpietro MG, et al (2005) Characterization of indomethacin-loaded lipid nanoparticles by differential scanning calorimetry. Int. J. Pharm. 304: 231–238.

138. Siekmann B, Westesen K (1992) Submicron-sized parenteral carrier systems based on solid lipids. Pharm. Pharmacol. Lett. 1: 123–126.

139. Manjunath K, Venkateswarlu V (2005) Pharmacokinetics, tissue distribution and bioavailability of clozapine solid lipid nanoparticles after intravenous and intraduodenal administration. J. Control. Rel. 107: 215–28.

140. Müller RH, Souto EB, Radtke M (2000) PCT application PCT/EP00/04111.

141. Attama AA, Momoh MA and Builders PF. Lipid Nanoparticulate Drug Delivery Systems: A Revolution in Dosage Form Design and Development. InTech. 2012:1-34. http://www.intechopen.com/subjects/pharmacology-toxicology-and-pharmaceutical-science.

142. Wissing SA, Müller RH (2002) The influence of the crystallinity of lipid nanoparticles on their occlusive properties. Int. J. Pharm. 242(1-2): 377-379.

143. Jenning V, Thünemann AF, Gohla SH (2000) Characterisation of a Novel Solid Lipid Nanoparticle Carrier System Based on Binary Mixtures of Liquid and Solid Lipids. Int. J. Pharm. 199(2): 167–177.

144. Villalobos-Hernandez JR, Müller-Goymann CC (2006) Sun protection enhancement of titanium dioxide crystals by the use of carnauba wax nanoparticles: The synergistic interaction between organic and inorganic sunscreens at nanoscale. Int. J. Pharm. 322(1-2): 161-170.

145. Guterres SS, Alves MP and Adriana R (2007) Pohlmann.Polymeric nanoparticles, nanospheres and nanocapsules, for cutaneous application. Drug Target Insights. 2: 147–157.

146. Xu ZP (2006) Inorganic nanoparticles as carriers for efficient cellular delivery. Chem. Eng. Sci. 61: 1027–40.

147. Scheindlin S (2004) Transdermal drug delivery: past, present, future. Mol. Interv. 4(6): 308–312.

148. Micromedex 1.0 (Healthcare Series) Thomson Reuters. [Accessed 27 December, 2013]. www.thomsonhc.com/home.

149. Yarosh DB (1992) Liposome-encapsulated enzymes for DNA repair. In: Braun-Falco O, Korting HC, Maibach H, editors. Liposome dermatics. Springer-Verlag; Heidelberg. pp. 258-69.

150. Barry BW (2001) Novel mechanisms and devices to enable successful transdermal drug delivery. Eur. J. Pharm. Sci. 14(2):101-14

151. Al-Obaidi H, Nasseri B, Florence AT (2010) Dynamics of lipid microparticles inside lipid vesicles: movement in confined spaces. J. Drug Target. 18(10): 821-830.

152. Chiranjib C, Souman P, George PDC, Zhi-Hong W, Chan-Shing L (2013) Nanoparticle as smart pharmaceutical delivery. Frontiers in Bioscience (Landmark Ed) 18: 1030-1050.

153. International Conference on Harmonisation Guidance on Nonclinical Safety Studies for the Conduct of Human Clinical Trials for Pharmaceuticals. Federal Register 1997; 62 (227): 62922–62925.

154. Makena H and Uday B. Kompella (2006) Nanotechnology and Nanoparticles: Clinical, Ethical, and Regulatory Issues. In Nanoparticle nanotechnology for drug delivery. vol. 159. Ed by Ram B Gupta AND Uday Kompella. pp. 381-393.

155. Florence AT, Attwood D (2009) Physicochemical principles of Pharmacy. 4th Ed. Pharmaceutical Press, UK. 329-390p.

156. Wertz PW, Miethke MC, Long SA, Strauss JS, Downing DT (1985) The composition of ceramides from human stratum corneum and from comedones.Journal of Investigative Dermatology 84: 410–12.

157. Devi VK, Nimisha J, Valli SK (2010) The importance of novel drug delivery systems in herbal medicines. Pharmacognosy Review. 4(7): 27-31.

158. Verma M, Gupta PK, Varsha BP, Purohit AP (2007) Development of transdermal drug dosage formulation for the anti-rheumatic ayurvedic medicinal plants. Ancient Sci. Life. 11: 66-69.

159. Kidd PM (2011) Phytosome: A technical revolution in phytomedicine. Available from: www.indena.com [Last accessed 10 September 2011].

160.   Ajazuddin SS (2010) Applications of novel drug delivery system for herbal formulations. Fitoterapia. 81: 680-9.
161.   Goyal A, Kumar S, Nagpal M, et al., (2011) Potential of novel drug delivery systems for herbal drugs. Ind. J. Pharm. Edu. Res. 45(3): 225-235.

CHAPTER 9

Synthesis and Biological Activity of Drug Delivery System Based on Chitosan Nanocapsules

Mohamed Gouda[1,2*], Usama Elayaan[1,3], Magdy M. Youssef[1,3]

[1]Department of Chemistry, College of Science, King Faisal University, Hofuf, Kingdom of Saudi Arabia
[2]Textile Research Division, National Research Center, Cairo, Egypt
[3]Department of Chemistry, Faculty of Science, Mansoura University, Mansoura, Egypt

ABSTRACT

Chitosan nanocapsules containing naproxen as an active ingredient were synthesized by ionic ge- lation method in presence of polyanion tripolyphosphate as a crosslinker. The morphology and diameter of the prepared chitosan nanoparticles was characterized using scanning electron mi- croscopy and transition electron microscopy. Different factors affecting on the size diameter of chitosan nanoparticles such as stirring time and temperature, pH values as well as chitosan concentration were studied. Different factors affecting on the immobilization of naproxen into chitosan nanoparticles such as time, temperature and pH values were optimized. Synthesized naproxen/chitosan nanocapsules were assessed against both Gram positive bacterial strain such as Bacillus subtilis and Staphylococcus aureus and Gram negative bacterial strain such as Pseudomonas aeruginosa and Escherichia coli. Also, the antifungal activity of the naproxen/chitosan nanocapsules against Saccharomyces cerevisiae was demonstrated. Super oxide dismutase like activity of naproxen/chitosan nanocapsules will be determined.

Keywords: Chitosan, Naproxen, Nanocapsules, Drug Delivery Systems, Antimicrobial, Antioxidant, DNA Degradation

1. INTRODUCTION

Chitosan, poly[β-(1-4)-linked-2-amino-2-deoxy-D-glucose], is the N-deacetylated product of chitin [1] . The difference between chitosan and chitin is only in the functional group situated at carbon-2 of the monomeric unit. Owing to the presence of free amino groups in chitosan, the solubility and reactivity of this polymer are greater than those of chitin [2]. In addition, chitosan has many significant biological and chemical properties: it is biodegradable, biocompatible, bioactive, and polycationic [3]. Thus, it has been widely used in many industrial and biomedical aspects, including in enzyme immobilization, and as carrier for controlled drug delivery [4] - [7]. In addition, chitosan is economically attractive because chitin is the most abundant natural polymer after cellulose. However, chitosan is macromolecular, which significantly marks its application. To overcome this drawback, the use of chitosan fabricated nano/submicron chitosan is effective. In recent years, nanotechnology has showed significant attractiveness for the preparation of drug carriers. Under the scale of nano, nanomaterials have characteristics such as magnetism and large surface area. These characteristics are favorable for drug immobilization. Many studies have mainly reported the preparation of chitosan nanoparticles and their applications in the carrier of drugs [8] [9].

Hollow nanocapsules in the fields of pharmaceutics, cosmetic, food, textile, adhesive, agricultural industry, artificial cells, and protection of proteins, enzymes, DNA, and catalysis have been widely used [10] [11]. All these based on their isolating property, large inner volume, and tunable permeability [12] [13]. To obtain the nanocapsules with multipurpose properties, new efforts are always tried to explore various techniques for fabrication. The template method as a common method to prepare nanocapsules often needs a template such as micelles [14] , calcium carbonate [15] , polyurea [16] , and emulsion droplets [17] . In the template method, the target material is precipitated or polymerized on the surface of the template. Then, the template is removed to form a cavity, leading to a hollow sphere structure [18] [19]. Another common method to fabricate nanocapsules by the layer by-layer assembly [20] [21] , multilayer nanocapsules with ultrathin wall thickness and tunable wall structures and properties has been fabricated [22] [23] . Furthermore, naproxen is a nonsteroidal antiinflammatory drugs. In the body, it works by reducing the hormones which causes inflammation and pain.

Naproxen (Figure 1) [24] , is one of the most commonly used propionic acid derivatives for the treatment of pain, joint swelling and symptoms of arthritis, it is worked by blocking the action of cyclooxygenase involved in the production of prostaglandins that are produced in response to injury or certain diseases and cause pain, swelling and inflammation. However it's causing some gastrointestinal side effects possibly caused by the free acidic group present. Because of their distinctive structural features and wide range of pharmacological activities, hydrazones have attracted enormous interest especially in medicinal chemistry [25]. Moreover, naproxen has shown anti-mycobacterial activities when tested in vitro [26] [27]. The objective of the present study is to synthesize and characterize chitosan nanocapsules using ionic gelation method in presence of polyanion tripolyphosphate as a cross-linker and to immobilize the naproxen as an active ingredient. The prepared nanocapsules will be characterized using ESM and TEM spectroscopy. Biological activities of prepared nanocapsules as antimicrobial activity against both gram positive and gram negative bacterial strains, DNA inhibition as well as super oxide radical inhibition will be evaluated.

2. EXPERIMENTAL METHODS

2.1. Materials

Chitosan low molecular weight (70,000 Daltons) with deacetylation degree of 85% and sodium triphosphate (TPP) were purchased from Aldrich Co., naproxen was obtained from Aldrich Co., acetic acid and ethanol (99.7%) were obtained from Merck Co.

2.2. Preparation of Chitosan Nanoparticles in Water Phase

Chitosan nanoparticles were prepared according to the ionotropic gelation process [28]. Various concentrations of chitosan (0.1%, 0.5%, 1.0%, 1.5% and 2%) were dissolved in acetic aqueous solution. At room temperature, 20 mL of various concentrations of TPP solution (0.1, 0.50, 1.0, 1.5 and 2.0 mg/mL) were added to 40 mL of chitosan solution at different pH, respectively. The mixtures were stirred with high speed ultrasonic stirrer at different time and temperatures. Prepared chitosan nanoparticles were filtered off from the solutions by centrifuge with 11,000 rpm. Filtered chitosan nanoparticles were dried in an oven at 50°C for 30 min.

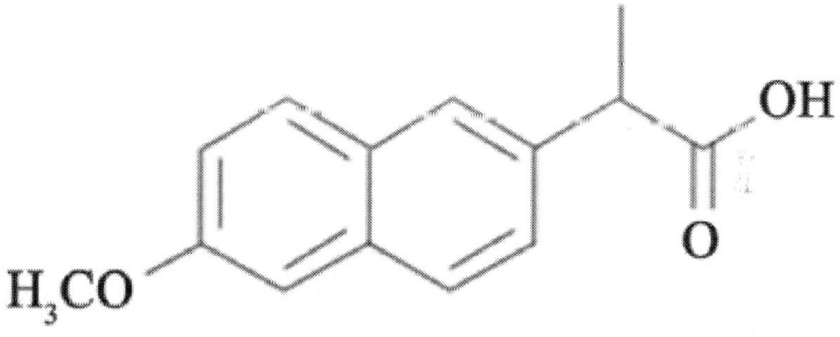

Figure 1. Structure of naproxen.

2.3. Characterization of Chitosan Nanoparticles

The morphologic characterization of chitosan nanoparticles was evaluated using a scanning electron microscopy (JSM-5610LV; JEOL, Japan). Particle size diameter of the prepared chitosan nanoparticles was evaluated using transmission electron microscopy (TEM) TEM (ZEISS-EM-10-GERMANY).

2.4. Immobilization of Naproxen

Ten milliliters of chitosan nanoparticle solution containing 1.0 mg of dry chitosan nanoparticles in the tube was incubated in a water bath for 30 min. Then 1.0 mL of 1.0 mg/mL naproxen was added to this tube, and the mixture was stirred with ultrasonic stirrer for different times (10, 20, 30, and 40 min) at different temperatures (40°C, 50°C, 60°C and 70°C) and different pH values (2.0, 3.0, 4.0, 5.0, 6.0 and 7.0). Naproxen immobilized on chitosan nanoparticles was removed by centrifugation at 11000 rpm, and the supernatant was collected to calculate the residual amount of naproxen.

2.5. Determination of Naproxen Concentration

Visible spectrophotometric method has been developed for the determination of naproxen concentration according to reported method [29] . The developed method is based on reaction of naproxen with phenol red. It was quantified spectrophotometrically at their absorption maximum at 422 nm.

2.5.1. Preparation of Standard Naproxen Solution
Standard solution of naproxen (1000 μg/ml) was prepared by dissolving 100 mg in methanol and diluting to the mark in a 100 ml volumetric flask. Working standard solution of 100 μg/ml was prepared by further dilution of the above standard stock solution

Phenol Red Method: Aliquots of the working standard solution of naproxen (60 - 80 μg/ml) were prepared and from that 1ml of sample was accurately measured and transferred into a series of volumetric flasks by means of a micro burette. To each of the above aliquots, 1 ml of 100 μg/ml of phenol red solution in methanol was added and mixed thoroughly. The volume was brought up to 5 ml mark with methanol, mixed thoroughly and after 10 min absorbance of each species was measured at 422 nm against reagent blank. A calibration curve was constructed by plotting the absorbance against the concentration of the drug.

2.6. Biological Activity of Naproxen/Chitosan Nanocapsules

2.6.1. Agarose Gels Electrophoresis
The chitosan, naproxen, and naproxen/chitosan nanocapsule (50 μg) were added individually to 1 μg of the DNA isolated from E. coli strain W3110 [30] . The samples were incubated for 1 h at 37°C. The DNA was analyzed by using horizontal agarose gels electrophoresis. The electrophoresis was performed using 0.7% (w/v) agarose gels in TAE buffer (5 mM sodium acetate, 1 mM EDTA and 0.04 M Tris-HCl pH 7.9). The agarose gels were stained with ethidium bromide (0.5 μg/ml) and the DNA was visualized on a UV transilluminator [31] .

2.6.2. Polyacrylamide Gel Electrophoresis
Egg Albumin (5 μg) was treated with the chitosan, naproxen, and naproxen/chitosan nanocapsule (10 μg) individually. The reaction mixtures were incubated for 1 h at 37°C. The protein samples were examined by using vertical one dimensional SDS-polyacrylamide gel electrophoresis according to the method of Laemmli [32] .

2.6.3. Antibacterial and Antifungal Activities

The antimicrobial investigation of chitosan, naproxen, and naproxen/chitosan nanocapsule was carried out using cup diffusion technique [33] . The test was carried out against the Gram-negative bacterial strains Pseudomonas aeruginosa (P. aeruginosa) and Escherichia coli (E. coli), the Gram-positive bacterial strains Bacillus subtilis (B. subtilis) and Staphylococcus aureus (S. aureus) and the fungal strain Saccharomyces cerevisiae (S. cerevisiae). The tested chitosan, naproxen and naproxen/chitosan nanocapsule were dissolved in Dimethyl sulfoxide (DMSO) at concentration I mg/ml. The Luria-Bertani Agar (LBA) Medium (10 g bacto-tryptone, 5 g yeast extract, 20 g agar, and 10 g NaCl in 1 Liter de-ionized water) was made for inoculation and bacterial growth. An aliquot of the solution of the tested naproxen/chitosan nanocapsule equivalent to 100 μg was placed separately in cups, cut in the agar. The LBA dishes were incubated for 24 hours at 37°C and the resulting inhibition zones were measured. From the inhibition zone diameter data analysis, the antimicrobial activity against the Gram-negative, Gram-positive bacteria and the fungal were determined.

2.6.4. Determination of Super Oxide Dismutase (SOD) Like Activity

The chitosan, naproxen and naproxen/chitosan nanocapsule were assayed for super oxide dismutase enzyme like activity [34] . SOD like activity of chitosan, naproxen and naproxen/chitosan nanocapsule was assayed by using phenazenmethosulphate (PMS) to generate a superoxide anion radicals at pH = 8.3 (phosphate buffer). Reduction of nitrobluetetrazolium (NBT) to form blue formazan was used as an indicator of superoxide production and followed spectrophotometrically at 560 nm. The addition of PMS (9.3×10^{-5} M) to a solution of NBT (3×10^{-3} M), NADH (4.7×10^{-4} M) and phosphate buffer (final volume of 1 ml) caused a change of OD ($\Delta1$) 560 nm per 4 min. The reactions in blank samples and in the presence of chitosan, naproxen and naproxen/chitosan nanocapsule were measured. For comparative purposes, the activity of native horseradish superoxide dismutase (HR SOD) has also been determined.

3. RESULTS AND DISCUSSION

3.1. Preparation of Chitosan Nanoparticles

3.1.1. Effect of Stirring Time on Size of Chitosan Nanoparticles

To study the influence of stirring time on the size of chitosan nanoparticles, stirring time was different from 10 to 40 min using ultrasonic stirrer at 21,000 rpm; the results are shown in Figure 2. It can be seen that, the particle size of prepared chitosan nanoparticles decreases with increasing of stirring time. Stirring time for 10 min, the particle size was 100 - 120 nm and by increasing of stirring time for 40 min the size of prepared nanoparticles became 30 - 60 nm. This trend could be explained by the energy transfer differences for different stirring times [17] [18] .

3.1.2. Effect of Temperature on Size of Chitosan Nanoparticles

Figure 3 shows the effect of temperature on the size of chitosan nanoparticles. The size of chitosan nanoparticles almost did not change from 30°C to 60°C. However, above 60°C, the solution became transparent, and chitosan nanoparticles were not fabricated. An explanation for this phenomenon is that the chitosan molecular chain might be cut short and fabrication was not done under the same conditions.

3.1.3. Effect of pH on Size of Chitosan Nanoparticles

Chitosan nanoparticles were successfully prepared by ionic gelation method. Figure 4 shows the effects of pH on particle size. The mean diameter of chitosan nanoparticles decrease with increases of pH. This could be related to different chitosan molecular conformation with changing pH before nanoparticle formation. Because the isoelectric point of chitosan was 6.8, the majority of amino groups were protonated to form an extended molecular chain in acid solution owing to strong repulsion existing among positively charged ammonium groups at low pH (3.0) [16] . A more extended spherical shape was formed on the addition of TPP solution. With an increase in pH (from 3.0 to 7.0), positive charges would be neutralized with gradual de-protonation of ammonium groups, resulting in a less extended molecular chain of chitosan to form uniform small nanoparticles.

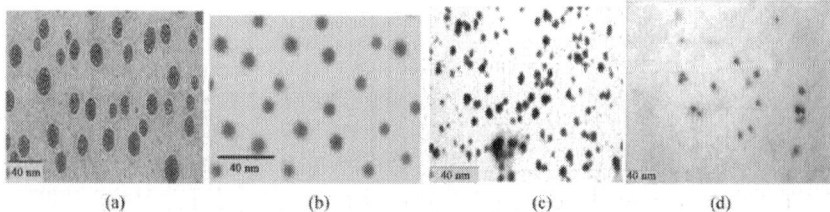

(a)   (b)   (c)   (d)

Figure 2. TEM micrograph of prepared chitosan nanoparticles at different stirring rate (a) 10 min; (b) 20 min; (c) 30 min and (d) 40 min. Using chitosan concentration; 0.1% with TPP concentration; 0.1 mg/mL, at pH; 7.0 and at temperature; 30°C.

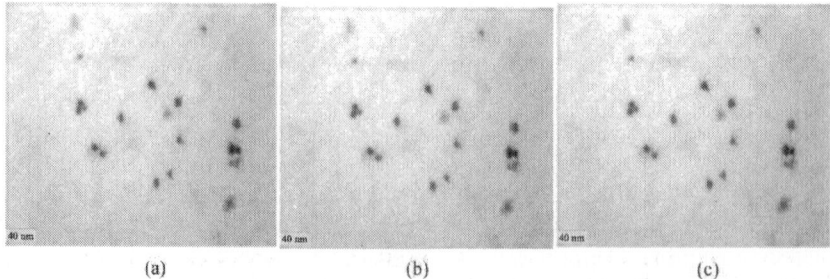

(a)   (b)   (c)

Figure 3. TEM micrograph of prepared chitosan nanoparticles at different temperature (a) 30°C; (b) 40°C; (c) 60°C. Using chitosan concentration; 0.1% with TPP concentration; 0.1 mg/mL, at pH; 7.0 and for stirring time; 40 min.

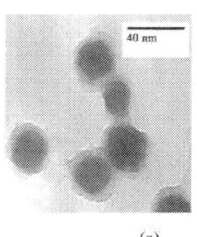 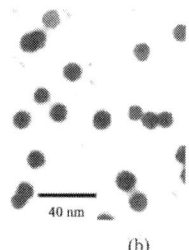 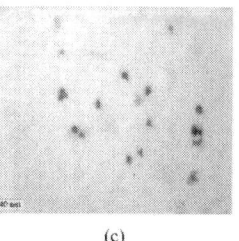

(a)   (b)   (c)

Figure 4. TEM micrograph of the prepared chitosan nanoparticles using different pH values (a) pH = 3.0; (b) pH = 5.0 and (c) pH = 7.0. Using chitosan concentration; 0.1% with TPP concentration; 0.1 mg/mL, at temperature 30°C and for stirring time; 40 min.

3.1.4. Effect of Chitosan Concentrations on the Size of Chitosan Nanoparticles

The results in Figure 5 show that the formation of nanoparticles at different chitosan concentrations (0.1%, 0.50%, 1.0%, 1.5% and 2.0%) was possible only within some moderate concentrations of chitosan and TPP. Some opalescent suspension was formed between TPP solution of 1.0 - 2.0 mg/ml. It has been observed that the highly viscous nature of the gelation medium will aggregate the formed nanoparticles [15] . Thus, it was supposed that a relatively lower viscosity of chitosan with a lower concentration promoted the formation of nanoparticles between chitosan and TPP.

3.2. Sem of Prepared Chitosan Nanoparticles

Figure 6 presents scanning electron micrographs of nanoparticles prepared by ionotropic gelation process. Spherical nanoparticles were formed spontaneously upon the incorporation of TPP solution to the chitosan solution under ultrasonic stirring. Chitosan nanoparticles are obtained by ionic gelation which is a simple process, where particles are formed by means of electrostatic interactions between the positively charged chitosan chains and polyanions (TPP) employed as cross linkers.

3.3. Conditions of Naproxen Immobilization

3.3.1. Effect of Time and Temperature on Immobilization

Naproxen was immobilized on chitosan nanoparticles at different times and temperature. The residual naproxen of the solution was determined using the method described above. As shown inTable 1, the optimal immobilization time and temperature was 15 min at 60°C. When the immobilization time was more than 15 min and temperature was higher than 60°C, the relative immobilization rate did not increase any more. This is attributed to that, at temperature higher than 60°C for time more than 15 min the prepared chitosan nanoparticles were aggregated and this may be effect on the immobilization velocity because the immobilization rate with chitosan nanoparticles was very quick because of their larger specific surface area of contact.

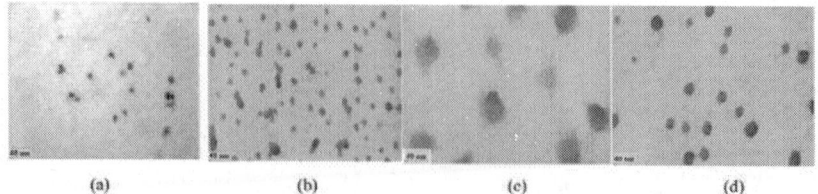

(a)   (b)   (c)   (d)

Figure 5. TEM micrograph of chitosan nanoparticles using different chitosan concentrations, (a) 0.1%; (b) 0.5%; (c) 1.0% and (d) 2.0%

Figure 6. SEM micrograph of chitosan nanoparticles prepared using 0.1% of chitosan with 0.1 mg/mL TPP at 30°C stirring time 40 min at pH 7.0.

Table 1. Residual concentration of naproxen (mg/gm. sample) at different time and temperature using 0.1 gm naproxen, 0.1 g% chitosan with 0.1 mg/ml TPP.

Time (min)	Naproxen concentration (mg/gm. sample)			
	40°C	50°C	60°C	70°C
5	0.18	0.16	0.13	0.15
10	0.15	0.12	0.1	0.11
15	0.1	0.08	0.05	0.07
20	0.1	0.08	0.04	0.07
25	0.1	0.08	0.04	0.07

3.3.2. Effect of pH on Immobilization

Naproxen was immobilized on chitosan nanoparticles at different pH values in phosphate buffer solutions (pH 7.2). The residual naproxen of solution was determined as the method described above. As shown in Figure 7, the relative immobilization rate increases with increasing the pH of the solution up to 5 and then decreases with increasing the pH value more than 5. This is may be attributed to that, when the solution pH was more than 5, the chitosan nanoparticles began to accumulate into larger chitosan microparticles which precipitated in the solution and the relative immobilization rate began to decrease.

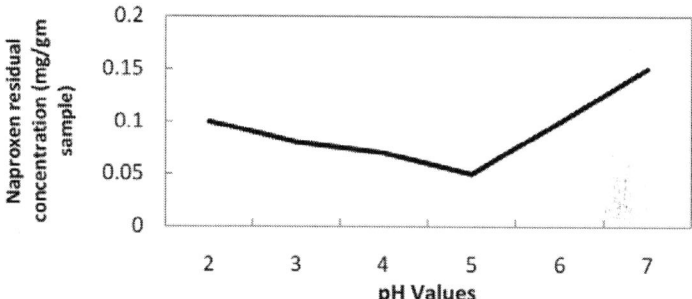

Figure 7. Effect of different pH values on immobilization of Naproxen into chitosan nanoparticles using 0.1 gm naproxen, 0.1 gm chitosan with 0.1 mg/ml TPP at 60°C for 15 min.

3.4. Biological Activity

3.4.1. Effect of Chitosan, Naproxen, and Chitosan Nanoparticles on DNA and Protein

The degradation effect of 50 µg of the chitosan, naproxen and naproxen/chitosan nanocapsule on the DNA in vitro is illustrated in Figure 8. The negative control (only DNA) and the positive control (DNA in DMSO) do not exhibit any degradation on DNA through the incubation period as illustrated in Figure 8 lanes 1 and 2, respectively. Chitosan has no degradation effect on the DNA as shown inFigure 8 lane 3. Naproxen has a significant degradation effect on the DNA as represented in Figure 8 lane 4. A complete degradation effect was exhibited by naproxen/chitosan nanocapsule as illustrated in Figure 8 lane 5. Therefore, the and naproxen/chitosan nanocapsule can be used as a promising anti-tumor agent in vivo to inhibit the DNA replication in the cancer cells and not allow the tumor for further growth. More work in vivo requirements to carry out to clarify their complete role. Further biochemical studies to reveal the effect of the chitosan, naproxen, and naproxen/chitosan nanocapsule on bovine serum albumin (BSA) as a high molecular weight biological compound was carried out. The result of the chitosan, naproxen, and naproxen/chitosan nanocapsule on the BSA was represented in Figure 9. The BSA and BSA in DMSO were utilized as controls as shown Figure 9 lanes 1 and 2, respectively. Chitosan has no effect on the BSA as compared to the control and represented in Figure 9 lane 3. While, naproxen exhibited a little degradation effect on the BSA as shown in Figure 9 lane 4. However, the naproxen/chitosan nanocapsule exhibited a partial degradation effect on the BSA as shown in Figure 9 lane 5.

The naproxen alone and naproxen/chitosan nanocapsule are able to interact with nucleophilic molecules including DNA. In the present study, naproxen/chitosan nanocapsule were demonstrated to degrade the in vitro DNA. The naproxen/chitosan nanocapsule interact with DNA forming inter-and intra-strand adducts, hindering DNA replication, leading to cell cycle arrest and apoptosis.

3.4.2. Antioxidant Activity of Chitosan, Naproxen, and Naproxen/Chitosan Nanocapsule

The SOD mimics like activity of chitosan, naproxen and naproxen/chitosan nanocapsule is represented in Table 2. Chitosan, naproxen, and naproxen/chitosan nanocapsule exhibit a significant SOD like activity and cause an inhibition percent of 78.01%, 66.61% and 72.07%, respectively. Chitosan, naproxen and naproxen/chitosan nanocapsule inhibited superoxide radical generation. Maintaining the equilibrium between the rate of radical generation and the rate of radical scavenging is an essential part of biological homeostasis. Therefore, it is suggested that the inhibition of superoxide radical generation by chitosan, naproxen and naproxen/chitosan nanocapsule is attributable to their free radical scavenging activity.

3.4.3. Antimicrobial Activity of Chitosan, Naproxen and Naproxen/Chitosan Nanocapsule

Ampicillin, a broad-spectrum antibiotic, was utilized as a positive control for this test. The results of the antimicrobial test of chitosan, naproxen and naproxen/chitosan nanocapsule against Gram-negative (P. aeruginosa and E. coli) and Gram-positive (B. subtilis and S. aureus) bacterial strains are summarized in Table 3. The naproxen/chitosan nanocapsule has maximal antimicrobial activity with inhibition zone diameter against E. coli 27 mm, P. aeruginosa 26 mm, B. subtilis 23 mm, S. aureus 27 mm and S. cerevisiae 24 mm as represented in Table 3. Also, Chitosan showed a wide spectrum of antimicrobial activity with inhibition zone diameter against E. coli 25 mm, P. aeruginosa 23 mm, B. subtilis 24 mm, S. aureus 24 mm and S. cerevisiae 20 mm as represented in

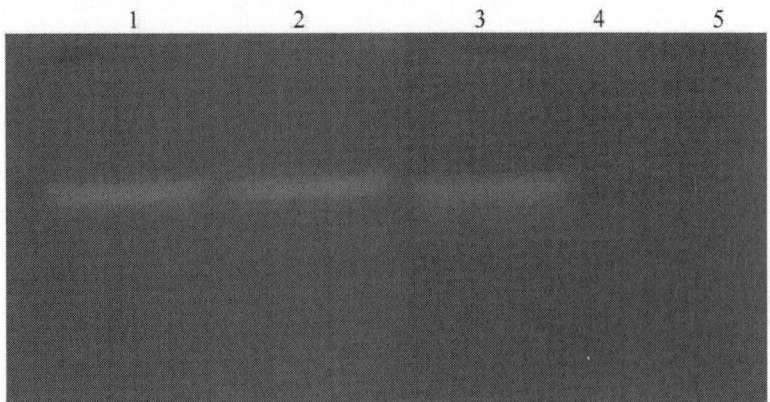

Figure 8. A figure showing the degradation effect of chitosan and naproxen chitosan nanoparticles on the DNA isolated from E. coli strain W3110.

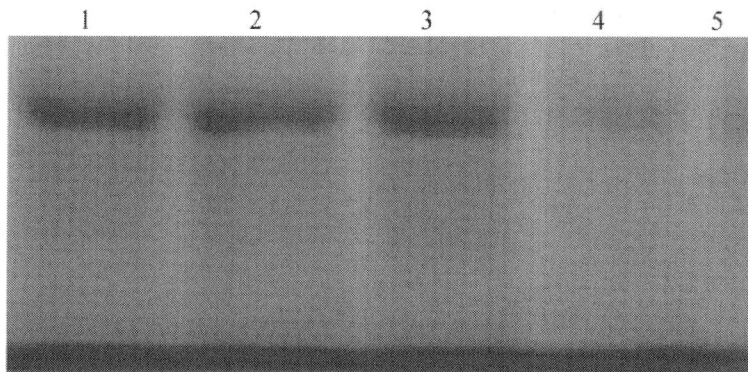

Figure 9. A figure showing the degradation effect of the chitosan, naproxen, and and naproxen/chitosan nanocapsule on the bovine serum protein.

Table 2. Superoxide (SOD) like activity of chitosan, naproxen and naproxen/chitosan nanocapsule as antioxidant enzyme.

	∆ through 4 min	% inhibition
Control	0.623	-
HR SOD	0.158	74.64
Chitosan	0.137	78.01
Naproxen	0.208	66.61
Naproxen/chitosan nanocapsule	0.174	72.07

% inhibition = (∆ Control − ∆ Test/∆ Control) × 100

Table 3. Effect of chitosan, naproxen, and chitosan nanoparticles on some microorganisms "the results expressed as zone inhibition in mm diameter".

	E. coli	P. aeruginosa	S. aureus	B. subtilis	S. cerevisiae
Ampicillin	22	18	23	19	17
Chitosan	25	23	24	24	20
Naproxen	14	11	9	13	11
Naproxen/chitosan nanocapsule	27	26	27	23	24

Table 3. However, the naproxen exhibited a moderate antimicrobial activity as the inhibition zone diameter against the tested microorganisms compared to the wide spectrum standard antibiotic ampicillin as illustrated in Table 3.

4. CONCLUSION

Ionic gelation technique was used as an efficient processing method to synthesize of chitosan nanocapsules containing naproxen as an active agent. Characterization and evaluation of these synthesized products were undertaken.

The synthesis involved preparation of chitosan nanoparticles cross linked with polyanion tripolyphosphate (TPP) and the factors affecting on the diameter of prepared chitosan nanoparticles were optimized and the major conclusion arrived at from these studies are given under: 1) The particle size of prepared chitosan nanoparticles decreases with increasing of stirring time. Stirring time for 10 min, the particle size was 100 - 120 nm and by increasing of stirring time for 40 min the size of prepared nanoparticles became 30 - 60 nm. 2) The size of chitosan nanoparticles almost did not change from 30°C to 60°C. However, above 60°C, the solution became transparent, and chitosan nanoparticles were not fabricated. 3) The mean diameter of chitosan nanoparticles decreases with increasing of pH, indicating an influence of pH on particle size. 4) The formation of nanoparticles at different chitosan concentrations (0.1%, 0.50%, 1.0%, 1.5% and 2.0%) was possible only within some moderate concentrations of chitosan and TPP. Some opalescent suspension was formed between TPP solution of 1.0 - 2.0 mg/ml. Different factors affecting on the immobilization of naproxen into chitosan nanoparticles were studied and the major conclusion arrived at from these studies are given as follows: 1) The optimal immobilization time and temperature was 15 min at 60°C. 2) The optimal immobilization pH was 5. The degradation effect of the chitosan, naproxen as well as naproxen/chitosan nanocapsule on the DNA in vitro. Chitosan as a macromolecule has no degradation effect on the DNA, naproxen alone has a significant degradation effect on the DNA while a complete degradation effect was exhibited by naproxen/chitosan nanocapsule. Chitosan has no effect on the BSA while, naproxen exhibited a little degradation effect on the BSA. However, the naproxen/chitosan nanocapsule exhibited a partial degradation effect on the BSA. The SOD mimics like activity of chitosan, naproxen, and naproxen/chitosan nanocapsule exhibit a significant SOD like activity and cause an inhibition percent of 78.01%, 66.61% and 72.07%, respectively. The naproxen/chitosan nanocapsule has maximal antimicrobial activity with inhibition zone diameter against E. coli 27 mm, P. aeruginosa 26 mm, B. subtilis 23 mm, S. aureus 27 mm and S. cerevisiae 24 mm. Also, Chitosan showed a wide spectrum of antimicrobial activity with inhibition zone diameter against E. coli 25 mm, P. aeruginosa 23 mm, B. subtilis 24 mm, S. aureus 24 mm and S. cerevisiae 20 mm. However, the naproxen exhibited a moderate antimicrobial activity as the inhibition zone diameter against the tested microorganisms compared to the wide spectrum standard antibiotic ampicillin.

REFERENCES

1.   Juang, R.S., Wu, F.C. and Tseng, R.L. (2001) Solute Adsorption and Enzyme Immobilizationon Chitosan Beads Prepared from Shrimp Shell Wastes. Bioresource Technology, 80, 187-193.

2.   Monteiro, O.A. and Airoldi, C. (1999) Some Studies of Crosslinking Chitosan Glutaraldehyde Interaction of Homogeneous System. International Journal of Biological Macromolecules, 26, 119-128.

3.   Denkbas, E.B., Kilicay, E., Birlikseven, C. and Ozturk, E. (2002) Magnetic Chitosan Microspheres: Preparation and Characterization. Reactive and Functional Polymers, 50, 225-232.

4.   Selmer-Olsen, E., Ratnaweera, H.C. and Pehrson, R. (1996) Novel Treatment Process for Dairy Wastewater with Chitosan Produced from Shrimp-Shell Waste. Water Science and Technology, 34, 33-40.

5.   Kucera, J. (2004) Fungal Myceliumthe Source of Chitosan for Chromatography. Journal of Chromatography B, 808, 69-73.

6.   Chiou, S.H. and Wu, W.T. (2004) Immobilization of Candida Rugosa Lipase on Chitosan with Activation of Hydroxyl Group. Biomaterials, 25, 197-204.

7.   Mi, F.L., Kuan, C.Y., Shyu, S.S., Lee, S.T. and Chang, S.F. (2000) The Study of Gelation Kinetics and Chain-Relaxa- tion Properties of Glutaraldehyde-Cross-Linked Chitosan Gel and Their Effects on Microspheres Preparation and Drug Release. Carbohydrate Polymers, 41, 389-396.

8.   Berthold, A., Cremer, K. and Kreuter, J. (1996) Preparation and Characterization of Chitosan Microspheres as Drug Carrier for Prednisolone Sodium Phosphate as Model for Anti-Inflammatory Drugs. Journal of Controlled Release, 39, 17-25.

9.   Tian, X.X. and Groves, M.J. (1999) Formulation and Biological Activity of Antineoplastic Proteoglycans Derived from Mycobacterium vaccae in Chitosan Nanoparticles. Journal of Pharmacology and Pharmacotherapeutics, 51, 151-157.

10. Du, J., Chen, Y., Han, C. and Schmidt, M. (2003) Organic/Inorganic Hybrid Vesicles Based on a Reactive Block Copolymer. Journal of the American Chemical Society, 123, 14710-14711.

11. Clark , C.G. and Wooley, K.L. (2001) Dendrimers and Other Dendritic Polymers. In: Tomalia, D .A ., Ed., Regioselectively-Crosslinked Nanostructures, Wiley, New York, 166-174.

12. Peyratout, C.S. and Dahne, L. (2004) Tailor-Made Polyelectrolyte Microcapsules: From Multilayers to Smart Containers. Angewandte Chemie International Edition, 43, 3762-3783.

13. Sukhorukov, G.B. , Rogach, A.L. , Zebli, B., Liedl, T., Skirtach, A.G. and Parak, W.J.(2005) Nanoengineered Polymer Capsules: Tools for Detection, Controlled Delivery, and Site-Specific Manipulation. Small, 1, 194-200.

14. Bédard, M.F. , De Geest , B.G. , Skirtach, A.G. , Möhwaldb, H. and Sukhorukov, G.B. (2010) Polymeric Microcapsules with Light Responsive Properties for Encapsulation and Release. Advances in Colloid and Interface Science, 158, 2-14.

15. Li, X.D. , Lu, T., Xu, J.J. , Hu, Q.L. and Shen, J.C. (2009) A Study of Properties of "Micelle-Enhanced" Polyelectrolyte Capsules: Structure, Encapsulation and in Vitro Release. Acta Biomaterialia, 5, 2122-2131.

16. Li, G., Feng, Y.Q. , Gao, P. and Li, X.G. (2008) Preparation of Mono-Dispersed Polyurea-Urea Formaldehyde Double Layered Microcapsules. Polymer Bulletin, 60, 725-731.

17.   Zhao, M.W. , Zheng, L.Q. , Bai, X.T. , Li, N. and Yu, L. (2009) Fabrication of Silica Nanoparticles and Hollow Spheres Using Ionic Liquid Microemulsion Droplets as Temdishes. Colloids and Surfaces A: Physicochemical and Engineering Aspects, 346, 229-236.

18.   Zhang, K., Zheng, L.L. , Zhang, X.H. , Chen , X. and Yang, B. (2006) Silica-PMMA Core-Shell and Hollow Nanospheres. Colloids and Surfaces A: Physicochemical and Engineering Aspects, 277, 145-150.

19.   Liu, H.X. , Wang, C.Y. , Gao, Q.X. , Chen , J.X. , Ren, B.Y. and Tong, Z. (2009) Facile Fabrication of Well-Defined Hydrogel Beads with Magnetic Nanocomposite Shells. International Journal of Pharmaceutics, 376, 92-98.

20.   Szarpak, A., Cui, D., Dubreuil, F., De Geest , B.G. , De Cock , L.J. , Picart, C. and Ly-Velty, R.A. (2010) Designing Hyaluronic Acid-Based Layer-by-Layer Capsules as a Carrier for Intracellular Drug Delivery. Biomacromolecules, 11, 713-720.

21.   Endo, Y., Sato , K. and Anzai, J.I. (2010) Preparation of Avidin-Containing Polyelectrolyte Microcapsules and Their Uptake and Release Properties. Polymer Bulletin, 66, 711-720.

22.   Taqieddin, E. and Amiji, M. (2004) Enzyme Immobilization in Novel Alginate-Chitosan Core-Shell Microcapsules. Biomaterials, 25, 1937-1942.

23.   Tang, Y.F. , Zhao, Y.Y. , Li, Y. and Du, Y.M. (2010) A Thermo-Sensitive Chitosan/Poly(Vinyl Alcohol) Hydrogel Containing Nanoparticles for Drug Delivery. Polymer Bulletin, 64, 791-804.

24.   Harrington , P.J. and Lodewijk, E. (1997) Large-Scale Synthetic Process for (S)-Naproxen by Syntex. Organic Process Research Development, 1, 72-76.

25.   Rollas, S. and Kucukguzel, S.G. (2007) Biological Activities of Hydrazone Derivatives. Molecules, 12, 1910-1939.

26.   Sriram, D., Yogeeswari, P. and Devakaram, R.V. (2006) Synthesis, in Vitro and in VivoAntimycobacterial Activities of Diclofenac Acid Hydrazones and Amides. Bioorganic Medicinal Chemistry, 14, 3113-3118.

27.   Munoz-Muniz,   O.   and   Juaristi,   E. (2003) Enantioselective Protonation of Prochiralenolates in the Asymmetric Synthesis of (S)-Naproxen. Tetrahedron Letters, 44, 2023-2026.

28.   Tang, Z .X . , Qian, J .Q . and Shi, L.E. (2007) Preparation of Chitosan Nanoparticles as Carrier for Immobilized Enzyme. Applied Biochemistry and Biotechnology, 136, 77-97.

29.   Syedakulsum, Padmalatha, M., Sandeep, K., Saptasila, B. and Vidyasagar, G. (2011) Spectrophotometric Methods for the Determination of Naproxen sodium in Pure and Pharmaceutical Dosage Forms. International Journal of Research in Pharmaceutical and Biomedical Sciences, 2, 1303-1307.

30.   Genthner, F.J. , Hook, L.A. and Strohl, W.R. (1985) Determination of the Molecular Mass of Bacterial Genomic DNA and Plasmid Copy

Number by High-Pressure Liquid Chromatography. Applied and Environmental Microbiology, 50, 1007-1013.

31. Sambrook, J., Fritsch, E.F. and Maniatis, T. (1989) Molecular Cloning: A Laboratory Manual. Cold Spring Harbor Laboratory Press, Cold Spring Harbor.

32. Laemmli, U.K. (1970) Cleavage of Structural Proteins during the Assembly of the Head of Bacteriophage T4. Nature, 227, 680-685.

33. Bauer, A.W. , Kirby, W.M. , Sherris , J.C. and Turck, M. (1966) Antibiotic Susceptibility Testing by a Standardized Single Disk Method. Journal of Clinical Pathology, 45, 493-496.

34. Bridges, S .M. and Salin, M.L. (1981) Distribution of Iron Containing Superoxide Dismutase in Vascular Plants. Plant Physiology, 68, 275-278.

CHAPTER 10

Nanoporous Platforms for Cellular Sensing and Delivery

Lara Leoni [1], Darlene Attiah [1] and Tejal A. Desai [1,2], *

[1]Department of Bioengineering, University of Illinois at Chicago;
[2]Department of Biomedical Engineering, Boston University, 44 Cummington Street, Boston, MA 02215, USA.

ABSTRACT

In recent years, rapid advancements have been made in the biomedical applications of micro and nanotechnology. While the focus of such technology has primarily been on in vitro analytical and diagnostic tools, more recently, in vivo therapeutic and sensing applications have gained attention. This paper describes the creation of monodisperse nanoporous, biocompatible, silicon membranes as a platform for the delivery of cells. Studies described herein focus on the interaction of silicon based substrates with cells of interest in terms of viability, proliferation, and functionality. Such microfabricated nanoporous membranes can be used both in vitro for cell-based assays and in vivo for immunoisolation and drug delivery applications.

Keywords: Nanotechnology; Drug delivery; Silicon; Immunoisolation; Cell-based assays

1. INTRODUCTION

The application of micro- and nanotechnology to the biomedical arena has tremendous potential in terms of developing new diagnostic and therapeutic modalities and has increasingly been used to solve complex problems at the molecular and cellular level. While the majority of research has focused on the development of miniaturized *diagnostic* tools such as electrophoretic, chromatographic, and cell micromanipulation systems [1,2,3,4,5], researchers have more recently concentrated on the development of microdevices and constructs for *therapeutic* applications. Micro- and nanofabrication techniques are currently being used to develop implants that can record from, sense, stimulate, and deliver to biological systems. Micromachined neural prostheses, drug delivery micropumps/needles, and microfabricated immunoisolation biocapsules [6,7,8,9] have all been fabricated using precision-based microtechnologies. Microfabrication methods have also been applied to biotechnology in areas such as DNA sequencing by hybridization, protein patterning, and functional cell sorting.

The interfacing of "chip" technology and cell biology has great potential for use in biomedical research. Moreover, the human body seems appropriate as a target of microtechnology since most structures in the body are in the micron to millimeter size range, the same size range as most micro and nanoscale constructs. Few other engineering technologies can so closely parallel the multidimensional size scale of the living cells and tissues, with both precision and accuracy, in the same fabrication process. The miniaturization and reproducibility of platform features greatly facilitates the use of these systems for cell-based applications.

Microfabricated substrates can provide unique advantages over traditional biomaterials used for biosensing and delivery due to the: 1) ability to control surface microarchitecture, topography, and feature size in the nanometer and micron size scale, and 2) control of surface chemistry in a precise manner through biochemical coupling or photopatterning processes. Microfabrication technologies, by their very nature, lend themselves to efficient, economic mass-scale replication, as convincingly demonstrated by the microelectronic industry. They also allow for precise control of feature size, chemistry, and topography. The long term integration of cells with inorganic materials such as silicon provides the basis for novel delivery and sensing platforms. Our recent work has focused on the ability to maintain cells long term in nanoporous silicon-based microenvironments.

2. RESULTS AND DISCUSSION

Silicon nanoporous membranes were fabricated with pore sizes ranging from 7 nm to 49 nm as described in Leoni and Desai, 2001 [10]. The pore size was controlled through a sacrificial oxide etching which allows one to create nanoscale features using conventional lithography. Figure 1 shows the overall process flow of the membrane fabrication and is discussed in details in the experimental methods section. Figure 2 shows a cross sectional SEM image of the membrane with the nanoscale channels visible. The membranes can be fabricated in arrays on a single wafer, allowing one to produce multiple nanoporous membranes for in vitro or in vivo applications.

2.1. Insulinoma Cells

Silicon nanoporous membranes, control petri dishes, and latex membranes were seeded with insulinoma cells as described. We found that the insulinoma cells grew without any marked changes on the silicon surfaces. In fact, viability of the cells in the silicon nanoporous environments was equivalent to conventional cell culture surfaces, ranging from 100 to 90 percent over an eight day period (figure 3). All cell types had normal

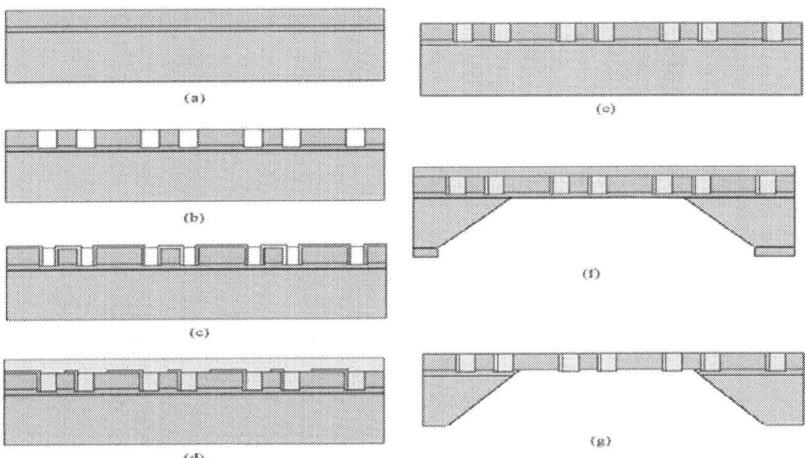

Figure 1. Process flow diagram for nanoporous membrane fabrication.

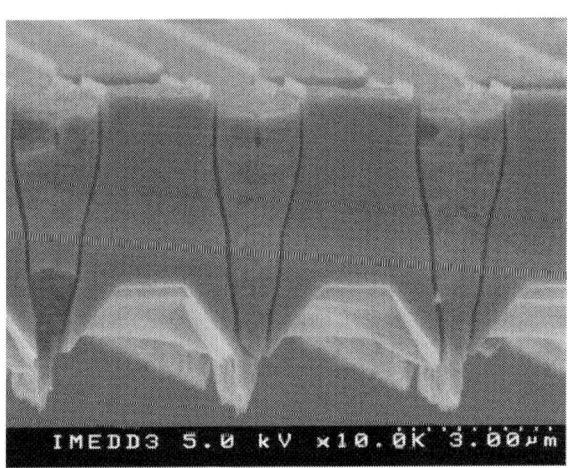

Figure 2. SEM micrograph of microfabricated nanoporous membrane: side view detail.

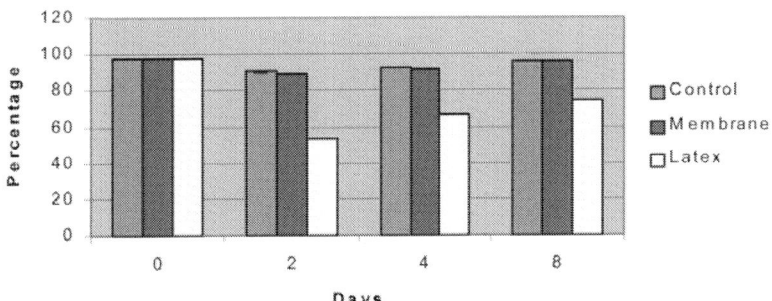

Figure 3. Viability of insulinoma cells over an 8-day period compared to control surface (petri dish) and negative control (latex).

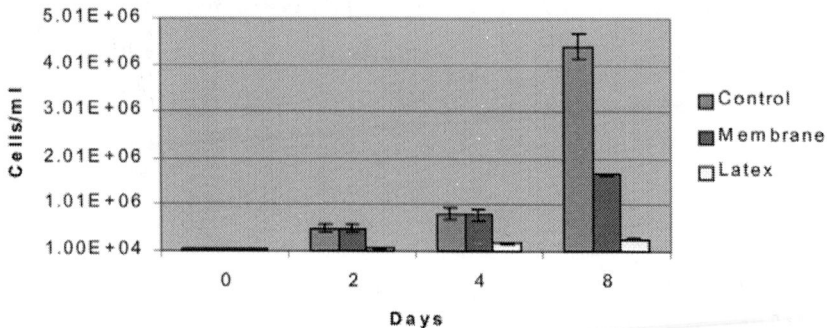

Figure 4. Proliferation of insulinoma cells over an 8-day period compared to control surface (petri dish) and negative control (latex).

morphology. In terms of proliferation, cells seeded in nanoporous microenvironments had similar levels of proliferation compared to control surfaces up to day 4 and then a decreased rate of proliferation at day 8 (figure 4). This behavior was presumably due to contact inhibition resulting from the limited nanoporous surface area that cells were seeded in as compared to the control surface area. Cells exhibited limited viability and proliferation on the negative control surface of latex. Figure 5 shows an image of insulinoma cells growing on a partially etched nanoporous membrane. It is interesting to note that the cells limit their attachment to the porous regions of the membrane. Such behavior could be due to the nanopores providing a greater surface area for cells to attach. In addition, the etched porous surface is more hydrophilic as its final etch step is in HF solution and therefore may promote greater cell attachment. Figure 6 shows that cells insulinoma cells seeded in silicon nanoporous microenvironments are indeed functional and can secrete insulin over time. Glucose-supplemented medium was allowed to diffuse to the cells, through the membrane, to stimulate insulin production and monitor cell functionality. Results indicate that the insulin secretion by cells and subsequent diffusion of the insulin through the nanoporous membrane channels is similar to that of cell grown in culture.

2.2. Diffusion Properties

When nutrients and time sensitive compounds are diffusing across a membrane, it is highly desirable to be able to precisely control the diffusion characteristics in order to retain the dynamic response of cells seeded on the membrane to external stimuli. The ability of the nanoporous microfabricated membranes to perform size-base exclusion and controlled diffusion of biomolecules has been studied. Membranes exhibited controlled diffusion of glucose and albumin based on membrane pore size. Such control over molecular diffusion was precise and reproducible (figure 7).

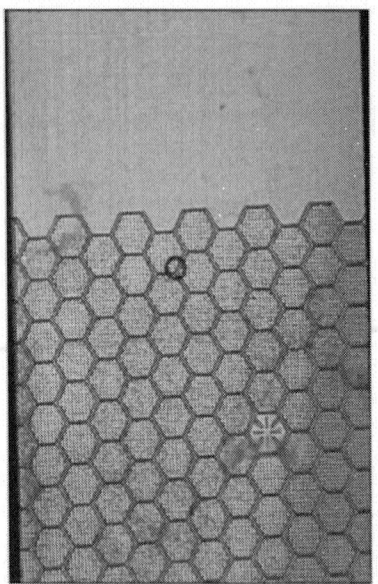 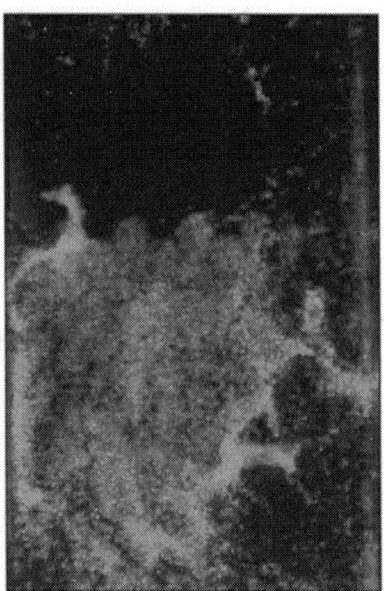

Figure 5. Micrograph of a) top surface of partially etched nanoporous membrane and b) nanoporous membrane seeded with fluorescently labeled insulinoma cells. Note the preferential adhesion of the cells to the etched membrane architecture.

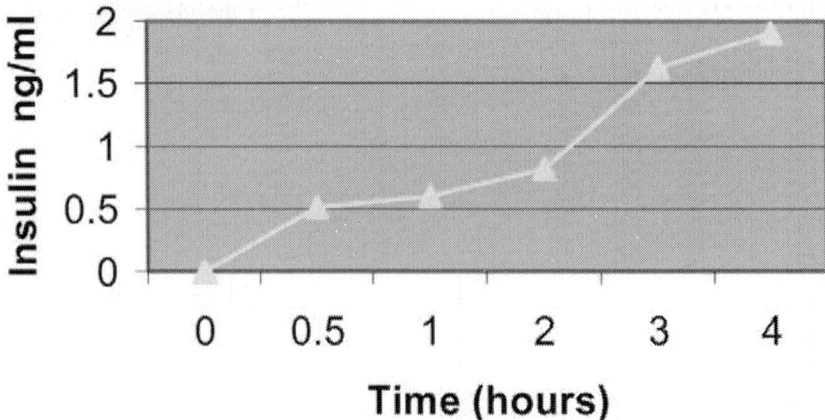

Figure 6. Insulin secretion profile from cells seeded on nanoporous membranes.

PC12 Cells

We found that PC12 cells also maintained approximately 100% viability as compared to control surfaces and actually proliferated to a greater extent in silicon nanoporous environments. Figure 8 shows empty nanoporous wells and wells seeded with PC12 cells at day 1 and day 7. The cells are able to

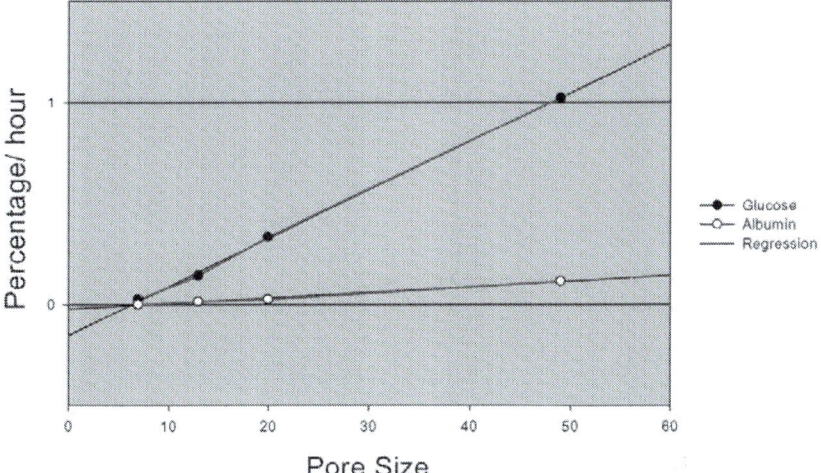

Figure 7. Glucose and Albumin diffusion rates.

become confluent and differentiated within the nanoporous wells. Cells seeded to the silicon nanoporous membranes experienced an approximate 245.31% ± 62.5% growth increase from day 1 to day 4 while those seeded to the polyurethane (control) capsules encountered only a 75.9% ° 21% growth increase (Figure 9). Moreover, cells attached to the silicon nanoporous membranes underwent high proliferation, increasing in number by approximately two fold every other day of the culture period. At day 1 of cell culture, the number of cells attached to the silicon based biocapsules and control capsules were observed approximately equal. By day 4, however, the silicon biocapsules exhibited a 110% ± 35.36% greater number of adherent cells. In contrast to the polyurethane capsules, cells that were attached to the silicon biocapsules were noticed to attach firmly to the membrane as they strongly resisted detachment upon trypsinizing.

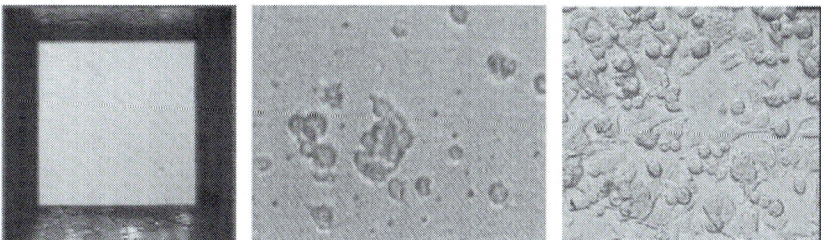

Figure 8. a) Empty nanoporous membrane and membrane seeded with PC12 cells seeded at b) day 1 and c) day 7. Cells show proliferation and differentiation within the nanoporous wells.

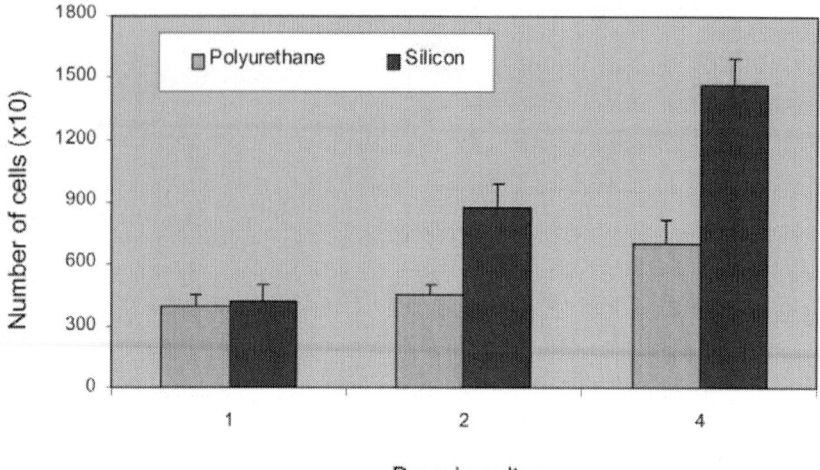

Figure 9. PC12 proliferation data in polyurethane (control) and silicon based capsules. Reported are average values of experiments performed with 3x standard deviations.

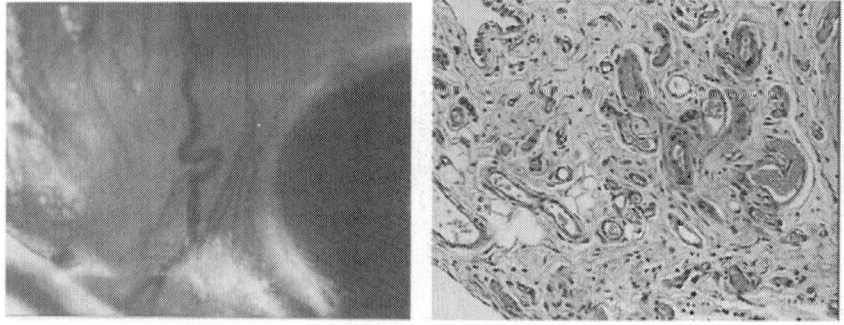

Figure 10. a) Nanoporous membrane retrieved from the peritoneal cavity: detail; b) H-E stained tissue (x20). Several blood vessels are visible (arrow).

In vivo Biocompatibility

At a gross examination, silicon nanoporous membranes seemed free of fibrotic tissue and clean. A rich network of blood vessels surrounded the microfabricated membrane in proximity of the diffusion area, minimizing possible limitations of glucose-insulin exchange due to the lack of a well developed vascular system surrounding the membrane (figure 10a). Microscopic analysis of tissue sampled from the membrane located in the omentum revealed a non-uniform structure, with prevalence of large round cells typical of adipose tissue (figure 10b). There was no evidence of macrophages or lymphocytes infiltration. Round structures of different diameters resembling ducts could be seen. Their lumen presented secretion of probable proteinaceous origin. Adenomers were also dispersed in the tissue,

as well as small vessels characterized by a thin layer of elongated cells, typical of the lining endothelium in capillaries.

3. CONCLUSION

A method to create precise nanoporous membranes via microfabrication technology has been described. Membranes can be fabricated to present uniform and well-controlled pore sizes as small as 7 nm, tailored surface chemistries, and precise microarchitecture. These platforms can be interfaced with living cells to allow for biomolecular separation and immunoisolation. Ideally a membrane in contact with cells should be biocompatible and allow for the free exchange of nutrients, waste products, and secreted therapeutic proteins. Furthermore, where nutrients and time sensitive compounds are diffusing across a membrane it is highly desirable to be able to control the diffusion characteristic precisely in order to retain the dynamic response of seeded cells to external stimuli. Membranes were shown to be sufficiently permeable to support the viability of insulinoma and PC12 cells. Applications of these nanoporous membranes range from cellular delivery to cell-based biosensing to in vitro cell-based assays (figure 11).

Figure 11. Arrays of nanoporous wells seeded with cell clusters that can potentially be used for high throughput cell-based assays.

In order to retain the same performance *in-vivo*, the biohybrid device must be fully biocompatible, which implies that the membrane should elicit little or

no foreign body response. The host response is a potentially serious problem to clinical implementation of the technology. The direct consequence of a nonbiocompatible membrane is a fibrotic overgrowth on the surface that interferes with diffusive transport of molecules and oxygenated blood supply. The microfabricated nanoporous membrane proved to be highly biocompatible. After 2 week implantation into rat peritoneal cavity, there was no or minimal fibrotic tissue, no significant host response was elicited, and a rich microvascular system was surrounding the device.

4. EXPERIMENTAL

The nanoporous membranes are achieved by applying fabrication techniques originally developed for Micro Electro Mechanical Systems (MEMS). Utilizing bulk and surface micromachining and microfabrication, silicon platforms can be engineered to have uniform and well-controlled pore sizes, channel lengths, and surface properties [8,9,10,11,12]. We have developed several variants of microfabricated diffusion barriers, containing pores with uniform dimensions as small as 7 nanometers [10]. One such variation is described below.

4.1. Fabrication

The process flow for fabrication of nanoporous membranes is depicted in Figure 1 below. The starting substrate is a 400 μm-thick, 100 mm-diameter, double side polished (100)-oriented silicon wafer. The first step is the etching of the support ridge structure into the substrate. A low stress silicone nitride layer (nitride), which functions as an etch-stop, is then deposited. A polysilicon film, that acts as the base structural layer (base layer) is deposited on top of the etch-stop layer (Fig. 1a). The next step is the etching of holes in the base layer, which defines the overall shape of the pores. The holes are etched through the polysilicon by chlorine plasma, with a thermally grown oxide layer used as a mask. After the pore holes are defined and etched through the base layer (Fig. 1b), the pore sacrificial oxide is grown on the base layer (Fig. 1c). The sacrificial oxide thickness determines the pore size in the final membrane, so control of this step is critical to reproducible pores in the membrane.

To mechanically connect the base polysilicon with the plug polysilicon, which necessary to maintain the pore spacing between layers, anchor points are defined in the sacrificial oxide layer (Fig. 1d). After the anchor points are etched through the sacrificial oxide, the plug polysilicon is deposited to fill in the holes. The plug layer is then planarized down to the base layer (Fig. 1e), leaving the final structure with the plug layer only in the base layer openings. A protective nitride layer is then deposited on the wafer, completely covering both sides of the wafer (Fig. 1f). This layer is completely impervious to KOH chemical etch used to release the membranes from the bulk silicon wafer in the desired areas, and the wafer is placed at 80°C KOH bath to etch. After the silicon is completely removed up to the membrane , the protective, sacrificial, and etch stop layers are removed by etching in HF (Fig. 1g). The pores fabricated on the membranes were characterized by SEM. The nanoporous membranes were seeded with islet cells, insulinoma cells, or PC12 cells and characterized in terms of viability, hormone secretion and/ proliferation.

4.2. Insulinoma Cells

Mouse insulinoma βTC3 cells (Efrat et al., 1998) have been obtained from the laboratory of Dr. Shimon Efrat, Department of Molecular Pharmacology, Albert Einstein College of Medicine, Bronx, NY. Cells are cultivated as monolayers in T-flasks in complete medium consisting of DMEM with 25 mM glucose and supplemented with 15% heat-inactivated horse serum (SIGMA) and 2.5% fetal bovine serum (SIGMA). Cultures are maintained at 37°C in a humidified 5% CO_2 /95% air atmosphere, and they are passed every 5-10 days. These cells maintain a stable phenotype of glucose responsiveness in the physiological range.

The membrane biocompatibility was characterized by direct contact tests looking at viability of cells in contact with silicon nanoporous membranes. The intracellular fluorescent dye carboxyfluorescein diacetate succinimidyl ester (CFDA, Molecular Probe) at a concentration of 5 M was used to label cells before in vitro culture. First cells were incubated in prewarmed CFDA labeling solution at 37 degrees C in a % CO_2 containing incubator for 45 minutes. The labeling solution was subsequently replaced by prewarmed DMEM medium supplemented with 10% FBS. Then, 10μl of 2×10^7 cells/ml were pipetted onto the membrane well. Cell viability was checked after 24 hours by counting with a hemacytometer under a light microscope. The viability was compared with cells grown on latex (negative control) and standard culture dishes. Cell attachment and proliferation to the various surfaces was also observed and measured.

4.3. Neurosecretory Cells (PC12)

PC12 cells (ATCC, Manassass, VA) of passages 3 to 5 were plated and at a density of $1-2 \times 10^6$ cells per 100 mm dish, maintained at 37 degrees C in 5% CO_2 and re-fed every 2-3 days. To determine the interaction between PC12 cells and silicon nanoporous membranes, the growth patterns of PC12 cells seeded at approximately $3-4 \times 10^3$ cells into micromachined membrane-based biocapsules and polyurethane (control) capsules (n=3) contained in 12 well culture dishes were monitored at days 1, 2, and 4. After the designated incubation times, the silicon and control capsules were transferred to empty culture wells and rinsed with PBS to remove loosely adherent cells. Samples were then incubated with 1x trypsin-EDTA solution to detach the adherent cells. Cell suspensions were then centrifuged at 850 rpm for five minutes. PC12 cell pellets were resuspended in fresh DMEM media and counted twice with a hemocytometer. PC12 viability was monitored by labeling with intracellular fluorescent dye as described above.

4.4. In-Vivo Biocompatibility

Implantation of nanoporous membranes were done according to the NIH guidelines for the care and use of laboratory animals. Male Lewis rats were anesthetized with inhalation of ether. A laparatomic incision was made and the biocapsules were either sutured on the adbominal wall or wrapped into the omentum and sutured to the same. Incision was closed by suture (polypropylene). Capsules were retrieved after two weeks. Tissue samples were fixed in 10% buffered formalin, paraffin embedded, sectioned and stained with Hematoxylin-Eosin.

ACKNOWLEDGEMENTS

Portions of this project were funded by NSF and The Whitaker Foundation. Special thanks to iMEDD,Inc. for technical support and fabrication of the membranes.

REFERENCES AND NOTES

1. Baxter, G.T.; Bousse, L.J.; Dawes, T.D.; Libby, J.M.; Modlin, D.N.; Owlck, J.C.; Parce, J.W. *Clinical Chemistry* **1994**, *40(9)*, 1800–1804.
2. Akin, T.; Najafi, K. *IEEE Transactions on Biomedical Engineering* **1994**, *4(4)*, 305–313.
3. Volkmuth, W.D.; Duke, T.; Austin, R.H.; Cox, E.C. Trapping of branched DNA in microfabricated structures. *Proc. Natl. Acad. Sci. USA* **1995**, *92(15)*, 6887–6891.
4. Gourley, P.L. *Nature Medicine* **1996**, *2(8)*, 942–944.
5. Anderson, D.J.; Najafi, K.; Tanghe, S.J.; Evans, D.A.; Levy, K.L.; Hetre, J.F.; Xue, X.; Zappia, J.J.; Wise, K.D. *IEEE Transactions on Biomedical Engineering* **1989**, *36(7)*, 693–704.
6. Santini, J.T., Jr.; Cima, M.J.; Langer, R. *Nature (London)* **1999**, 335–338.
7. Henry, S.; McAllister, D.V.; Allen, M.G.; Prausnitz, M.R. *J. Pharm. Sci.* **1998**, 922–925.
8. Desai, T.A.; Hansford, D.; Ferrari, M. *J. Membr. Sci.* **1999**, 221–231.
9. Desai, T.A.; Chu, W.H.; Tu, J.K.; Beattie, G.M.; Hayek, A.; Ferrari, M. *Biotechnol. Bioeng.* **1998**, 118–120.
10. Leoni, L.; Desai, T.A. Nanoporous Biocapsules for the Encapsulation of Insulinoma Cells: Biotransport and Biocompatibility

Considerations. *IEEE Transactions in Biomedical Engineering* **2001**, *Vol. 48 (11)*.

11. Desai, T. A.; Hansford, D. J.; Leoni, L.; Essenpreis, M.; Ferrari, M. Nanoporous Anti-fouling Silicon Membranes for Implantable Biosensor Applications. *Biosensors and Bioelectronics* **2000**, *15(9-10)*, 453–462.

12. Desai, T.A.; Hansford, D.J.; Kulinsky, L.; Nashat, A.H.; Rasi, G.; Tu, J.; Wang, Y.; Zhang, M.; Ferrari, M. *Biomed. Microdevices.* **1999**, 11–40.

CHAPTER 11

Nanoparticle Insulin Drug Delivery — Applications and New Aspects

Hatice Kübra Elçioğlu[1] and Ali Demir Sezer[2]

[1]Department of Pharmacology, Faculty of Pharmacy, Marmara University, Istanbul, Turkey
[2]Department of Pharmaceutical Biotechnology, Faculty of Pharmacy, Marmara University, Istanbul, Turkey

1. INTRODUCTION

Insulin is a hormone secreted from the β cells of the islets of Langerhans, specific groups of cells in the pancreas. Insulin is a protein consisting of two polypeptide chains, one of 21 amino acid residues and the other of 30, joined by two disulfide bridges. It was isolated in 1921 with its first clinical use in 1922 [1]. Insulin is prepared different techniques; One of these isolated from animals and the other is biotechnological preparation using with the recombinant DNA techniques [2, 3].

Insulin is a important player in the control of intermediary metabolism and profound effects on both carbohydrate and lipid metabolism. It has significant influence on protein and mineral metabolism [4,5].

The traditional and most predictable method for the administration of insulin is by subcutaneous injections. This method is often painful and hence, deterrent to patient compliance especially for those requiring multiple dose injections of four times a day. Also, there have been reports of hypoglycemic episodes following multi dose injections of insulin [6, 7]. Several new approaches to the method have been adopted to decrease the suffering of the diabetic patients including the use of supersonic injector, infusion pump, sharp needles and pens. Some insulin delivery routs so problematic way for example oral administration; Oral delivery eliminates the pain caused by injection, psychological barriers associated with multiple daily injections. Oral delivery of insulin as a non-invasive therapy for Diabetes Mellitus is still a challenge to the drug delivery technology, because insulin is degraded by the enzymes in the acidic environment of stomach. Otherwise insulin delivery via transdermal delivery is so popular way of insulin administration but there are some disadvantages of this route, for example insulin molecular size and application problems etc. While some of them eased the pain encountered by the diabetic patients, they offer incomplete convenience. Even though the ultimate goal would be to eliminate the need to deliver insulin exogenously and regaining the ability of patients to produce and use own insulin, new concepts are currently explored to deliver insulin using oral, pulmonary, nasal, ocular and rectal routes [8, 9].

The success of the route of administration is judged on the basis of its ability to elicit effective and predictable lowering of blood glucose level and therefore minimizing the risk of diabetic complications. It is clear that several difficulties have to overcome with the use of formulation and application devices technology [10, 11]. The various explored routes are reviewed in this chapter. On the other hand, the chapter is an attempt to illustrate the use of insulin drug delivery and their body route in diabetes management benefiting many diabetic patients with promising patient compliance.

Diabetes mellitus (DM) which is a metabolic disorder characterized by chronic hyperglycemia (increased blood and hepatic glucose levels) with disturbances in carbohydrate, fat and protein metabolism, resulted by diminished insulin secretion, impaired insulin action or both. It's expected to increase from 171 million in 2000 to 366 million by the year 2030 as predicted by the WHO so it continues to increase in prevalence and will become a serious threat of mankind health [12]. Insulin injections remain to be preferred approach for the treatment of insulin-dependent diabetes mellitus (T1DM) and for many patients non-insulin-dependent diabetes mellitus (T2DM) also. People with type 1 diabetes mellitus have an autoimmun mediated destruction of pancreatic islet beta-cells and insulin deficiency. T1DM usually occurs in children and young adults and require daily insulin administration by injection or an insulin pump for survival. On the other hand, insulin resistance (which is associated with excessive glucose production by the liver and impaired glucose utilization by peripheral tissue, especially muscle) is observed in T2DM. They have an impaired endogenous insulin secretion to deal with the increased blood glucose level and majority needs oral antidiabetic drugs. As the disease progresses, the pancreas looses its ability to produce insulin and necessity of insulin therapy increases [12, 13, 14].

Hyperglycemia, recurrence of ample fluctuation of blood glucose levels and insulin resistance can lead to long term complications such as micro and macrovascular. It is well known that improved metabolic control significantly reduces both microvascular (ie, retinopathy, nephropathy and neuropathy) or macrovascular [ie, cardiovascular disease (CVD), cerebrovascular accidents and peripheral vascular disease] complications in diabetes. The development of complications is a cause of considerable morbidity and increases disability and mortality for the individual with diabetes [15].

The conventional pharmacotherapies currently available for the treatment of type-2 diabetes include insulin sensitisers (metformin and thiazolidinediones), insulin secretagogues (sulphonylureas and glinides), alpha-glucosidase inhibitors, insulin and insulin analogues. Glucagon like peptide (GLP)-1 agonists and dipeptidyl peptidase (DPP)-4 inhibitors are the new therapies; that improve glycemic control have recently been developed [16, 17, 18, 19]. These therapies are proposed to treat the key metabolic abnormalities associated with T1DM and T2DM and minimize the side effects noted with conventional therapies. Also in development there are additional therapies that have effects on the kidney to promote glucose excretion [15]. SGLT-2 (proximal renal tubule) has high transport capacity for reabsorption of approximately 90% of primarily filtered glucose. SGLT-2 inhibitors inhibit glucose reabsorption in proximal renal tubule. It results glycosuria leads to a decline in plasma glucose level. A wide variety of SGLT-2 inhibitors are currently under development with Dapagliflozin, Canagliflozin, Empagliflozin being the most advanced substances. Excretion of approximately 40% of primarily filtered glucose translates to a loss of 50–100 g glucose every day. The consequential decline in fasting and postprandial glucose leads to an HbA1c reduction of approximately 0.8%. The loss of energy substrate reduces body weight approximately 3 kg.

Current therapy for diabetes mellitus through oral anti-diabetic drugs and subcutaneous administration of insulin suffers from serious disadvantages, such as patient noncompliance and occasional hypoglycemia. Moreover, these approaches don't mimic the normal physiological fate of insulin release and doesn't provide better glucose homeostasis. In normal human physiology when the blood glucose level increases insulin releases from the pancreas, reaches to the hepatic portal vein and goes to liver which is its primary site of action. Subcutaneous administration of insulin moves firstly peripheral tissues and can produce peripheric hyperinsulinemia. In order to overcome the problems associated with parenteral administration of insulin, substantial progress has been made for insulin route such as ocular, vaginal, rectal, oral, pulmonary, transdermal, intranasal, and other routes (Figure 1) [20]. The barriers to reaching the bloodstream are either physical, such as poor absorption at barrier surfaces, or chemical, such as pH inactivation and enzymatic degradation. Delivery of insulin via the ocular route was tested in animal models in combination with different absorption enhancers, with particular attention given to toxicity as polymers were added to overcome low absorption. Vaginal and rectal routes of insulin have also been evaluated but the absorption rate and bioavailability are poor due to the thick mucosal layers in these tissues. Lots of absorption enhancers (bile salts, chelating agents, surfactants, cyclodextrins, and dihydrofusidate) used but they couldn't prevent local reactions with severe complications.

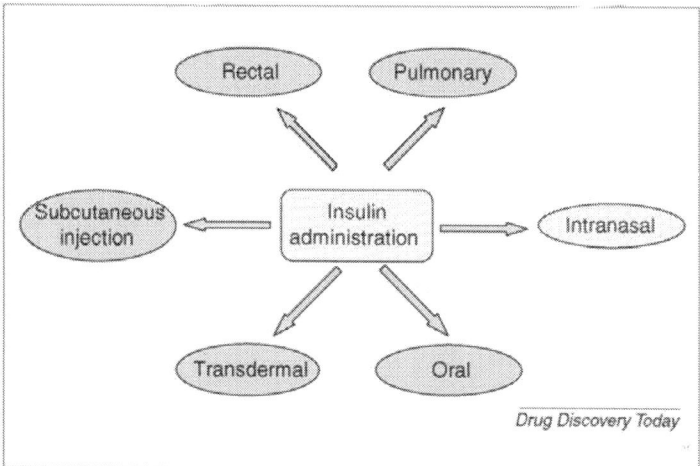

Figure 1. The main administration routes for insulin delivery (Reproduced with permission from Ref. [20], Copyright 2013 Elsevier)

Nasal delivery has also been evaluated because of the easy access, high vascularity and large absorption area associated with this route. Unfortunately, highly active mucociliary clearance in the nose hindered prolonged drug action resulting in poor bioavailability. Buccal and sublingual insulin administration provide better results due to the low levels of proteolytic enzyme activity, the high vascularization of the tissue, the large surface area for absorption and the ease of administration. Unlike other delivery routes, the gut is the natural route of nutrient absorption into the circulation. The fact that the gut presents the largest absorption surface of all routes provides better efficacy. However, the multiple layers of oral epithelial cells represent a significant GI barrier to drug penetration, which, coupled with the continuous flow of saliva, leads to poor efficacy.

Taking all of this account oral administration is considered to be the most safest and convenient which delivers the drug directly into the liver through portal circulation, where it inhibits hepatic glucose production. Hence by oral delivery to a greater extent the natural physiological route of insulin can be mimicked (Figure 2) [21]. The highly acidic environment in the stomach and the presence of proteolytic enzymes cause structural instability of the oral delivery of protein and peptide drugs including in the harsh environment of the gastrointestinal system [22, 23, 24]. These drugs should overcome some various GI barriers such as chemical, enzymatic and absorption barriers to obtain adequate bioavailability [25]. Different formulation of polymers for insulin delivery such as liposomes, microspheres, microemulsion and nanoparticles (NPs) have been investigated to circumvent these GI barriers [26, 27].

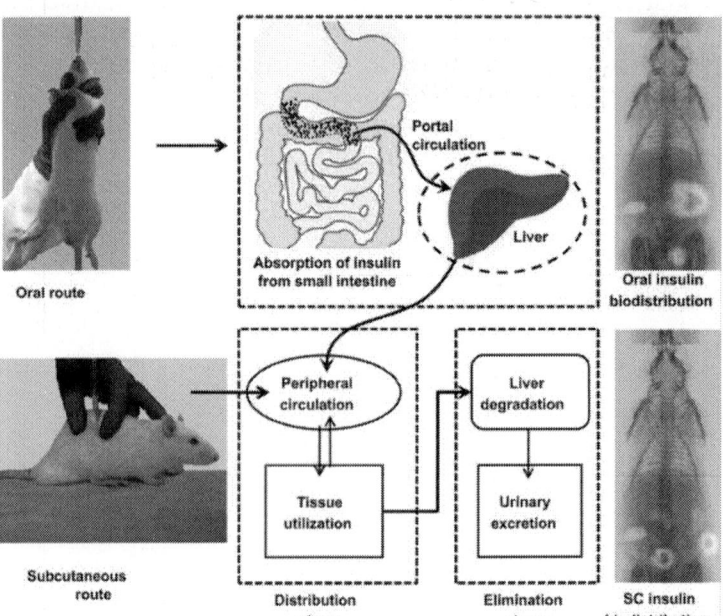

Figure 2. Schematic diagrams illustrating the absorption, distribution and elimination of aspart insulin following oral or subcutaneous (s.c.) administration to rats (Reproduced with permission from Ref. [21], Copyright 2013 Elsevier).

Among these approaches nanoparticular sysytems have attracted special interest because of providing the protection to the highly acidic medium in the stomach (preventing enzymatic degradation), prolonging intestinal residence time, increasing the permeability of drugs to systemic circulation (increasing absorption) and providing controlled-release properties for encapsulated drug [12, 28]. For the conventional medicine it is well understood the nanosize along with other characteristics does play an important role as evident from the improved bioavailability/pharmacological availability [29, 30]. Owing to the high surface area to volume ratio of NPs the window of absorption is also high in comparison with microparticles, this is an added advantage in improving the bioavailability of the administered drug [31, 32].

2. DELIVERY ROUTE OF INSULIN

2.1. Oral Delivery

Insulin therapy is effectively used in treatment of diabetes mellitus. Insulin is a key player in lowering blood glucose levels for type 1 diabetes and also required at later stages in type 2 diabetes patients. The widely accepted route for delivery of insulin is by parenteral administration but this delivery of insulin usually requires at least three or four daily insulin injections for good glycemic control. Consequently more acceptable different routes of insulin delivery have been searched to decrease suffering from discomfort, local pain, irritation, infection, immune reactions and lipoatrophy at the injection site of insulin. Oral delivery of insulin would deliver the drug directly into the liver through portal circulation and could mimic the physiological fate of endogenously secreted insulin [33, 34, 35]. However polypeptides, like insulin are degraded in the stomach pH and undergo proteolysis by enzymes in the gastrointestinal tract [22, 36]. Moreover the gastrointestinal mucosa has low permeability for large hydrophilic peptides.

In order to overcome the problems associated with parenteral administration of insulin several strategies that are based on nanotechnology has been developed to enhance the intestinal absorption of different protein and peptides. NPs consist of naturally occurring biodegradable polymers are widely investigated in this regard. They have emerged as potential carriers of several therapeutic agents for controlled drug delivery as well as the oral route of insulin. Various natural hydrophilic and hydrophobic polymers used as carrier of oral insulin such as chitosan, alginate, dextran sulphate, etc. are commonly used to prepare NPs.

2.1.1. Polymers Used As Matrices for Oral Insulin Delivery
Over the past few decades, enhancing attention has been paid to the use of polymeric NPs either hydrophilic or hydrophobic as carriers for insulin delivery. Hydrophilic polymers are of particular interest due to their non-toxic, biocompatible, biodegradable and natural polymers. Among them, chitosan is

widely used because of its ease of chemical modification and promising biological properties.

2.1.1.1. Hydrophilic Polymers

Chitosan (CS): CS is well known naturally occuring copolymer of beta [1-4] linked and N-acetyl glucosamine and have been generally found in crustacean (crabs, shrimps and lobsters) shell and in some fungi or yeast. It is a biodegradable, biocompatible, non toxic, non-allergic easily absorbable natural hydrophilic polymer properties that have resulted in a wide array of applications in biomedical and drug delivery research [29, 30, 37]. Moreover it prolongs the intestinal residence time that shows its mucoadhesive property [38] (Figure 3). It has also been shown as a paracellular permeability enhancer by interacting with the TJ proteins occluding ZO-1 and opens the tight junctions between epithelial cells [34, 39, 40]. In addition to these properties it increases the stability of nanospheres and facilates effective encapsulation of proteins and drugs that make it as a suitable carrier material [38,41, 42]. CS have been extensively used to develope new chitosan derivatized polymers. CS combined with poly(γ-glutamic acid) (γ-PGA) based insulin NPs are used as hydrophilic polymers for oral insulin delivery. *In vivo* preclinical studies of this formulations at a dose of 30 IU/kg in streptozotocin (STZ) induced diabetic rat models showed increased intestinal absorption of insulin from γ-PGA NPs. It has got long lasting hypoglycemic effect and 15% relative bioavailability compared to subcutaneous (sc) injection [43]. The same formulation filled in enteric coated capsules was even better at the same dose, showing 20% oral bioavailability. Also aspart insulin (=monomeric, 3 times faster than regular) is encapsulated in the same CS-γ-PGA; has got 15.7 % oral bioavailability [21, 44].

Moreover insulin loaded NPs with carboxylated chitosan and Poly-methyl methacrylate (PMMA) were developed to improve the insulin delivery via oral route. One of the most widely investigated polymer towards peptide delivery is acrylates which have high interest because of its pH sensitivity and carboxyl groups to enhance the bioadhesivity, alter the tight junction, chelate the Ca^{2+} there by inhibiting the proteolytic activity of proteases, etc. They evaluated their ability to reduce blood glucose levels in diabetic rats. *In vivo* experiments resulted in the reduction of blood glucose levels by 67% at a dose of 100 IU/kg and the pharmacological bioavailability of the 25 IU/kg at a dose of PMMA NPs was 9.7% [45, 46].

Chitosan with sodium alginate is being prepared another insulin loaded nanoparticle product which is used to improve the loading capacity and activity maintenance. It's observed that when insulin-loaded nanospheres (25, 50, 100 IU/kg) administered orally to diabetic rats they reduced glycaemia in a dose dependent manner. Their pharmacological availabilities are found 7.1, 6.8 and 3.4 %, respectively [33,47, 48].

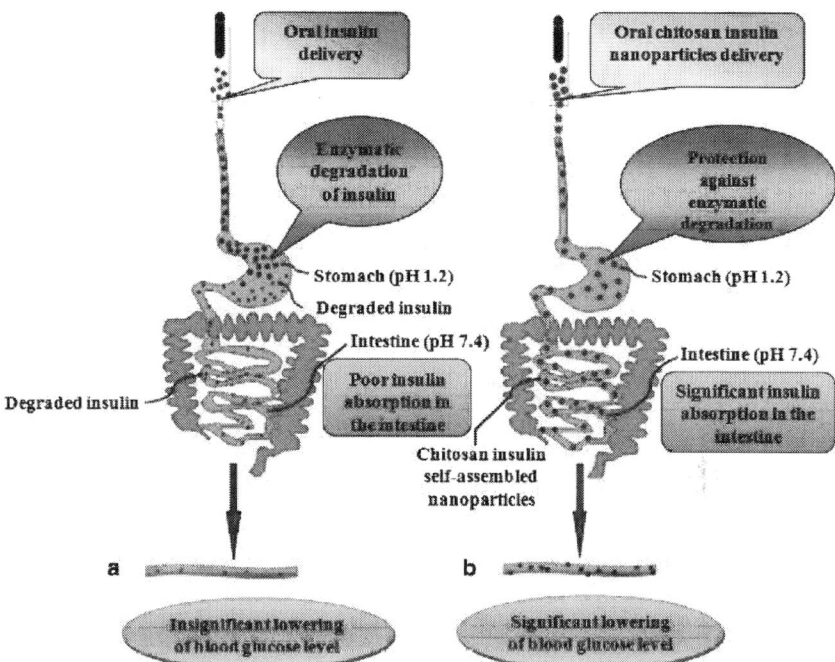

Figure 3. In-vivo efficiency of orally delivered insulin and chitosan/insulin self-assembled NPs (Reproduced with permission from Ref. [38], Copyright 2013 Elsevier).

In addition hydroxypropyl methylcellulose phthalate (HPMCP) is a pH-sensitive polymer designed as an enteric coating material. It reduces the release of drug in acidic conditions and also to improve the colloidal stability of the particles. The release of insulin from CS/HPMCP NPs were significantly reduced at acidic pH and even after 6 h it was less about 25% only. Insulin was protected from enzymatic degradation in the case of CS/HPMCP in comparison with native chitosan particles. Insulin loaded chitosan and HPMCP NPs were orally administered to diabetic wistar rats. The pharmacological availability was 3.02% and 8.47%, respectively, for the chitosan and the modified NPs. In comparison to oral insulin solution the hypoglycaemic effect was increased by 2.8 and 9.8-folds for the chitosan and the modified NPs, respectively [49, 50, 51].

Dextran sulphate-vitamin B12: Dextran sulphate is a non-toxic and highly water soluble different polymer used as matrices for oral delivery of insulin. Vitamin-B_{12} demonstrated as a ligand to enhance the uptake of the dextran NPs and their translocation across the gastrointestinal tract for high bioavailability. Insulin conjugated to dextran-vitamin B_{12} NPs to diabetic rats that had the least amount of cross linking were found to be most effective at lowering blood glucose levels (70-75%) in STZ induced diabetic rats. In addition the hypoglycemic effect lasted for 54 h. This modification showed the greatest hypoglycemic effect with a pharmacological availability of 29.4% (Table 1) [49,52, 53].

Table 1. Hydrophilic Polymers.

Particle	Size (nm)	Dose (IU/kg)	Pharmacological Activity (PA %)	Reference
CS-γPGA	218	30	15.1	
Entericcoated(EC)	233	30	20.1	21, 43, 44
Aspart Insulin	245	30	15.7	
CS-PMMA-PEG	1000	50	1.8	45
CS-PMMA	200-300	25	9.7	46
CS-NP	25-400	7, 14, 21	14, 15.6, 15.3	54
CS-dextran sulfate	500	50, 100	5.6, 3.4	55
CS/HPMCP	255	12.5	8.47	56
CS-Na alginate	748	25, 50, 100	7.1, 6.8, 3.4	57
B_{12}vitamin-dextran	192	20	29.4, 26.5	52, 53

2.1.1.2. Hydrophobic Polymers

Poly (lactide-co-glycolide) (PLGA): Particles consisting of PLGA have been widely studied as therapeutic delivery vehicles owing to their biodegradabile and biocompatibile particles. The hydrophobic nature of PLGA matrices generally makes them incapable of entrapping water-soluble insulin. Intragastric administration of the insulin-loaded PLGA NPs (20 IU/kg) to diabetic rats reduced fasting plasma glucose levels to 57.4% within the first 8h of administration. The relative bioavailability of insulin following oral administration of NPs was 7.7% compared to subcutaneous injection of its solution. Star-branched PLGA (β-cyclodextrin-PLGA) NPs are highly promising for mitigating the burst effect and prolonging the release of insulin. Another study attempted to prevent the burst release of insulin in the stomach by using a cellulose derivative (hydroxypropyl methylcellulose phthalate, HPMCP) to prepare PLGA NPs. This modification reduced the initial release of PLGA NPs in simulated gastric fluid from 50% to 20%, and their relative bioavailability in diabetic rats was approximately 6.2% [49, 58, 59].

Poly lactide acide (PLA): PLA exhibit a strong affinity toward the small intestine due to their polyethylene oxide (PEO) blocks and a high permeation capability toward the cell membrane owing to their amphiphilic property. When orally treated with vesicular PLA NPs loaded with insulin to diabetic mice (50 IU/kg), the highest blood glucose reduction was achieved at 4.5 h. Although this effect lasted at least an additional 18.5 h, increasing the insulin concentration to 100 IU/kg did not enhance this hypoglycemic effect (hypoglycemic effect lasted for 23 hr) [60].

Poly-ε-caprolactone (PCL): NPs prepared with PCL and a monomeric form of insulin analog (aspart-insulin). Their results demonstrated that this formulation allows for preservation of biological activities of insulin, increase of serum insulin levels and improvement of the glycemic response. The maximum effect of reduction in hyperglycemia was found at 8 h after oral administration, which was more pronounced with aspart-insulin-loaded NPs (52%) at the dose of 50 IU/kg [61].

Lipidic polymers [Solid lipid NPs (SLN)]: Previous studies have demonstrated that nanoencapsulation of proteins in SLNs prolongs their blood residence time, modifies their biodistribution and improves their bioavailability [32]. Oral insulin delivery with SLNs administered to diabetic rats, their relative pharmacological bioavailability was 5.1% in comparison to SC injection of insulin; a considerable hypoglycemic effect was also observed during 24 h. To facilitate the transport of particles across the cellular barriers, in an another study the relative bioavailability increased to 7.1%. That study also suggested that increasing the drug entrapment efficiency and utilizing protease inhibitors in SLNs may further enhance the bioavailability of insulin (Table 2).

Table 2. Hydrophobic Polymers.

Particle	Size (nm)	Dose (IU/kg)	BA (relative bioavailability)%	Reference
PLGA	150	20	40 7.7	46
PLGA-HPMCP	169	20	60 6.27	59
PLA		50	Hypoglycemic effect lasted for 23 h	60
PCL, Aspart insulin loaded	700	50	52/ 12-24 hr	61

2.2. Nasal Delivery

Nasal administration has attracted a lot of interest as a highly efficient route for the systemic delivery of insulin. It has been well known that the pharmacokinetic profile of intranasal insulin resembles the pulsatile pattern of endogenous insulin secretion in healthy volunteers during meal times [62]. In addition it's considered as a promising route for the following reasons: the nose has relatively large surface area [150 cm^2] of absorption because of numerous microvilli, high vascularized subepithelial layer that passes directly into the systemic circulation, thereby avoiding the loss of drug by first pass metabolism in the liver, high permeability of the nasal epithelial membrane and lower enzymatic activity than the gastrointestinal tract. Although nasal administration of insulin has many advantages, there are also some barriers that limit the intranasal absorption of insulin. Macociliary clearance of formulations from the nasal cavity, low permeability of nasal mucosa to large molecules and the low bioavailability of insulin act as barriers to intranasal absorption. To overcome the various barriers by the nasal route, researchers have studied many extensive range of enhancers such as bile salts and derivatives, sodium lauryl sulfate, laureth-9, phospholipids, cyclodextrins, chitosan and enzyme inhibitors.

At first CS NPs is seemed to be the safest and most effective as a carrier for the nasal delivery of insulin. It protected insulin from degradation in the nasal cavity and increased intranasal absorption of insulin with its positive charge [20, 63]. Also PEG-grafted(g) have been used to enhance the solubility and improve the biocompatibility of CS. Insulin PEG-g with CS NPs administered intranasally to transport insulin across the nasal mucosa in rabbits [64]. However recent studies showed that insulin-CS solution formulation was more effective than the intranasal NP complex (Bioavailability 17 %, 3.6 % respectively) [65]. Because of NPs couldn't enhance the uptake of insulin; PEGylated trimethyl CS NP results was also not found significantly different from the insulin-CS solution formulation. Besides chitosan reduced gold NPs could enhance insulin transport into cells effectively. After insulin loaded gold Nps administered to diabetic rats by intranasally blood glucose concentration was decreased by 20.27 % [66]. Moreover intranasal route of insulin loaded starch NPs containing sodium glycocholate which's used as mucoadhesive carrier, caused 70 % reduction of plasma glucose levels and significant hypoglycemia untill 6h in the STZ induced diabetic rats [67].

2.3. Pulmonary Delivery

Pulmonary administration is one of the most promising alternative route of insulin delivery. The lungs offer a large and highly vascularised surface area for drug absorption approximately 80-140 m^2. Alveoles are covered by a very thin (0.1-0.2 mm) monolayer epithelium, that permits rapid drug absorption. The alveoli can be effectively targeted for drug absorption by drug delivery as an aerosol with a mass median aerodynamic diameter of less than 5μm. First pass metabolism in this administration avoids gastrointestinal system metabolism. Although metabolic enzymes are found in the lungs, their activities and pathways may be different from those found in the GIT and this makes the pulmonary route of many therapeutic proteins and peptides very promising [68, 69, 70].

There are variety of inhalation devices such as metered-dose inhalers or drug powder inhalers. Such as AERx®. Insulin Diabetes Management system developed by Novo Nordisk which delivers aerosol of human insulin; Exubera® developed by Nektar/Pfizer which uses a dry powder formulation. [36].

Dry powder inhalers are currently the most commonly used devices because of their stability and sterility to develope pulmonary insulin. The surfactants, bile salts and fatty acids have been evaluated as absorption enhancers which increase the permeability of drugs through the epithelial membranes. However polyoxyethylene (PE) oleyl ether showed good enhancement sorbitan trioleate exhibited moderate enhancing ability. The enhancing effects of glycerol trioleate, ethyl oleate, oleyl alcohol, palmitic acid and stearic acid were very low. In contrast liposomes are very effective pulmonary absorption enhancers for peptide and protein drugs. They've biogenic phospholipids and biocompatible, biodegradable and non immunogenic natural properties.

Experimental studies investigated that insulin could be efficiently encapsulated in liposomes which has approximately 1μm particle size. Liposome mediated pulmonary drug causes enhancement in drug retention time in the lungs and decreases side effects which results increased therapeutic effects. When aerolized insulin liposomes delivered by the inhalation route in mice caused significantly reduction in plasma glucose concentrations [71]. Insulin-calcium phosphate (CAP) and polyethylene glycol (PEG) particles were administered to the lungs by a route of administration of respiratory tract and these particles positively affected the disposition of the insulin in the lungs of rats [72]. Poly-lactide-co-glycolide (PLGA) particles have used to improve insulin loaded particles which has a mean diameter of 400 nm. After the pulmonary administration of insulin with PLGA nanospheres, blood glucose levels were significantly decreased and has got prolonged hypoglycemic response over 48 h in guinea pigs [73]. In another related study poly butyl cyanoacrylate NPs have been used given by pulmonary inhalation of insulin in the lungs resulted stable and prolonged pharmacological effect. A significant reduction in glucose levels were found and the relative pharmacological bioavailability was 57.2 % [70]. Besides insulin/1, 2 dipalmitoyl phosphatidylcholine (DPPC) physical mixture used to enhance insulin absorption in pulmonary route of inhalation. This mixture caused higher blood glucose decrease because of their potentially effective, non-toxic and natural absorption enhancer property [74].

2.4. Buccal Delivery

Insulin delivered by buccal route is through an aerosol spray into the oral cavity. It's absorbed through the inside of the cheeks and in the back of the mouth. The buccal mucosa is excellently accessible with surface area approximately 100-200 cm^2, lower risk to be traumatized and a relatively good permeability and perfusion [36, 75, 76, 77]. Several formulations and factors alone or in combination can influence release properties of buccal insulin delivery system. These formulations should contain absorption enhancers (such as surfactants, bile salts, chelators, sodium lauryl sulfate or fatty acids) to increase membrane permeability, enzyme inhibitors to protect the drug from degradation, protease inhibitors (aprotinin and sodium glycocholate) to function drug permeation across mucosa, lipophilicity modifications (conjugation with polymers) bioadhesive delivery systems (gels, films, patchs) and liposomal formulations [36, 76, 78]. Lysalbinic acid which is applied as an absorption enhancer was shown to enhance significantly buccal mucosa permeability for insulin. They investigated that it's a product of the alkaline hydrolysis of egg albumin and has no irritating or sensibilizing effect upon buccal use. Co-administration of lysalbinic acid and relatively small proteins such as insulin can increase insulin's permeability from the cheek mucosa of hamster [79].

In the last years a new innovative system has been developed by Generex Biotechnology Corporation (Toronto, Canada). It's based on a liquid formulation (Oral-Lyn®) of recombinant human insulin, absorption enhancers (which encapsulate and protect the Insulin molecules) and Rapid Mist® device (advanced buccal drug delivery technology). This device sends fastly small particles from an aqueous spray into the oral cavity. This allows rapid insulin absorption.

Oral-Lyn® has been evaluated in healthy persons and type I diabetes. It appears in the circulation within 10 min, the time to peak insulin concentration is is around 25 min. It has observed a more fast onset of action and less prolonged hypoglycemic action. Several studies in patients both type 1 and 2 diabetes demonstrated that this oral insulin can be efficient in controlling postprandial glucose levels. This new buccal insuin system needs further investigations in diabetic patients [35, 36, 76, 78].

2.5. Transdermal Delivery

Transdermal insulin delivery is an appealing alternative to the invasive parenteral route of administration and other alternative routes of insulin such as pulmonary and nasal routes because the skin offers the advantages of an easy access and a very large surface area (1-2 m^2). It improves patient compliance and avoids both liver's first pass metabolism and degradation of drugs in gastrointestinal tract. The skin also represents an important painless interface for systemic drug administration. Despite these advantages the human skin limits permeation of foreign compounds especially large hydrophilic molecules like insulin. The stratum corneum; which is the upper layer causes impermeability of the skin by its lipid-rich matrix. Several attemps have been made to overcome the skin barrier and to allow the transfer of large drugs such as insulin. They can be divided into chemical (liposome and chemical enhancers) and physical methods (mainly iontophoresis and sonophoresis).

2.5.1. Transdermal Delivery Methods

Chemical enhancers such as surfactants, fatty acids, fatty esters and azone-like compounds alter the lipid structure of the stratum corneum. They reduce its barrier properties and enhance its permeability for large molecule drugs that would not pass through the skin.

Iontophoresis is a non invasive technique used to increase transdermal insulin penetration through the skin by the application of a small electric current potential. Large drug molecules can be delivered in a shorter time with the help of this method and it increases drug's mobility.

Another non invasive technique *sonophoresis (ultrasound, phonophoresis)* which has been used to enhance (and or or delivery and activity of drugs) skin permeability to various low and high molecules weight drugs such as inulin. Low frequency ultrasound (20-160 kHz) decreases blood glucose levels both in animal and human studies [36, 72, 75].

Microneedles are minimally invasive painless and promising technology to deliver drugs into the skin without disruption of nerve endings. This technology create micronsized channels which interstitial fluid fills up the channels in the skin. It makes hydrophilic transport pathway, faciliates the stratum corneum barrier and increases skin permeability to large molecules [35, 36, 62, 75].

Also other methods have been investigated like microdermabrasion, pressure waves and electroporation but they're still in at a preliminary stage. Altogether chemical and physical methods they all need further investigations.

2.6. Ocular Delivery

Ocular delivery is another the most promising and challenging delivery of ophthalmologically active peptides and proteins for the treatment of ocular diseases. The advantages of the ocular delivery are; less development of immunological reactions in eye tissues, less side effects, no tolerance and avoidance of hepatic first pass metabolism. While the enhancers such as saponin, dodecylmaltoside, tetradecylmaltoside, fusidic acid and glycocholate increases the systemic absorption of insulin in animals they may also increase the eye toxicity [36, 80, 81]. A series of alkylglycosides including tetradecyl-, tridecyl-and dodecylmaltoside and dodecylsucrose were potent stimulators of insulin absorption after topical ocular delivery in anesthetized rats when used at concentrations as low as 0.125 %. These are the most hydrophobic alkylglycoside reagents and were the most effective at enhancing systemic insulin absorption [82]. Moreover sucrose cocoate, a pharmacological excipient of cosmetic and dermatologic preparation was used to determine its possible absorption enhancer in ocular drug delivery. When insulin was delivered ocularly in the presence of 0.5 % sucrose cocoate, plasma insulin levels were significantly enhanced and blood glucose levels were reduced [83]. Because of this observation insulin-containing liposome was prepared to prolong the retention time of the formulation in the precorneal area [84]. This positively charged formulation decreased the blood glucose levels 65-70%.

More recently Gelfoam® an absorbable gelatin sponge ocular devices have been developed as insulin carriers for systemic administration of insulin. Although Gelfoam® containing 0.2 mg insulin has been showed prolonged systemic absorption of insulin within the desired therapeutic levels it may also cause long term toxicity such as slowing the tear production. Because of this toxicity sodium insulin and zinc insulin Gelfoam ocular devices have been developed and these devices were sufficient to control blood glucose levels (60 % of initial) for over 8 hours [62].

2.7. Vaginal Delivery

In recent years numerous studies prove that vagina has got rich blood supply and large surface area that means good permeability and can be a potential route for systemic delivery to a wide range of compounds. The main advantages of vaginal drug route are avoidance of first pass metabolism, ease of administration and good permeability for low molecular weight drugs. For systemic delivery bile salts, dihydrofusidate, cyclodextrins, surfactants and chelating agents have been tested as enhancers to facilitate the rate of vaginal absorption but sometimes they induced several local reactions [36, 75].

2.8. Rectal Delivery

Rectal route of delivery have been tested soon after the discovery of insulin but several investigators have met absorption problems through the mucosa. This administration's promising advantage is the possibility of avoiding, to some extent, the hepatic first-pass metabolism. Absorption promoters and surfactants were used to provide highest hypoglycemic effect in rectal insulin delivery. The most effective rectal absorption enhancer polyoxyethylene-9-lauryl ether (POELE) or sodium salicylate were used in insulin suppositories on diabetic dogs. It was investigated that hypoglycemic effect can be achieved about 50-55 % [85].

3. CONCLUSION

Over the last years numerous studies summerised polymeric NPs focused on different routes of insulin delivery. The association of insulin with NP formulations designed to protect insulin from degradation and enhance its uptake in the ileum. However, more research in this area is needed to achieve the goal that has plagued researchers for many decades. At any rate, polymeric NPs for routes of insulin delivery seems to be the better alternative compared to others.

REFERENCE

1. Rosenfeld L. Insulin: Discovery and controversy. Clinical Chemistry (48):2270-88, 2002.
2. Verge D. Biotechnological and administration innovations in insulin therapy. Médecine sciences (Paris) (20):986-98, 2004.
3. Walsh G. Therapeutic insulins and their large-scale manufacture. Applied Microbiology and Biotechnology (67):151-9, 2005.
4. Saltiel AR, Khan CR. Insulin signaling and regulation of glucose and lipid metabolism. Nature (414):799-806, 2001.
5. Rotte M, Baerecke C, Pottag G, Klose S, Kanneberg E, Heinze HJ, et al. Insulin affects the neuronal response in the medial temporal lobe in humans. Neuroendocrinology. (81):49-55, 2005.
6. Hermansen K, Rönnemaa T, Petersen AH, Bellaire S, Adamson U. Intensive therapy with inhaled insulin via the AERx® insulin diabetes management system: A 12-week proof-of-concept trial in patients with type 2 diabetes. Diabetes Care. (27):162-7, 2004.
7. Rolla AR, Rakel RE. Practical approaches to insulin therapy for type 2 diabetes mellitus with premixed insulin analogues. Clinical Therapeutics (27):1113-25, 2005.
8. Hermansen K, Rönnemaa T, Petersen AH, Bellaire S, Adamson U. Intensive therapy with inhaled insulin via the AERx® insulin diabetes

management system: A 12-week proof-of-concept trial in patients with type 2 diabetes. Diabetes Care. (27):162-7, 2004.

9. Alemzadeh R, Palma-Sisto P, Parton EA, Holzum MK. Continuous subcutaneous insulin infusion and multiple dose of insulin regimen display similar patterns of blood glucose excursions in pediatric type 1 diabetes. Diabetes technology & therapeutics (7):587-96, 2005.

10. Wild S, Roglic G, Green A, Sicree R, King H. Global prevalence of diabetes, estimates for the year 2000 and projections for 2030. Diabetes Care. (27):1047-53, 2004.

11. Owens DR, Zinman B, Bolli G. Alternative routes of insulin delivery. Diabetic Medicine (20):886-98, 2003.

12. Krol S., Ellis-Behnke R., Marchetti P., Nanomedicine for treatment of diabetes in an aging population: state-of-the-art and future developments, Nanomedicine: Nanotechnology, Biology, and Medicine, (8):S69-S76, 2012.

13. Sonia T.A., Sharma C.P., An overview of natural polymers for oral insulin delivery, Drug Discovery Today 17(13/14):784-792, 2012.

14. Golden SH., Sapir T., Methods for Insulin Delivery and Glucose Monitoring in Diabetes: Summary of a Comparative Effectiveness Review, Journal of Managed Care Pharmacy, 18(6):S3-S17, 2012.

15. Cefalu W.T., Evolving treatment strategies fort he management of type 2 diabetes, The American Journal of the Medical sciences, 343(1):21-26, 2012.

16. Jain S., Saraf S., Influence of processing variables and in vitro characterization of glipizide loaded biodegradable NPs, Diabetes & Metabolic Syndrome: Clinical Research & Reviews, (3):113-117, 2009.

17. Russel-Jones D., The safety and tolerability of GLP-1 receptor agonists in the treatment of type 2 diabetes, International Journal of Clinical Practice 64(10):1402-1414, 2010.

18. Jean M., Alameh M., De Jesus D., Thibault M., Lavertu M., Darras V., Nelea M., Buschmann MD., Merzouki A., Chitosan-based therapeutic NPs for combination gene therapy and gene silencing of in vitro cell lines relevant to type 2 diabetes, European Journal of Pharmaceutical Sciences, (45):138-149, 2012.

19. Rekha MR., Sharma CP., Oral delivery of therapeutic protein/peptide for diabetes-Future perspectives, International Journal of Pharmaceutics, (440):48-62, 2013.

20. Duan X., Mao S., New Strategies to improve the intranasal absorption of insulin, Drug Discovery Today, 15(11/12):416-427, 2010.

21. Sonaje K., Lin KJ., Wey SP., Lin CK., Yeh TH., Nguyen HN., Hsu CW., Yen TC., Juang JH., Sung HW., Biodistribution, pharmacodynamics and pharmacokinetics of insulin analogues in a rat model: Oral delivery using pH-responsive NPs vs. subcutaneous injection, Biomaterials, (31):6849-6858, 2010.

22. Pickup JC., Zhi ZL., Khan F., Saxl T., Birch DJS., Nanomedicine and its potential in diabetes research and practice, Diabetes/Metabolism Research and Reviews (24):604-610, 2008.

23. Chaturverdi K., Ganguly K., Nadagouda MN., Aminabhavi TM., Polymeric hydrogels for oral insulin delivery, Journal of Controlled Release, (165):129-138, 2013.

24. Chen M.C., Sonaje K., Chen K.J., Sung H.W.: A review of the prospects for polymeric nanoparticle platforms in oral insulin delivery. Biomaterials, 32:9826-9838, 2011.

25. Maroni A., Zema L., Del Curto MD., Foppoli A., Gazzaniga A., Oral colon delivery of insulin with aid of functional adjuvants, Advanced Drug Delivery Reviews, (64):540-556, 2012.

26. Packhaeuser CB., Kissel T., On the design of in situ forming biodegrable parenteral depot systems based on insulin loaded dialkylaminoalkyl-amine-poly(vinyl alcohol)-g-poly(lactide-co-glycolide) NPs, Journal of Controlled Release, (123):131-140, 2007.

27. Elsayed A., Remawi MA., Qinna N., Farouk A., Badwan A., Formulation and characterization of an oily-based system for oral delivery of insulin, European Journal of Pharmaceutics and Biopharmaceutics, (73):269-279, 2009.

28. Durazo SA., Kompella UB., Functionalized nanosystems for targeted mitochondrial delivery, Mitechondrion, (12):190-201, 2011.

29. Avadi MR., Sadeghi AMM., Mohammadpour N., Abedin S., Atyabi F., Dinarvand R., Rafiee-Tehrani M., Preparation and characterization of insulin NPs using chitosan and Arabic gum with ionic gelation method, Nanomedicine: Nanotechnology, Biology, and Medicine, (6):58-63, 2010.

30. Sung HW., Sonaje K., Liao ZX., Hsu LW., Chuang EY., pH-Responsive NPs Shelled with Chitosan for Oral Delivery of Insulin: From Mechanism to Therapeutic Applications, Accounts of Chemical Research, 45(4):619-629, 2012.

31. Gupta AS., Nanomedicine approaches in vascular disease: a review, Nanomedicine: Nanotechnology, Biology, and Medicine, (7):763-779, 2011.

32. Liu J., Gong T., Wang C., Zhong Z., Zhang Z., Solid lipid NPs loaded with insulin by sodium cholate-phosphatidylcholine –based mixed micelles: Preparation and characterization, International Journal of Pharmaceutics, (340):153-162, 2007.

33. Woitiski CB., Neufeld RJ., Veiga F., Carvalho RA., Figueiredo IV., Pharmacological effect of orally delivered insulin facilitated b multilayered stable NPs, European Journal of Pharmaceutical Sciences, (41):556-563, 2010.

34. Woitiski CB., Sarmento B., Carvalho RA., Neufeld RJ., Veiga F., Facilitated nanoscale delivery of insulin across intestinal membrane models, International Journal of Pharmaceutics, (412):123-131, 2011.

35. Yaturu S., Insulin therapies: Current and future trends at dawn, World Journal of Diabetes, 4(1):1-7, 2013.

36. Lassmann-Vague V., Raccah D., Alternatives routes of insulin delivery, Diabetes & Metabolism, (32):513-522, 2006.

37. Bayat A., Dorkoosh FA., Dehpour AR., Moezi L., Larijani B., Junginger HE., Rafiee-Tehrani M., NPs of quaternized chitosan derivatives as a carrier

for colon delivery of insulin: Ex vivo and in vivo studies, International Journal of Pharmaceutics, (356):259-266, 2008.

38. Mukhopadhyay P., Sarkar K., Chakraborty M., Bhattacharya S., Mishra R., Kundu PP., Oral insulin delivery by self-assembled chitosan NPs: In vitro and in vivo studies in diabetic animal model, Materials Science and Engineering C, (33):376-382, 2013.

39. Jintapattanakit A., Junyaprasert VB., Mao S., Sitterberg J., Bakowsky U., Kissel T., Peroral delivery of insulin using chitosan derivatives: A comparative study of polyelectrolyte nanocomplexes and NPs, International Journal of Pharmaceutics, (342):240-249, 2007.

40. Ma Z., Lim TM., Lim LY., Pharmacological activity of peroral chitosan-insulin NPs in diabetic rats, International Journal of Pharmaceutics, (293):271-280, 2005.

41. Mukhopadhyay P., Mishra R., Rana D., Kundu P.P., Strategies for effective oral insulin delivery with modified chitosan NPs, Progress in Poymer Science, (37):1457-1475, 2012.

42. Jin Y., Song Y., Zhu X., Zhou D., Chen C., Zhang Z., Huang Y., Goblet cell-targeting NPs for oral insulin delivery and the influence of mucus on insulin transport, Biomaterials, (33):1573-1582, 2012.

43. Sonaje K., Lin YH., Juang JH., Wey Sp., Chen CT., Sung HW., In vivo evaluation of safety and efficacy of self-assembled NPs for oral insulin delivery, Biomaterials, (30):2329-2339, 2009.

44. Sonaje K., Chen YJ., Chen HL., Wey SP., Juang JH., Nguyen HN., Hsu CW., Lin KJ., Sung HW., Enteric-coated capsules filled with freeze-dried chitosan/poly(γ-glutamic acid)NPs for oral insulin delivery, Biomaterials, (31):3384-3394, 2010.

45. Sajeesh, S, Vauthier C., Gueutin C., Ponchel G., Sharma C.P., 2010. Thiol functionalised polymethacyrlic acid-based hydrogel microparticles for oral insulin delivery. Acta Biomater. (6):3072-3080, 2010..

46. Cui F., Qian F., Zhao Z., Yin L., Tang C., Yin C., Preparation, characterization, and oral delivery of insulin loaded carboxylated chitosan grafted poly(methyl methacrylate) NPs, Biomacromolecules, (10):1253-1258, 2009.

47. Woitiski CB., Veiga F., Ribeiro A., Neufeld R., Design for optimization of NPs integrating biomaterials for orally dosed insulin, European Journal of Pharmaceutics and Biopharmaceutics, (73):25-33, 2009.

48. Sarmento B., Riberio A., Veiga F., Ferreira D., Development and characterization of new insulin containing polysaccharide NPs, Colloids and Surfaces, (53):193-202, 2006.

49. Card JW., magnuson BA., A review of the efficacy and safety of nanoparticle-based oral insulin delivery systems, American Journal of Physiology – Gastrointestinal and Liver Physiology, (301):G956-G967, 2011.

50. Li MG., Lu WL., Wang JC., Zhang X., Wang XQ., Zheng AP., Zhang Q., Distribution, transition, adhesion and release of insulin loaded NPs in the gut of rats, International Journal of Pharmaceutics, (329):182-191, 2007.

51. Jin HE., Xia F., Zhao YP., Preparation of hydroxypropyl methyl cellulose phthalate NPs with mixed solvent using supercritical antisolvent process

and its application in co-precipitation of insulin, Advanced Powder Technology, (23):157-163, 2012.

52. Chalasani KB., Russell-Jones GJ., Jain AK., Diwan PV., Jain SK., Effective oral delivery of insulin in animal models using vitamin B12-coated dextran NPs, Journal of Controlled Release, (112):141-150, 2007.

53. Chalasani KB., Russell-Jones GJ., Yandrapu SK., Diwan PV., Jain SK., A novel vitamin B12-nanosphere conjugate carrier system for peroral delivery of insulin, Journal of Controlled Release, (117):421-429, 2007.

54. Pan Y., Li YJ., Zhao HY., Zheng JM., Xu H., Wei G., Hao JS., Cui FD., Bioadhesive polysaccharide in protein delivery system: chitosan nanoparticles improve the intestinal absorption of insulin in vivo. International Journal of Pharmaceutics (249):139-147, 2002.

55. Sarmento B., Ribeiro A., Veiga F., Ferreira D., Neufelt R., Oral bioavailability of insulin contained in polysaccharide nanoparticles, Biomacromolecules, 8(10): 3054-60, 2007.

56. Makhlof A., Tozukaa Y., Takeuchia H., Design end evaluation of novel pH-sensitive chitosan nanoparticles for oral insulin delivery. Eur. J. Pharm. Sci.42, 445-451, 2011.

57. Sarmento B., Riberio A., Veiga F, Sampaio P., Neufeld R., Ferreira D. Alginate/chitosan nanoparticles are are effective for oral insulin delivery. Pharmaceutical Research, (24):2198-206, 2007.

58. Wu ZM., Zhou L., Guo XD., Jiang W., Ling L., Qian Y., Luo KQ., Zhang LJ., HP55-coated capsule containing PLGA/RS NPs for oral delivery of insulin, International Journal of Pharmaceutics, (425):1-8, 2012.

59. Cui FD., Tao AJ., Cun DM., Zhang LQ., Shi K., Preparation of insulin loaded PLGA-HP55 NPs for oral delivery, Journal of Pharmaceutical Sciences, (96):421-427, 2007.

60. Xiong XY, Li YP, Li ZL, Zhou CL, Tam KC, Liu ZY et al., Vesicles from pluronic/poly(lactic acid) block copolymers as new carriers for oral insulin delivery. Jornal of Control Release (120):11-7, 2007.

61. Damge C., Maincent P., Ubrich N., Oral delivery of insulin associated to polymeric NPs in diabetic rats, Journal of Controlled Release, (117):163-170, 2007.

62. Khafagy ES., Morishita M., Onuki Y., Tkayama K., Current Challenges in non-invasive insulin delivery systems: A comparative review, Advanced Drug Delivery Reviews, (59):1521-1546, 2007.

63. Ferna´ndez-Urrusuno, R. et al. (1999) Enhancement of nasal absorption of insulin using chitosan NPs. Pharm. Res. 16, 1576–1581 (Duan X, 82)

64. Zhang, X. et al. (2008) Nasal absorption enhancement of insulin using PEG-grafted chitosan NPs. European Journal of Pharmaceutics and Biopharmaceutics (68):526-534, 2008.

65. Dyer, A.M. et al. (2002) Nasal delivery of insulin using novel chitosan based formulations: a comparative study in two animal models between simple chitosan formulations and chitosan NPs. Pharmaceutical Research (24):1415-1426, 2007.

66. Bhumkar, D.R. et al. (2007) Chitosan reduced gold NPs as novel carriers for transmucosal delivery of insulin. Pharmaceutical Research (24):1415-1426, 2007.

67. Jain, A.K. et al. (2008) Effective insulin delivery using starch NPs as a potential transnasal mucoadhesive carrier. European Journal of Pharmaceutics and Biopharmaceutics (69):426-435, 2008.

68. Henkin RI., Inhaled insulin-Intrapulmonary, intranasal and other routes of administration: Mechanisms of action, Nutrition (26):33-39, 2010.

69. Liu J., Gong T., Fu H., Wang C., Wang X., Chen Q., Zhang Q., He Q., Zhang Z.: Solid Lipid NPs for pulmonary delivery of insulin, International Journal of Pharmaceutics, (356):333-344, 2008.

70. Zhang Q., Shen Z., Nagai T., Prolonged hypoglycemic effect of insulin-loaded polybutylcyanoacrylate NPs after pulmonary administration to normal rats, International Journal of Pharmaceutics, (218):75-80, 2001.

71. Huang Y.Y., Wang C.H., Pulmonary delivery of insulin by liposomal carriers, Journal of Control Release (113):9-14, 2006.

72. Garcia-Contreras L., Morçöl T., Bell S.J.D., Hickey A.J., Evaluation of novel particles as pulmonary delivery systems for insulin in rats, AAPS PharmSciTech (5): (Article 9), 2003.

73. Kawashima Y., Yamamoto H., Takeuchi H., Fujioka S., Hino T., Pulmonary delivery of insulin with nebulized DL-lactide/glycolide copolymer (PLGA) nanospheres to prolong hypoglycemic effect. Journal of Control Release (62):279-287, 1999.

74. Mitra R., Pezron I., Li Y., Mitra A.K., Enhanced pulmonary delivery of insulin by lung lavage fluid and phospholipids, International Journal of Pharmaceutics (217):25-31, 2001.

75. Sharma JPK., Bansal S., Banlk A., Noninvasive routes of proteins and peptids drug delivery, Indian Journal of Pharmaceutical Sciences, 73(4):367-375, 2011.

76. Palermo A., Napoli N., Manfrini S., Lauria A., Strollo R., Pozzili P., Buccal spray insulin in subjects with impaired glucose tolerance: the prevoral study, Diabetes, Obesity and Metabolism, 13:42-46, 2011.

77. Palermo A., Maddaloni E., Pozzili P., Buccal Spray Insulin (Oralgen) for type 2 diabetes:what evidence? Expert Opinion on Biological Therapy 12(6):767-772, 2012.

78. Heinemann L., Jacques Y., Oral Insulin and Buccal Insulin, Journal of Diabetes Science and Technology, 3(3):568-584, 2009.

79. Starokadomskyy P.L., Dubey IY., New absorption promoter for the buccal delivery: preparation and characterization of lysalbinic acid, International Journal of Pharmaceutics (308):149-154, 2006.

80. Lee Y.C., Yalkowsky S.H., Effect of formulation on the systemic absorption of insulin from enhancer-free ocular devices, International Journal of Pharmaceutics (185):199-204, 1999.

81. YamamotoA., Luo A.M., Dodda-Kashi S., Lee V.H., The ocular route for systemic insulin delivery in the albino rabbit, Journal of Pharmacology and Experimental Therapeutics (249):249-255, 1989.

82. Pillion D.J., Atchison J.A., Wang R.X., Meezan E., Alkylglycosides enhance systemic absorption of insulin applied topically to the rat eye, Journal of Pharmacology and Experimental Therapeutics (271):1274-1280, 1994.

83. Ahsan F., Arnold J.J., Meezan E., Pillion D.J., Sucrose cocoate, a component of cosmetic preparations, enhances nasal and ocular peptide absorption, International Journal of Pharmaceutics (251):195-203, 2004.

84. Adikwu M.U., Evaluation of snail mucin motifs as rectal absorption enhancer for insulin in non-diabetic rat models, Biological and Pharmaceutical Bulletin (28):1801-1804, 2005.

85. Hosny E., Al-Shora H.I., Elmazar M.M, Relative hypoglycemic effect of insulin suppositories in diabetic beagle dogs: optimization of various concentrations of sodium salicylate and polyoxyethylene-9-lauryl ether, Biological and Pharmaceutical Bulletin (24):1294-1297, 2001.

CHAPTER 12

Multidrug Delivery Systems with Single Formulation

Tatsuya Okuda, Satoru Kidoaki

Division of Biomolecular Chemistry, Institute for Materials Chemistry and
Engineering, Kyushu University, Fukuoka, Japan.

ABSTRACT

Development of new way and system for multidrug delivery has recently
attracted much attention and became one of major issue in drug delivery
research. Although this research field is still immature compared to the single
drug delivery system, intensive efforts have recently been devoted by
researchers in order to realize more efficient, functional, and safe combination
therapy using multiple drugs or agents. In this review article, we outline several
targets in terms of application for biochemical modulation together with various
concrete attempts of simultaneous and sequential delivery of multiple drugs or
agents with single formulation. Finally, we will also summarize the possible
contribution of biomaterial sciences and nanobiotechnology for improvement of
future multidrug delivery system.

Keywords: Multidrug Delivery System; Single Formulation; Combination
Therapy; Biochemical Modulation

1. INTRODUCTION

Almost all diseases, especially intractable diseases such as cancer and human
immunodeficiency virus infection, cannot cure by treatment with only one type
of drug because of pathological complexity. The combinatorial use of multiple
drugs having different pharmaceutical action mechanism often redound to better
outcome compared with monochemotherapy, because the combination of
adequately selected drugs might bring the synergistic or additive effects at lower
dose. Furthermore, the low dose combination therapy may concurrently lead to
suppression of severe adverse effect. Therefore, the combination therapy by
multiple drug treatment has been commonly used in clinical chemotherapy.

Biochemical modulation (BCM) is defined as one of a methodology of
combination therapy that aimed to enhance pharmacological effect and/or
suppress serious side effect by taking advantage of modulator drug(s) to

modifying pharmacokinetics of effector (main) drug [1,2]. In the BCM-based chemotherapy, the drugs are administered sequentially or simultaneously according to action mechanism of individual drugs. Since Bertino et al. reported in 1977 [3], the theory of BCM has attracted much attention especially in clinical medication and now constructed a part of basis of current cancer chemotherapy. To date, several regimens have been developed and applied to clinical cancer chemotherapy by improving drug combination, administration dose, order, timing, and so on (**Table 1**). Refinement of BCM regimens has surely yielded better outcome of chemotherapy. Since most of regimens, however, need temporally controlled bolus injection, infusion, and continuous infusion, the patients are forced to restrain to the bedside for a long time. Therefore, the development of multi drug delivery systems (M DDSs) with single formulation that matched for BCM regimens has been expected.

The development of nanotechnology and subsequent fusion with biotechnology, i.e., the emergence of nanobiotechnology, has made significant impact on life science, especially medical and pharmaceutical sciences. In addition to development of nanobiotechnology, the rapid advances in materials science also contributes to improvement of quality of medicine. Although the excellent drugs often have some inconvenient natures and/or characters, such as low solubility, unfavorable biodistribution, and serious side effect, in return for its therapeutic potential, several drug delivery systems (DDSs) based on approaches in nanobiotechnology, such as liposomes [8], polymeric micelles [9,10], polymer particulate [11], and so on, have been developed so far. Some DDS formulations have already been reached to the market [12-15], and consequently the established DDSs, especially for cancer chemotherapy, have successfully improved not only therapeutic outcome of conventional drugs but also the patient's quality of life.

Although more functional and smart DDSs have been reported with recent remarkable advance of nanobiotechnology, the development of the MDDSs that are suitable for controlling of pharmacokinetics of multiple drugs have been still insufficient compared to DDSs for single drug delivery. Therefore, development of new modes and systems, which can effectively and functionally regulate release profiles of multiple drugs from a single formulation, is now important issue with significant interest in current and future DDS research. In past decade, many researchers have made huge effort to develop MDDS, and consequently various MDDSs, which can adopt for combination therapy, have been established. These attempts have been well summarized in some previous review articles that are described based on the types of formulation [16-18]. Meanwhile, we will especially focus on the MD DSs related to biochemical modulation (BCM), and review the recent attempts from this point of view in this article. The concrete examples of MDDS that can simultaneously deliver multiple agents will be respectively introduced according to individual applications (Section 2). Several approaches for development of sequential multiple agent delivery from single formulation are shown in Section 3. Finally, some future perspectives and conclusions will given in Section 4.

Table 1. Representative BCM based regimens for clinical cancer chemotherapy.

| Regimen name | Utilizing drugs* | | | | Cycle | Ref.# |
| | Order and way of administration | | | | | |
	First	Second	Third	Fourth		
FOLFOX6	I-LV 200 mg/m² and L-OHP 100 mg/m² Infusion for 2 h	5-FU 400 mg/m² Bolus injection	5-FU 2400 ~ 3000 mg/m² Infusion for 46 h	-	Every 2 weeks	[4]
FOLFOX7	I-LV 200 mg/m² and L-OHP 130 mg/m² Infusion for 2 h	5-FU 400 mg/m² Bolus injection	5-FU 2400 mg/m² Infusion for 46 h	-	Every 2 weeks	[5]
FOLFIRI	I-LV 200 mg/m² and CPT-11 180 mg/m² Infusion for 2 h	5-FU 400 mg/m² Bolus injection	5-FU 2400 ~ 3000 mg/m² Infusion for 46 h	-	Every 2 weeks	[4]
mFOLFOX6 + Panitumumab	Panitumumab 6 mg/kg Infusion for 1 h	I-LV 200 mg/m² and L-OHP 85 mg/m² Infusion for 2 h	5-FU 400 mg/m² Bolus injection	5-FU 2400 mg/m² Infusion for 46 h	Every 2 weeks	[6]
FOLFIRI + Panitumumab	Panitumumab 6 mg/kg Infusion for 1 h	I-LV 200 mg/m² and CPT-11 180 mg/m² Infusion for 2 h	5-FU 400 mg/m² Bolus injection	5-FU 2400 mg/m² Infusion for 46 h	Every 2 weeks	[7]

* Abbreviated of drug names are respectively following; I-LV: Levofolinate, L-OHP: Oxaliplatin, CPT-11: Camptothecin 11; 5-FU: 5-Fluorouracil.

2. APPLICATIONS OF MDDS FOR BCM-BASED CHEMOTHERAPIES

2.1. Serious Side-Effect Reduction

Anticancer drugs often have intolerable side effects, such as cardiotoxicity, myelosuppression, and gastrointestinal dysfunction. Although anticancer drugs bring enough therapeutic efficiency, there are many cases in which administration of anticancer drug is forced to withdraw due to serious adverse effect. For instance, doxorubicin (DOX) that is one of the most effective and widely used anticancer drugs has crucial cardiotoxicity. Current major strategies to prevent heart injury caused by this drug are encapsulation of DOX into carrier molecules, such as poly (ethylene glycol) (PEG) modified liposome and polymeric micelles, or covalent conjugation to biocompatible polymer. These approaches have succeeded in suppression of side effect and enhancement of therapeutic outcome, however, it seems that more active approach against drug-induced cardiomyopathy is needed for more safe and efficient cancer chemotherapy.

Santucci et al. have developed simultaneous anticancer drug epirubicin (EPI) and nitric oxide (NO) carrying system (EPI-PEG-NO), in which NO (NO releasable moiety, butandiol mononitrate) and EPI were covalently conjugated to each terminal of PEG [19]. In their system, NO act as not only protecting reagent against anthracyclineinduced cardiomyopathy but also sensitizer of anticancer drug treatment. In order to increase anticancer efficacy and enhance cardiocyte-protecting ability of EPI-PEGNO system, the improved system used branched PEG as polymer backbone instead of linear one has been developed. The newly constructed PEG conjugates have one EPI and four or eight NO releasable moieties (EPI-bPEG- $(NO)_4$, EPI-bPEG-$(NO)_8$) [20]. The ternary conjugate showed higher anticancer activity for carcinoma cells (Caco-2), whereas it decreased cytotoxicity against endothelial cells and cardiomyocytes, with respect to free EPI treatment.

2.2. Modulation of Cancer Surviving Mechanism

A better understanding of cancer pathophysiology is very important on planning of chemotherapy. Mutated molecules and upor down-regulated molecules in cancer cells compared with normal cells must be good target of medication. Recently, several researches focused on these molecules as target of therapy have been actively progressed.

It is well known that the tumor suppressor gene, p53, is mutated in a wide variety of human cancer cells. The translational product of p53 gene plays an important role as a transcription factor. There are many target genes related to cell cycle arrest, senescence, and apoptosis, under the regulation of p53. Since the p53 protects cells from several damages by regulating programmed cell death mechanism in the downstream of p53 pathway, the mutation of p53 promotes the surviving of cancer cells. Therefore, it is expected that the recovery of p53 function lead to better outcome of cancer therapy. Wu et al. have reported that chemosensitivity of cancer against several anticancer drugs, such as DOX, cisplatin, paclitaxel (PTX), and mytomycin C, have been enhanced by adenoviral transfection of wild-type p53 gene [21]. Wiradharma et al. have developed the MDDS focused on this p53 mediated chemosensitizing mechanism [22]. In their co-delivery system of DOX and p53-encoding plasmid DNA (pDNA), the three blocks oligopeptide, Ac-(AF)$_6$-H$_5$-K$_{15}$-NH$_2$(FA32), which can simultaneously assemble and form surface cationic core-shell structure at above the critical micelle concentration, was used as carrier molecule. DOX was encapsulated into hydrophobic core and p53-encoding pDNA was complexed with cationic surface of FA32 via electrostatic interaction. The synergistic effect on suppression of Hep-G2 cell proliferation was observed by co-delivery of DOX and p53- encoding pDNA using FA32, whereas no synergism was observed by individual treatment of free DOX and FA32/p53-encoding pDNA complex.

Survivin is an unique apoptosis suppressor molecule, which suppress the programmed cell death by inhibit caspase family in several cell types [23,24]. Since survivin molecules are expressed in almost of all major cancer cells whereas are not expressed in other differentiated cells [25], it is one of another good target molecule on tumor chemotherapy. To date, many attempts have been made to offset survivin function in cancer cells by dominant-negative mutant, antisense oligonucleotide (ASO), RNA interference, and so on. Recently, two successful examples of MDDS, in which abolishment of survivin function and anticancer drug treatment are combined, have been reported. Xiao et al. have constructed the liposomal co-delivery system of anticancer drug DOX and dominant negative survivin mutant (Msurvivin T34A)-encoding pDNA [26]. This survivin mutant has antiproliferative potential and caspase-dependent apoptosis induction ability for tumor cells. In this system, DOX was loaded into internal aqueous phase of truncated bFGF peptide-modified cationic liposome taking advantage of pH gradient remote loading method, and then Msurvivin T34A-encoding pDNA was complexed with DOX-loaded cationic liposome via electrostatic interaction. This MDDS has achieved dose reduction of DOX to 3-fold lower level compared to free DOX treatment in vitro. Furthermore, significant tumor growth suppression effect was observed in the Lewis lung carcinoma-bearing C57BL/6 mice by co-delivery of DOX and Msurvivin T34A-encoding pDNA compared to liposomal DOX or lipoplex alone. Another example of MDDS related to the combination of anticancer drug treatment and survivin counteraction has been reported from Xu et al [27]. In this system, the survivin-targeted RNAi encoding pDNA (iSur-pDNA)/protamine complexes and DCTX were encapsulated by folate-modified lipid-based envelope via the lipid film hydration technique. This newly constructed formulation showed lower cytotoxicity compared to commercially available liposomal transfection reagent and enhanced internalizing ability due to folatemodification. As the result of in vitro cytotoxicity evaluation against for hepatocellular carcinoma cell line BEL 7402, the MDDS showed much higher cytotoxicity than the single treatment by DCTX or iSur-pDNA monoloaded folate-modified lipid-based envelope. In addition, the cytotoxicity provided by co-delivery of DCTX and iSur-pDNA using the MDDS was also higher than the combination of free

DCTX and iSur-pDNA mono-loaded folate-modified lipid-based envelope. These results suggest that the simultaneous delivery (i.e., the MDDS) is more effective on the combination of cancer chemotherapy and survivin targeted therapy.

2.3. Tumor Starving Therapy

The tumor growth and angiogenesis are closely related to each other. The primary tumors and metastatic tumors upregulate angiogenesis by secreting of several angiogenesis related factors, such as vascular endothelial growth factor (VEGF), matrix metalloproteases (MMPs), and so on, and acquire nutrition needed for own growth from the induced neovessels. If suppression of the tumor-induced angiogenesis and/or disruption of neovessels that provide nutrition to tumor can be achieved, tumor growth inhibition therapy will be realized. The angiogenesis-targeted cancer therapy is especially called as "Tumor starving therapy", because nutrition supply to tumor is blocked by suppression of angiogenesis or disruption of neovasculature.

Combrestatin A4 (CA4) is dephosphorylated form of combrestatin A4 phophate, which is natural product isolated from the South African tree Combretum caffrum. CA4 acts as tubulin inhibitor in endothelial cells, and consequently induce neovascularture disruption. Owing to its ability, CA4 gets much attention as antivasculature (vascular disrupting) agents. Even though the single CA4 treatment shows some promising results in clinical trial, it is expected that the combination of CA4 and another drug(s) will redound superior therapeutic outcome [28].

Some groups have developed the MDDS aimed for co-delivery of CA4 and anticancer drug [29-31]. Wang et al. have developed the MDDS, which can co-deliver CA4 and PTX [30]. In their system, nanocapsule that consists of mPEG$_{2000}$-PLA$_{2000}$ block copolymer was used as carrier molecule. PTX was covalently conjugated to the terminal of PLA side via ester bond, and then CA4 was encapsulated into hydrophobic core during nanocapsules formation. In the in vivo Matrigel plug assay, the dual loaded nanocapsule showed significant antiangiogenesis effect compared to treatment with single drug loaded nanocapsule. Moreover, in the Lewis lung carcinoma tumorbearing mice, the most significant tumor growth suppression effect was observed by dual drug loaded nanocapsule treatment. The improved co-delivery system of CA4 and DOX has been reported from Yang et al. [31]. The cyclic pentapeptide, cyclic arginine-glycine-aspartic acidtyrosine-lysine (cRGDyK), which is well known as recognition motif for integrins, was conjugated to the PEGside terminal of PEG-b-PLA block copolymer. Since expression of integrins, especially $\alpha_v\beta_3$ and $\alpha_v\beta_5$, are upregulating in sprouting neovasculature as well as various cancer cells, the targeting ability to angiogenic endothelial cells and tumor have provided to this system by introducing RGD motif (cRGDyK). The cRGDyK-modified nanomicelle system, in which DOX and CA4 were loaded by chemical conjugation and simultaneous encapsulation, respectively, showed significantly enhanced uptake of the drugs by cancer and endothelial cells via receptor-mediated endocytosis. Furthermore, in the in vivo experiment, this cRGDyk-modified and dual-loaded system achieved superior tumor growth inhibition, antineovasculature, and apoptosis induction compared to each single-drug loaded nanomicelles.

TNP-470 is a synthetic analogue of fumagillin with potent antiangiogenic ability [32]. Although TNP-470 has strong potency as anticancer drug, the use of TNP-470 is limited due to its dose limiting neurotoxicity. Since serious adverse effect of TNP-470 is ascribed to its unfavorable biodistribution characteristic and low solubility, the improvement of bioavailability has been achieved by conjugation to highly biocompatible N-(2-hydroxypropyl)methacrylamide (HPMA) copolymer [33]. In order to acquire further therapeutic outcome, Segal et al. have constructed the MDDS based on the HPMA-TNP-470 conjugate [34]. In this work, both alendronate (ALN) and TNP-470 were covalently conjugated to HPMA copolymer via short peptide linker (Gly-Gly-Pro-Nle), which can be cleaved by cathepsin K specifically overexpressed in bone tissue. The drug ALN has not only bone targeting ability but also antitumor activity and antiangiogenic potential. Thus this co-delivery system utilizing HPMA copolymer-ALN-TNP-470 conjugate exerted passive/activetargeting ability and synergistic antineovasculature and anticancer activity in vivo. The HPMA copolymer-ALNTNP-470 conjugate remarkably suppressed MG-63-ras human osteosarcoma growth in xenograft mice by 96%, whereas the only 45% growth inhibition was observed by combination treatment of free drugs. Simultaneously, immunohistochemistry revealed 74% reduction of neovascularization in mice treated with HPMA copolymerALNTNP-470 conjugate, meanwhile the dual treatment with free drugs showed only 39% reduction.

VEGF is one of glycoproteins, which is related to vasculogenesis and angiogenesis. VEGF binds to vascular endothelial cells via specific VEGF receptor on cell surface, and induces cell mitosis, migration, and differentiation. Since many cancer cells are upregulating the production and secretion of VEGF for own survival and growth, suppression of the VEGF expression in tumor cells and overriding of the secreted VEGF will be good strategies for cancer therapy. In several clinical trials, addition of VEGF specific monoclonal antibody (Avastin), which can void VEGF function, to pre-existing combination chemotherapy has lead to better outcome.

As far as the MDDS for VEGF targeted combination therapy is concerned, the combination between conventional anticancer drug treatment and knock down of VEGF production by RNAi has been attempted [35,36]. For instance, Huang et al. have developed the MDDS of DOX and VEGF targeted siRNA by using cationic polymer micelle that consists of stearic acid-grafted polyethyleneimine (PEI-SA) [36]. Since the carrier molecule PEISA can spontaneously form micelle structure having hydrophobic core and cationic hydrophilic shell, hydrophobic PTX was encapsulated into core region and then and anionic siRNA was complexed with cationic PEI-SA polymer micelle via electrostatic interaction. In the in vitro experiment, dual-loaded PEI-SA micelles treatment showed significant reduction of VEGF production in Hep G-2 (human hepatocellular carcinoma) cells. At day 30 of postintratumoral administration, remarkable tumor growth suppression (13.0% of control) by simultaneous delivery of DOX and VEGF targeted siRNA using PEI-SA micelles was observed, whereas siRNA or DOX single loaded PEISA micelles treatment displayed slightly lower effect (33.7% and 56.7% of control, respectively).

2.4. Multidrug Resistance Reversal of Cancer

Multidrug resistance (MDR) of malignant neoplasm is the survival ability of cancer cells under the treatment with structurally and functionally diverse anticancer drugs. Since MDR is a major obstacle in clinical cancer chemotherapy, establishment of treatment strategies to overcome the cancer MDR is now becoming the great issue to be resolved for success of tumor therapy. The underlying mechanism of cancer MDR has been intensively researched and revealed to date. MDR phenotype in various type of cancer is frequently associated with upregulation of the permeability-glycoprotein (P-gp). P-gp, encoded by the human multidrug resistance gene, MDR1, is a member of ATP-binding cassette (ABC) transporter superfamily, which can act as active efflux pump at plasma membrane. Cancer cells possessing MDR ability are exerting chemoresistance by reducing intracellular drug concentration utilizing the overexpressed P-gp. Therefore, P-gp has attracted much attention as a promising target for overcoming cancer chemoresistance, and many researches about cancer chemotherapy that combined with P-gp targeted MDR reversal and conventional anticancer drug treatment have been intensively progressed. As far as P-gp-targeted MDR reversal is concerned, two strategies, i.e., inhibition of P-gp function by chemosensitizers, and MDR1 gene silencing utilizing ASO or RNAi, have been employed.

Chemosentitizers, such as verapamil (VRP), Elacridar, Tariquidar, and so on, mainly act as antagonist for P-gp and suppress drug efflux, and consequently recover chemosensitivity of MDR cancer cells. Since required dose (2 μM - 6 μM) of VRP for P-gp inhibition is, however, significantly higher than the clinical dose for arrhythmia treatment and may induce cardiotoxicity, adoption of DDS, especially MDDS, to this system will be appropriate way to reduce unfavorable side effect and to enhance therapeutic outcome. Indeed, various liposomal or nanoparticulate MDDSs that can co-deliver both chemosensitizer and anticancer drug have been reported. Wang et al. developed the liposomal MDDS, in which both VRP and DOX were simultaneously loaded into internal aqueous phase of Stealth liposomes (PEGylated liposomes) by pHgradient remote loading method. [37]. This system had demonstrated high MDR reversal ability against the multidrug resistant rat prostate cancer cell line (Mat-LyLuB2 cells). Based on the preceding work, Wu et al. established improved liposomal MDDS possessing cancer-targeting ability [38]. In their system, transferrin (Tf), a ligand for transferrin receptor frequently overexpressed in tumor and leukemia cells, was conjugated to surface of Stealth liposome. Tf-modified dual-loaded Stealth liposome (Tf-L-DOX/VRP) showed Tf-mediated targeting ability and high efficiency of MDR reversal to DOX-resistant K562 leukemia cells. The MDR reversal efficiency of Tf-L-DOX/VRP was 5.61 times superior to free DOX treatment, whereas non-targeted Stealth liposome treatment was almost equivalent to free DOX.

In the case of nanoparticulate MDDSs aimed for MDR reversal, various types of carriers, such as ionic polysaccharide microsphere [39], polyalkylcyanoacrylate nanoparticle [40], nanoparticles consist of highly biocompatible polymers [41-43], and polymer-lipid hybrid nanoparticle [44,45], are utilized for dual-loading of chemosensitizer and anticancer drug. The successful example has been reported from Patil et al. [43]. In their system, PTX and tariquidar, the third generation P-gp inhibitor, were encapsulated into biotin-modified PLGA nanoparticle during preparation by oil-in-water emulsion solvent evaporation method. Dual loaded nanoparticles showed significantly higher cytotoxicity in vitro compared to PTX single-loaded nanoparticles. Enhanced cytotoxicity caused by dual loaded nanoparticles treatment had been in good agreement with increased PTX-accumulation in drug-resistant cancer cells. The remarkable tumor growth inhibition was observed in drug-resistant tumor-bearing mice that treated with biotinylated nanoparticles encapsulating both PTX and tariquidar. Noteworthy, this therapeutic effect had been achieved at lower PTX dose that would be ineffective in the absence of tariquidar.

MDR1 gene silencing by ASO or siRNA are also attempted to suppress overexpression of P-gp and increase chemosensitivity of drug-resistant cancer [46,47]. For example, three agents (DOX, MDR1 and Bcl-2 targeted ASO) loaded Stealth liposome was prepared and applied to MDR reversal [46]. The distinctive feature of this system is that simultaneously aiming for suppression of DOX efflux by P-gp and enhancement of DOX-induced apoptosis. Three agents loaded liposomal MDDS achieved almost five-order reduction of IC_{50} value against for multidrug-resistant A2870/AD human ovarian carcinoma cells compared to free DOX treatment. The PEGylated liposome encapsulating DOX and ASOs showed significant tumor growth inhibition that correlated to enhanced apoptosis observed in tumor tissue of multidrug-resistant A2870 xenograft mouse model.

3. SEQUENTIAL MULTI-AGENT DELIVERY BY BULK SYSTEM

In the preceding section, concrete examples of various BCM-based MDDS that can simultaneously deliver multiple agents are introduced. On the other hand, there are many cases, in which not only the drug combination but also administration order and/or time lag will be important. Development of sequential delivery system that can achieve differential release control of individual agent from single formulation is another major task of DDS research. From this point of view, many researchers have made much effort to establish novel sequential multidrug delivery system. In particulate systems, the core/shell structure has been employed for the sequential multidrug delivery [30,48,49]. Meanwhile, the type of developed system with bulk matrices could be roughly classified into two categories, i.e., multilayered and composite system.

3.1. Multilayered System

The structural feature of this system is the multilayered structure that composed of hydrogels, polymer-form, polymer-film, and so on. In the multilayered systems, the release sequence of individual agents can control by lamination of agent-loaded layer in adequate and required order. Exceptionally the monolayer system of alginate hydrogel, in which release order of VEGF and PDGF is controlled by difference of diffusibility from gel matrix, has been reported from Hao et al. [50].

Some multilayered systems have been reported to date [51-53]. For example, the two-layered heterogeneously loaded and crosslinked gelatin gel system was prepared for sequential delivery of two bone growth factors, BMP- 2 and IGF-I, by Raiche and Puleo [51]. In their system, the burst release of growth factors was suppressed by crosslink ratio of individual layer. Pluripotent C3H10T1/2 cells exhibited significantly higher levels of osteoblastic phenotype, such as alkaline phosphatase activity and mineralized matrix formation, at early phase through cultivation on the two-layered gel matrix, which was releaseable BMP-2 and IGF-I in appropriate order (i.e., BMP-2 followed sequentially by IGF-I or BMP-2 plus IGF-I). More practical use directed system has been developed by Strobel et al. [53]. In their system, titanium Kirschner wire (K-wire) commonly used for fracture treatment was multiply coated with poly(D,L-lactide) (PDLLA) by dipping method. To form multiple coating on the surface, K-wire was sequentially immersed into three solutions of PDLLA that individually included gentamicin (antibiotics), IGF-I, or BMP-2. Prevention of infections on wound region was also aimed in addition to the stimulation of bone healing in this system. The multilayer coated K-wire displayed distinctitve release profile: 1) a burst release of gentamicin; 2) a burst and sustained release of IGF-I; and 3) a slow sustained release of BMP-2.

3.2. Composite System

Composite systems are commonly constructed by doping particulate DDS into bulk materials, such as hydrogel, polymer-form, and so on. The agents should be delivered are separately incorporated into particulate DDS and bulk material, respectively. Or the agents are loaded respectively into different particulate DDSs, and the bulk material is used only as a support of doped particulate. In this system, release profiles of individual agents are controlled by diffusibility from bulk material and/or releasing property of the utilized particulate DDS.

Hasirci and co-workers have attempted to develop sequential multi-agents delivery system for bone tissue engineering utilizing composite system. They first reported the poly(4-vinyl pyridine) (P_4VP)/alginate microsphere doped PLGA-form scaffold for sequential delivery of BMP-2 and BMP-7 [54]. In this system, the release profiles of two growth factors can be controlled by P_4VP/-alginic acid concentration at microsphere preparation. They also reported another BMP-2/BMP-7 sequential delivery system [55]. Two different polymers, PLGA and poly(3-hydroxybutyrate-co-3-hydroxyvalerate) (PHBV), were used for preparation of fastand slow-releasing nanocapsules, respectively, and these nanoparticles having different release rate were loaded to the chitosan-based wet-spun fiber scaffold. In each system, the sequential delivery of BMP-2 and BMP-7 from single formulation synergistically enhanced osteogenic differentiation against rat bone marrow derived stem cells. On the other hand, the injectable sequential multi-agents delivery system, which consists of PLG microsphere loaded alginate hydrogel, has been developed by Sun et al. [56]. In their system, two angiogenesis related growth factors, VEGF and PDGF, were respectively incorporated into alginate hydrogel and PLG microsphere for efficient neovasculalization on ischemic lesion site. The local concentrations of VEGF and PDGF on injected site reached peak level at week 2 and 4, respectively, and sustained PDGF release lasted to week 6. Sequential VEGF/PDGF delivery significantly promoted angiogenesis and induced reperfusion compared to single VEGF delivery or control.

3.3. Approaches for More Rigorous Temporal Release Control

The importance of sequential administration with pre-determined order and/or timing has been demonstrated in several systems, such as BCM-based cancer chemotherapy [3,57,58], bone regeneration [59], angiogenesis [60], and so on. According to these reports, it is expected that the precise control of release characteristics, especially administration order and timing, can be led more improved therapeutic outcome. However, all systems that developed to date have simultaneous releasing period. To realize more functional and effective combination therapy, more rigorous temporally controlled sequential release system has to be developed. It seems that this is one of major issue should be attempted. To address this issue, some approaches have been reported.

For instance, sequential delivery systems, which are capable of controlled release behaviors in response to external environmental changes such as pH and temperature, have been established. Xia et al. have developed the pH-controlled selective gentamicin/naproxen release system utilizing mesoporous bioactive glass (MBG)/poly(γ- benzyl-L-glutamate)-poly(ethylene glycol) graft copolymer (PBGL-g-PEG) nanomicelle composite materials [61]. The water-soluble gentamicin and lipophilic naproxen were respectively encapsulated into interior of MBG via hydrogen bond and core of PBGL-g-PEG nanomicelle via hydrophobic interaction. At low pH condition, the interaction between gentamicin and MBG surface was weakened with increasing of H+, and consequently gentamicin was quickly released, whereas the L-glutamate side chain was protonated and naproxen was stably encapsulated in PBGL-g-PEG nanomicelle core. Meanwhile, the opposite phenomenon was caused at high pH condition. The pH-controlled release of this system was realized due to opposite pH-dependency of release behavior of each material. The further functionalized dual-drug delivery system that can be response to both pH and temperature has been reported from Wei et al. [62]. In their system, pH-responsive chitosan/poly(vinyl alchol) hydrogel and pH/temperature dual responsive poly(L-glutamic acid)-b-poly(propylene oxide) -b-poly(L-glutamic acid) (GPG) micelle were employed and combined to encapsulate aspirin and DOX, respectively. With increasing pH value, the release rate of aspirin from chitosan/PVA hydrogel was accelerated, whereas DOX release from GPG micelle was facilitated with pH value lowering and/or temperature elevation.

Although the above-mentioned approaches are very interesting and useful from the standpoint of the rigorous release timing control, the temporal regulation is still insufficient. To address this issue, we have developed the multilayered nanofiber mesh system for time-programmed dual-drug delivery utilizing the sequential electrospinning method [63]. Electrospinning is a key technology that can easily fabricate nano-/micro-fiber mesh from various polymers. Since electrospun products have some distinctive properties such as high surface-area-to-volume ratio, flexible designability of morphology, and extracellular matrix-like structure, they have attracted much attention in the research fields of DDS and tissue engineering. The multilayered drug-loaded fiber mat that consists of biodegradable PLCL was designed with the following construction from top to bottom: 1) the first drug-loaded mesh; 2) plain mesh act as barrier; 3) the second drug-loaded mesh; and 4) the basement plain mesh. In our system, to prevent simultaneous release of the second drug during the first drug release, the basement and barrier mesh were prepared as having enough thickness and the second drug loaded mesh was sealed between the basement and barrier mesh. The release suppression period of the second drug could be prolonged in the barrier mesh thickness dependent manner. Based on four-layered structure, introduction of the complete suppression period of the seconddrug release at the end of the first-drug release was achieved by setting of barrier mesh with appropriate thickness, and subsequently the second-drug release was initiated and lasted. Moreover, since the release rate of individual drugs from electrospun fiber was accelerated with decreasing of fiber diameter, the release profile of whole system could be altered by controlling of fiber diameter. This is the first example of time-programmed dual-drug sustained release utilizing multilayered electrospun fiber mat formulation. Namely, we have successfully developed the sequential MDDS that would be able to realize more functional and effective combination therapy.

4. FUTURE PERSPECTIVE

As we introduced and mentioned in preceding sections, huge efforts have been devoted by researchers to establish the MDDSs with single formulation in the past decade. With recent remarkable evolution of nanobiotechnology and material sciences, not only exploitation of novel system but also improvement of pre-existing system have been actively attempted, and consequently various promising results have been demonstrated. However, it seems that there are still some rooms for further improvement. Especially, the development of more efficient active targeting system, more functionalized system, and more rigorously time controlled multi-drug/agent release system would be important issue in current and future DDS research.

The development of novel ligand would be essentially important for development of more efficient active targeting system. The contribution of nanobiotechnology would be indispensable for exploring of novel targeting molecules by high-throughput screening. Since DNA aptamers, which has antibody-compatible specific binding ability for several kinds of antigen, has been found out day by day [64], DNA aptamers would be one of promising molecule for more efficient active targeting. Furthermore, in recent years, development of multi-functional DDSs possessing not only drug delivery but also bioimaging ability has attracted much attention in the research field of DDS. Supplementation of bioimaging capability is very useful attempt in terms of visualization of the therapeutic effect. Indeed, development of bioimaging compatible MDDSs have been started [65-67]. The addition of extra function such as bioimaging has also been expected.

Meanwhile, site-specific drug release can be more easily achieved by bulk system in exchange for restriction of administration site. Furthermore, the individual release profiles of loaded drugs can also be easily controlled in the bulk system compared to particulate systems. As we mentioned in preceding section 3, rigorous release control of individual drugs has not sufficiently been achieved yet. However, we successfully demonstrated that the rigorously temporal controlled MDDS could be developed by appropriate material design and fine processing utilizing electrospinning. Therefore, both exploitation of novel functional materials and technical advance of material processing would be essentially needed for fine release control. Concretely, smart material that can change own characteristics responded to various environmental stimuli such as temperature, pH, and so on, and nanoscale precise processing technology such as layering, alignment, and configuration will be extensively required. In addition, although pre-constructed bulk formulation is potentially possessing restriction on selection of administration site, this issue will be circumvented by development of more flexible system such as injectable hydrogel that can perform in situ sol-gel phase transition or chemical polymerization.

In conclusion, although development of MDDS with single formulation having simultaneous or precisely-controlled sequential delivery capability is still immature and challenging field in DDS research, it is expected that establishment of more improved and functional MDDS would expand the potential of combination therapy based on BCM. For realization of more efficient, functional, and safe BCM-based multiple medications with the MDDS, the contributions of biomaterials science and nanobiotechnology to every steps of MDDS construction are essentially required. Moreover, since investigation and selection of effective drug combinations is an important factor for BCM-based chemotherapy, strategic alliance among experts of different fields such as medical doctors, paramedics, biologists, and engineers, would be also needed. We have expected early realization of combination therapy by MDDS with single formulation in clinical medication by synergistic cooperation of researchers.

REFERENCES

1. D. S. Martin, R. L. Stolfi, R. C. Sawyer and C. W. Young, "Application of biochemical modulation with a therapeutically inactive modulating agent in clinical trials of cancer chemotherapy," Cancer Treatment Report, Vol. 69, No. 4, 1985, pp. 421-423.

2. B. Leyland-Jones and P. J. O'Dwyer, "Biochemical modulation: application of labolatory models to the clinic," Cancer Treatment Report, Vol. 70, No. 1, 1986, pp. 219- 229.

3. J. R. Bertino, W. L. Sawicki, C. A. Lindquist and V. S. Gupta, "Schedule-Dependent Antitumor Effects of Methotrexate and 5-Fluorouracil," Cancer Research, Vol. 37, No. 1, 1977, pp. 327-328.

4. C. Tournigand, et al., "FOLFIRI Followed by FOLFOX6 or the reverse sequence in advanced colorectal cancer: a randomized GERCOR study," Journal of Clinical Oncology, Vol. 22, No. 2, 2004, pp. 229-237.

5. F. Maindrault-Goebe, et al., "High-dose intensity oxaliplatin added to the simplified bimonthly leucovorin and 5-fluorouracil regimen as second-line therapy for metastatic colorectal cancer (FOLFOX 7)," European Journal of Cancer, Vol. 37, No. 8, 2001, pp. 1000-1005.

6. J. C. Bendell, et al., "Randomized phase II study of perifosine in Combination with capecitabine (P-CAP) versus capecitabine plus placebo (CAP) in patients with secondor third-line metastatic colon cancer (mCRC): Updated results," Gastrointestinal Cancers Symposium, Orland, 22-24 January 2010.

7. M. Peeters, et al., "Randomized phase III study of panitumumab (pmab) with FOLFIRI versus FOLFIRI Alone as second-line treatment (tx) in patients (pts) with metastatic colorectal cancer (mCRC): Patient-Reported Outcomes (PRO)," Gastrointestinal Cancers Symposium, Orland, 22-24 January 2010.

8. V. Torchilin, "Recent advances with liposomes as pharmaceutical carriers," Nature reviews Drug discovery, Vol. 4, No. 2, 2005, pp. 145-160.

9. Y. Kakizawa and K. Kataoka, "Block copolymer micelles for delivery of gene and related compounds," Advanced Drug Delivery Reviews, Vol. 54, No. 2, 2002, pp. 203-222.

10. M. Yokoyama, "Polymeric micelles as a new drug carrier system and their required considerations for clinical trials," Expert opinion on drug delivery, Vol. 7, No. 2, 2010, pp. 145-158.

11. C. Amin, N. Mackman and N. S. Key, "Microparticles and Cancer," Pathophysiology of haemostasis and thrombosis, Vol. 36, No. 3-4, 2008, pp. 177-183.

12. H. Maeda, T. Sawa and T. Konno, "Mechanism of tumortargeted delivery of macromolecular drugs, including the EPR effect in solid tumor and clinical overview of the prototype polymeric Drug SMANCS," Journal of Controlled Release, Vol. 74, No. 1-3, 2001, pp. 47-61.

13. N. D. James, et al., "Liposomal doxorubicin (Doxil): an effective new treatment for Kaposi's sarcoma in AIDS," Clinical oncology, Vol. 6, No. 5, 1994, pp. 294-296.

14. S. de Marie, R. Janknegt, and I. A. Bakker-Woudenberg, "Clinical use of liposomal and lipid-complexed amphotericin B," The Journal of antimicrobial chemotherapy, Vol. 33, No. 5, 1994, pp. 907-916.

15. M. Harries, P. Ellis and P. Harper, "Nanoparticle albuminbound paclitaxel for metastatic breast cancer," Journal of clinical oncology, Vol. 23, No. 31, 2005, pp. 7768-7771.

16. F. Greco and M. J. Vicent, "Combination therapy: opportunities and challenges for polymer-drug conjugates as anticancer nanomedicines," Advanced Drug Delivery Reviews, Vol. 61, No. 13, 2009, pp. 1203-1213.

17. C.-M. J. Hu, S. Aryal and L. Zhang, "Nanoparticle-Assisted combination therapies for effective cancer treatment," Therapeutic Delivery, Vol. 1, No. 2, 2010, pp. 323-334.

18. H. Zhang, G. Wang, and H. Yang, "Drug delivery systems for differential release in combination therapy," Expert opinion on drug delivery, Vol. 8, No. 2, 2011, pp. 171-190.

19. L. Santucci, et al., "Nitric Oxide Modulates Proapoptotic and Antiapoptotic Properties of Chemotherapy Agents: The Case of NO-Pegylated Epirubicin," The FASEB Journal, Vol. 20, No. 6, 2006, pp. 765-767.

20. G. Pasut, et al., "Polymer-drug conjugates for combination anticancer therapy: investigating the mechanism of action," Journal of Medicinal Chemistry, Vol. 52, No. 20, 2009, pp. 6499-6502.

21. Q. Wu, R. Kreienberg and I. B. Runnebaum, "Growth suppression of human ovarian carcinoma OV-MZ-2a and OV-MZ-32 cells mediated by gene transfer of wildtype p53 enhanced by chemotherapy in vitro," Journal of Cancer Research and Clinical Oncology, Vol. 126, No. 3, 2000, pp. 139-144.

22. N. Wiradharma, Y. W. Tong and Y. Y. Yang, "Self-assembled oligopeptide nanostructures for co-delivery of drug and gene with synergistic therapeutic effect," Biomaterials, Vol. 30, No. 17, 2009, pp. 3100-3109.

23. S. Shin, et al., "An Anti-Apoptotic Protein Human Survivin Is a Direct Inhibitor of Caspase-3 and -7," Biochemistry, Vol. 40, No. 4, 2001, pp. 1117-1123.

24. A. Chandele, V. Prasad, J. C. Jagtap, R. Shukla and P. R. Shastry, "Upregulation of survivin in G2/M cells and inhibition of caspase 9 activity enhances resistance in staurosporine-induced apoptosis," Neoplasia, Vol. 6, No. 1, 2004, pp. 29-40.

25. J. C. Reed, "The Survivin Saga Goes in Vivo," The Journal of Clinical Investigation, Vol. 108, No. 7, 2001, pp. 965-969.

26. W. Xiao, X. Chen, L. Yang, Y. Mao, Y. Wei and L. Chen, "Co-delivery of doxorubicin and plasmid by a novel FGFR-mediated cationic liposome," International Journal of Pharmaceutics, Vol. 393, No. 1-2, 2010, pp. 119-126.

27. Z. Xu, Z. Zhang, Y. Chen, L. Chen, L. Lin and Y. Li, "The characteristics and performance of a multifunctional nanoassembly system for the co-delivery of docetaxel and iSur-pDNA in a mouse hepatocellular carcinoma model," Biomaterials, Vol. 31, No. 5, 2010, pp. 916-922.

28. G. M. Tozer, C. Kanthou and B. C. Baguley, "Disrupting Tumour Blood Vessels," Nature Reviews Cancer, Vol. 5, No. 6, 2005, pp. 423-435.

29. Y.-F. Zhang, J.-C. Wang, D.-Y. Bian, X. Zhang and Q. Zhang, "Targeted delivery of RGD-modified liposomes encapsulating both combretastatin A-4 and doxorubicin for tumor therapy: in vitro and in vivo studies," European journal of pharmaceutics and biopharmaceutics, Vol. 74, No. 3, 2010, pp. 467-473.

30. Z. Wang and P. C. Ho, "A nanocapsular combinatorial sequential drug delivery system for antiangiogenesis and anticancer activities," Biomaterials, Vol. 31, No. 27, 2010, pp. 7115-7123.

31. T. Yang, et al., "Targeted delivery of a combination therapy consisting of combretastatin A4 and low-dose doxorubicin against tumor neovasculature," Nanomedicine: Nanotechnology, Biology, and Medicine, 2011, in press.

32. D. Ingber, et al., "Synthetic analogues of fumagillin that inhibit angiogenesis and suppress tumour growth," Nature, Vol. 348, No. 6301, 1990, pp. 555-557.

33. R. Satchi-Fainaro, et al., "Targeting Angiogenesis with a Conjugate of HPMA Copolymer and TNP-470," Nature Medicine, Vol. 10, No. 3, 2004, pp. 255-261.

34. E. Segal, et al., "Targeting angiogenesis-dependent calcified neoplasms using combined polymer therapeutics," PloS One, Vol. 4, No. 4, 2009, p. e5233.

35. C. Zhu, et al., "Co-delivery of siRNA and paclitaxel into cancer cells by biodegradable cationic micelles based on PDMAEMA-PCL-PDMAEMA triblock copolymers," Biomaterials, Vol. 31, No. 8, 2010, pp. 2408-2416.

36. H.-Y. Huang, W.-T. Kuo, M.-J. Chou and Y.-Y. Huang, "Co-delivery of anti-vascular endothelial growth factor siRNA and doxorubicin by multifunctional polymeric micelle for tumor growth suppression," Journal of biomedical materials research Part A, Vol. 97, No. 3, 2011, pp. 330-338.

37. J. Wang, et al., "In vitro cytotoxicity of Stealth liposomes co-encapsulating doxorubicin and verapamil on doxorubicin-resistant tumor cells," Biological & Pharmaceutical Bulletin, Vol. 28, No. 5, 2005, pp. 822-828.

38. J. Wu, et al., "Reversal of multidrug resistance by transferrin-conjugated liposomes co-encapsulating doxorubicin and verapamil," Journal of pharmacy & pharmaceutical sciences, Vol. 10, No. 3, 2007, pp. 350-357.

39. Z. Liu, X. Y. Wu and R. Bendayan, "In vitro investigation of ionic polysaccharide microspheres for simultaneous delivery of chemosensitizer and antineoplastic agent to multidrug-resistant cells," Journal of Pharmaceutical Sciences, Vol. 88, No. 4, 1999, pp. 412-418.

40. C. E. Soma, C. Dubernet, D. Bentolila, S. Benita and P. Couvreur, "Reversion of multidrug resistance by co-encapsulation of doxorubicin and cyclosporin A in polyalkylcyanoacrylate nanoparticles," Biomaterials, Vol. 21, No. 1, 2000, pp. 1-7.

41. X. Song, et al., "PLGA nanoparticles simultaneously loaded with vincristine sulfate and verapamil hydrochloride: systematic study of particle size and drug entrapment efficiency," International Journal of Pharmaceutics, Vol. 350, No. 1-2, 2008, pp. 320-329.

42. X. R. Song, et al., "Reversion of multidrug resistance by co-encapsulation of vincristine and verapamil in PLGA nanoparticles," European journal of pharmaceutical sciences, Vol. 37, No. 3-4, 2009, pp. 300-305.

43. Y. Patil, T. Sadhukha, L. Ma and J. Panyam, "Nanoparticle-mediated simultaneous and targeted delivery of paclitaxel and tariquidar overcomes tumor drug resistance," Journal of Controlled Release, Vol. 136, No. 1, 2009, pp. 21-29.

44. H. L. Wong, R. Bendayan, A. M. Rauth and X. Y. Wu, "Development of solid lipid nanoparticles containing ionically complexed chemotherapeutic drugs and chemosensitizers," Journal of Pharmaceutical Sciences, Vol. 93, No. 8, 2004, pp. 1993-2008.

45. H. L. Wong, R. Bendayan, A. M. Rauth and X. Y. Wu, "Simultaneous delivery of doxorubicin and GG918 (Elacridar) by new polymer-lipid hybrid nanoparticles (PLN) for enhanced treatment of multidrug-resistant

breast cancer," Journal of Controlled Release, Vol. 116, No. 3, 2006, pp. 275-284.

46. R. I. Pakunlu, Y. Wang, M. Saad, J. J. Khandare, V. Starovoytov and T. Minko, "In vitro and in vivo intracellular liposomal delivery of antisense oligonucleotides and anticancer drug," Journal of Controlled Release, Vol. 114, No. 2, 2006, pp. 153-162.

47. X.-B. Xiong and A. Lavasanifar, "Traceable multifunctional micellar nanocarriers for cancer-targeted co-delivery of MDR-1 siRNA and doxorubicin," ACS Nano, Vol. 5, No. 6, 2011, pp. 5202-5213.

48. S. Sengupta, et al., "Temporal targeting of tumour cells and neovasculature with a nanoscale delivery system," Nature, Vol. 436, No. 7050, 2005, pp. 568-572.

49. H. Nie, Z. Dong, D. Y. Arifin, Y. Hu and C.-H. Wang, "Core/shell microspheres via coaxial electrohydrodynamic atomization for sequential and parallel release of drugs," Journal of Biomedical Materials Research Part A, Vol. 95, No. 3, 2010, pp. 709-716.

50. X. Hao, et al., "Angiogenic effects of sequential release of VEGF-A165 and PDGF-BB with alginate hydrogels after myocardial infarction," Cardiovascular Research, Vol. 75, No. 1, 2007, pp. 178-185.

51. A. T. Raiche and D. A. Puleo, "In vitro effects of combined and sequential delivery of two bone growth factors," Biomaterials, Vol. 25, No. 4, 2004, pp. 677-685.

52. R. R. Chen, E. A. Silva, W. W. Yuen and D. J. Mooney, "Spatio-temporal VEGF and PDGF delivery patterns blood vessel formation and maturation," Pharmaceutical Research, Vol. 24, No. 2, 2007, pp. 258-264.

53. C. Strobel, N. Bormann, A. Kadow-Romacker, G. Schmidmaier and B. Wildemann, "Sequential release kinetics of two (gentamicin and BMP-2) or three (gentamicin, IGF-I and BMP-2) substances from a one-component polymeric coating on implants," Journal of Controlled Release, Vol. 156, No. 1, 2011, pp. 37-45.

54. F. B. Basmanav, G. T. Kose and V. Hasirci, "Sequential growth factor delivery from complexed microspheres for bone tissue engineering," Biomaterials, Vol. 29, No. 31, 2008, pp. 4195-4204.

55. P. Yilgor, K. Tuzlakoglu, R. L. Reis, N. Hasirci and V. Hasirci, "Incorporation of a sequential BMP-2/BMP-7 delivery system into chitosan-based scaffolds for bone tissue engineering," Biomaterials, Vol. 30, No. 21, 2009, pp. 3551-3559.

56. Q. Sun, et al., "Sustained release of multiple growth factors from injectable polymeric system as a novel therapeutic approach towards angiogenesis," Pharmaceutical Research, Vol. 27, No. 2, 2010, pp. 264-271.

57. J. A. MacDiarmid, et al., "Sequential treatment of drugresistant tumors with targeted minicells containing siRNA or a cytotoxic drug," Nature Biotechnology, Vol. 27, No. 7, 2009, pp. 643-651.

58. J. Jiangm, et al., "Sequential treatment of drug-resistant tumors with RGD-modified liposomes containing siRNA or doxorubicin," European journal of pharmaceutics and biopharmaceutics, Vol. 76, No. 2, 2010, pp. 170-178.

59. P. Yilgor, N. Hasirci and V. Hasirci, "Sequential BMP-2/ BMP-7 delivery from polyester nanocapsules," Journal of Biomedical Materials Research Part A, Vol. 93, No. 2, 2010, pp. 528-536.

60. J. E. Tengood, K. M. Kovach, P. E. Vescovi, A. J. Russell and S. R. Little, "Sequential delivery of vascular endothelial growth factor and sphingosine 1-phosphate for angiogenesis," Biomaterials, Vol. 31, No. 30, 2010, pp. 7805-7812.

61. W. Xia, J. Chang, J. Lin and J. Zhu, "The pH-controlled dual-drug release from mesoporous bioactive glass/ polypeptide graft copolymer nanomicelle composites," European journal of pharmaceutics and biopharmaceutics, Vol. 69, No. 2, 2008, pp. 546-552.

62. L. Wei, C. Cai, J. Lin and T. Chen, "Dual-drug delivery system based on hydrogel/micelle composites," Biomaterials, Vol. 30, No. 13, 2009, pp. 2606-2613.

63. T. Okuda, K. Tominaga and S. Kidoaki, "Time-programmed dual release formulation by multilayered drug-loaded nanofiber meshes," Journal of Controlled Release, Vol. 143, No. 2, 2010, pp. 258-264.

64. A. S. Barbas, J. Mi, B. M. Clary and R. R. White, "Aptamer applications for targeted cancer therapy," Future Oncology, Vol. 6, No. 7, 2010, pp. 1117-1126.

65. X.-B. Xiong and A. Lavasanifar, "Traceable multifunctional micellar nanocarriers for cancer-targeted co-delivery of MDR-1 siRNA and doxorubicin," ACS Nano, Vol. 5, No. 6, 2011, pp. 5202-5213.

66. A. Singh, F. Dilnawaz, S. Mewar, U. Sharma, N. R. Jagannathan and S. K. Sahoo, "Composite polymeric magnetic nanoparticles for co-delivery of hydrophobic and hydrophilic anticancer drugs and MRI imaging for cancer therapy," ACS Applied Materials & Interfaces, Vol. 3, No. 3, 2011, pp. 842-856.

67. C.-C. Huang, W. Huang and C.-S. Yeh, "Shell-by-shell synthesis of multi-shelled mesoporous silica nanospheres for optical imaging and drug delivery," Biomaterials, Vol. 32, No. 2, 2011, pp. 556-564.

Index